《模糊数学与系统及其应用丛书》编委会

主　　编　罗懋康

副 主 编　陈国青　李永明

编　　委　(以姓氏笔画为序)

史福贵　李庆国　李洪兴　吴伟志

张德学　赵　彬　胡宝清　徐泽水

徐晓泉　曹永知　寇　辉　裴道武

薛小平

模糊数学与系统及其应用丛书 9

格值模糊凸结构与格值模糊代数

史福贵 著

科学出版社

北 京

内 容 简 介

本书是作者及其团队多年来部分研究成果的总结. 本书给出了模糊代数中的模糊子(半)群度、模糊子环度、模糊理想度、模糊子域度、模糊向量子空间度、模糊子格度和模糊效应子代数度等概念, 并建立了它们和模糊凸空间之间的联系.

本书可作为数学专业的硕士研究生和博士研究生的教学用书, 也可以供相关科研工作者参考.

图书在版编目(CIP)数据

格值模糊凸结构与格值模糊代数 / 史福贵著. -- 北京：科学出版社, 2024.11. (模糊数学与系统及其应用丛书; 9) -- ISBN 978-7-03-079889-3

I. O159

中国国家版本馆 CIP 数据核字第 2024N5S696 号

责任编辑：李静科　李　萍 / 责任校对：彭珍珍
责任印制：赵　博 / 封面设计：无极书装

科学出版社 出版
北京东黄城根北街 16 号
邮政编码：100717
http://www.sciencep.com
固安县铭成印刷有限公司印刷
科学出版社发行　各地新华书店经销
*

2024 年 11 月第　一　版　开本：720×1000　1/16
2025 年 1 月第二次印刷　印张：14 1/4
字数：280 000
定价：108.00 元
(如有印装质量问题, 我社负责调换)

《模糊数学与系统及其应用丛书》序

自然科学和工程技术，表现的是人类对客观世界有意识的认识和作用，甚至表现了这些认识和作用之间的相互影响，例如，微观层面上量子力学的观测问题.

当然，人类对客观世界最主要的认识和作用，仍然在人类最直接感受、感知的介观层面发生，虽然往往需要以微观层面的认识和作用为基础，以宏观层面的认识和作用为延拓.

而人类在介观层面认识和作用的行为和效果，可以说基本上都是力图在意识、存在及其相互作用关系中，对减少不确定性，增加确定性的一个不可达极限的逼近过程；即使那些目的在于利用不确定性的认识和作用行为，也仍然以对不确定性的具有更多确定性的认识和作用为基础.

正如确定性以形式逻辑的同一律、因果律、排中律、矛盾律、充足理由律为形同公理的准则而界定和产生一样，不确定性本质上也是对偶地以这五条准则的分别缺损而界定和产生. 特别地，最为人们所经常面对的，是因果律缺损所导致的随机性和排中律缺损所导致的模糊性.

与随机性被导入规范的定性、定量数学研究对象范围已有数百年的情况不同，人们对模糊性进行规范性认识的主观需求和研究体现，仅仅开始于半个世纪前 1965 年 Zadeh 具有划时代意义的 *Fuzzy sets* 一文.

模糊性与随机性都具有难以准确把握或界定的共同特性，而从 Zadeh 开始延续下来的 "以赋值方式量化模糊性强弱程度" 的模糊性表现方式，又与已经发展数百年而高度成熟的 "以赋值方式量化可能性强弱程度" 的随机性表现方式，在基本形式上平行——毕竟，模糊性所针对的 "性质"，与随机性所针对的 "行为"，在基本的逻辑形式上是对偶的. 这也就使得 "模糊性与随机性并无本质差别" "模糊性不过是随机性的另一表现" 等疑虑甚至争议，在较长时间和较大范围内持续.

然而时至今日，应该说不仅如上由确定性的本质所导出的不确定性定义已经表明模糊性与随机性在本质上的不同，而且人们也已逐渐意识到，表现事物本身性质的强弱程度而不关乎其发生与否的模糊性，与表现事物性质发生的可能性而不关乎其强弱程度的随机性，在现实中的影响和作用也是不同的.

例如，当情势所迫而必须在 "于人体有害的可能为万分之一" 和 "于人体有害的程度为万分之一" 这两种不同性质的 150 克饮料中进行选择时，结论就是不言而喻的，毕竟前者对 "万一有害，害处多大" 没有丝毫保证，而后者所表明的 "虽然有害，但极微小" 还是更能让人放心得多. 而这里，前一种情况就是 "有害" 的随机性表现，后一种情况就是 "有害" 的模糊性表现.

模糊性能在比自身领域更为广泛的科技领域内得到今天这一步的认识，的确不是一件容易的事。到今天，模糊理论和应用的研究所涉及和影响的范围也已几乎无远弗届。这里有一个非常基本的原因：模糊性与随机性一样，是几种基本不确定性中，最能被人类思维直接感受，也是最能对人类思维产生直接影响的。

对于研究而言，易感知、影响广本来是一个便利之处，特别是在当前以本质上更加逼近甚至超越人类思维的方式而重新崛起的人工智能的发展已经必定势不可挡的形势下。然而也正因为如此，我们也都能注意到，相较于广度上的发展，模糊性研究在理论、应用的深度和广度上的发展，还有很大的空间；或者更直接地说，还有很大的发展需求。

例如，在理论方面，思维中模糊性与直感、直观、直觉是什么样的关系？与深度学习已首次形式化实现的抽象过程有什么样的关系？模糊性的本质是在于作为思维基本元素的单体概念，还是在于作为思维基本关联的相对关系，还是在于作为两者统一体的思维基本结构，这种本质特性和作用机制以什么样的数学形式予以刻画和如何刻画才能更为本质深刻和关联广泛？

又例如，在应用方面，人类是如何思考和解决在性质强弱程度方面难以确定的实际问题的？是否都是以条件、过程的更强定量来寻求结果的更强定量？是否可能如同深度学习对抽象过程的算法形式化一样，建立模糊定性的算法形式化？在比现在已经达到过的状态、已经处理过的问题更复杂、更精细的实际问题中，如何更有效地区分和结合"性质强弱"与"发生可能"这两类本质不同的情况？从而更有效、更有力地在实际问题中发挥模糊性研究本来应有的强大效能？

这些都是模糊领域当前还需要进一步解决的重要问题；而这也就是作为国际模糊界主要力量之一的中国模糊界研究人员所应该、所需要倾注更多精力和投入的问题。

针对相关领域高等院校师生和科技工作者，推出这套《模糊数学与系统及其应用丛书》，以介绍国内外模糊数学与模糊系统领域的前沿热点方向和最新研究成果，从上述角度来看，是具有重大的价值和意义的，相信能在推动我国模糊数学与模糊系统乃至科学技术的跨越发展上，产生显著的作用。

为此，应邀为该丛书作序，借此将自己的一些粗略的看法和想法提出，供中国模糊界同仁参考。

罗懋康

国际模糊系统协会 (IFSA) 副主席 (前任)

国际模糊系统协会中国分会代表

中国系统工程学会模糊数学与模糊系统专业委员会主任委员

2018 年 1 月 15 日

前　　言

本书是在作者及其团队多年研究成果的基础上整理完成的. 从 2020 年起, 作者开始为牡丹江师范学院的硕士生开设模糊方法论选讲课程, 目的是让选择此课程的硕士研究生尽快掌握模糊代数和模糊凸空间的一些基本思想和方法, 同时也让他们能够尽可能快地定下毕业论文选题. 其间正有新冠疫情, 为了网上授课的需要, 我不得不把一些已发表的相关论文整理成有动画效果的 PDF 文件放映模式. 几年下来, 积攒的这些文件也渐渐多了, 遂产生了整理出书的念头, 在学生们的帮助下, 终于完成此书初稿.

在 Zadeh L A 引入模糊集不久, 就有了模糊子群、模糊子环和模糊子域等方面的研究工作, 至今已有大量相关模糊代数的论文, 详见 [95, 96, 103, 111, 127]. 实际上, 在作者的早期工作中, 鲜有模糊代数的工作, 主要工作是关于模糊拓扑的研究. 由于模糊拟阵与模糊拓扑有着某种相似性, 所以后来自然又开始了模糊拟阵的研究工作.

拟阵的模糊化工作最早是由 Goetschel R 和 Voxman W 率先开始的, Goetschel R 和 Voxman W 使用模糊集取代分明集引入了模糊独立集的概念, 从而定义了一种模糊拟阵, 我们现在称之为 L-拟阵, 它的独立集是模糊的, 但是独立集族是分明的. 从另外一个角度, 作者引入了模糊化拟阵的概念, 这个模糊化拟阵中的独立集仍然是分明的, 但是由独立集构成的集族是模糊的, 从而开始了模糊化拟阵的研究工作. 之后作者又同时考虑独立集和独立集族两者同时是模糊的情形, 引入了 (L, M)-模糊拟阵的概念.

在模糊拟阵的研究过程中, 我们发现, 拟阵可以看作一个特殊的凸空间, 从而模糊拟阵理论又可看作模糊凸结构理论的一种特殊情况, 因此我们又开始了有关模糊凸空间与凸结构的研究, 先后提出了模糊化凸空间与 (L, M)-模糊凸空间理论. 作者团队从事模糊凸空间的工作时间并不长, 但是取得的成果却是很丰富的, 这一点从参考文献可见, 就不详细介绍了. 之所以近期开始重视模糊代数的研究, 起源于我们在模糊凸结构方面的工作. 在从事模糊凸结构的研究中, 我们发现模糊凸结构在模糊代数中有相当多的理论应用, 这些应用很多都是前所未有的. 在文献 [137, 140] 中, 我们把模糊拓扑学中的研究方法应用到了模糊代数中, 引入了子群度和正规子群度的概念, 但当时并没有发现这种子群度和模糊凸结构的直接联系, 直到 2016 年, 李娟首次发现了模糊子格度可以诱导模糊凸结构. 这样, 我们便开始注意模糊凸结构在模糊代数中的具体表现问题.

迄今为止, 我们发现, 模糊凸结构不仅仅在模糊代数中有诸多理论应用, 在模糊拓扑、模糊拟阵、模糊图、模糊向量空间、模糊博弈论等分支中都有重要的应用. 因此, 模糊凸结构的研究就越来越显示出它巨大的生命力, 尚未发现的应用还有待于广大读者继续深入探讨. 希望本书能够起到抛砖引玉的作用.

在研究这些模糊子代数时, 用的基本工具就是格上的两种集合套理论, 因此我们在第 2 章先介绍了两种集合套和它们的一些应用. 由于凸空间和拓扑空间既相似又不同, 为了建立模糊子代数和模糊凸空间的关系, 我们在第 3 章中介绍了模糊拓扑空间和模糊凸空间的一些比较基本的概念和相关结果, 一些结果可看作两种集合套的理论应用. 第 4 章是模糊子群度理论, 其可看作已有模糊子群的推广, 这种模糊子群度很自然地就诱导出了模糊凸结构. 第 5 章是模糊子环度理论, 其可看作已有模糊子环的推广. 第 6 章是模糊向量子空间理论, 介绍了模糊子空间维数的新提法. 第 7 章是模糊子格理论, 除了给出模糊子格的刻画外, 还建立了它和模糊凸结构的联系. 第 8 章是模糊效应子代数度理论, 模糊效应子代数度同样可以诱导模糊凸结构. 从范畴的角度来说, 各种模糊子代数范畴均可看作模糊凸空间范畴的子范畴, 因此从范畴论角度可以继续研究各种模糊子代数, 限于篇幅, 我们没有再过多介绍, 相关研究者可以继续深入挖掘新成果.

本书可作为模糊数学方向的研究生教材, 尤其是模糊代数、模糊拓扑和模糊凸空间这些研究方向的研究生的入门教材. 因为书中的部分结果是作者所带研究生完成的研究成果, 所以对于这些方向上的研究生来说是很容易理解和掌握的. 对于想在模糊拓扑学方向从事研究的研究生来说, 可以重点阅读第 2, 4 章内容; 而对于想在模糊代数方向从事研究的研究生来说, 可以略过第 3 章的阅读.

本书之所以能够完成, 首先应该感谢我的团队成员信秀、王岚、王冰、庞斌、李娟、韩元良、温宇飞、董彦彦、魏晓伟、曾铭仪、安英英等. 他们有的是作者论文的合作者, 有的在书稿排版和校对的过程中投入了很多精力.

另外, 本书的出版还得到了国家自然科学基金项目 "模糊凸代数理论的研究" (2023—2026, 项目编号: 12271036) 和国家自然科学基金项目 "模糊凸空间理论的研究" (2019—2022, 项目编号: 11871097) 的支持, 在此也要一并感谢.

由于我们的水平有限, 加上准备仓促, 书中疏漏在所难免, 希望读者不吝指教.

作者史福贵现为泉州师范学院数学与计算机科学学院和广西民族大学数学与物理学院的特聘教授.

史福贵

2023 年 11 月 26 日于北京

目 录

《模糊数学与系统及其应用丛书》序
前言
第 1 章　预备知识 ··· 1
　　1.1　集合与映射 ·· 1
　　1.2　完全分配格 ·· 3
　　1.3　代数系统 ·· 6
　　习题 1 ·· 10
第 2 章　L-模糊集的表现定理 ··· 11
　　2.1　L-模糊集的分解定理 ·· 11
　　2.2　表现定理 I ··· 17
　　2.3　表现定理 II ·· 20
　　2.4　L-模糊关系 ·· 22
　　2.5　L 值 Zadeh 型函数 ··· 27
　　2.6　L-模糊偏序 ·· 30
　　2.7　模糊映射 ·· 31
　　习题 2 ·· 35
第 3 章　(L,M)-模糊拓扑与 (L,M)-模糊凸结构 ······················ 37
　　3.1　拓扑的模糊化 ·· 37
　　3.2　(L,M)-模糊内部算子 ·· 39
　　3.3　(L,M)-模糊闭包算子 ·· 43
　　3.4　诱导拓扑空间 ·· 46
　　3.5　Hutton 一致空间和 Erceg 伪度量空间 ························· 50
　　3.6　凸结构 ··· 53
　　3.7　凸结构的模糊化 ··· 54
　　3.8　(L,M)-模糊凸包算子 ·· 58
　　3.9　由 M-模糊化凸结构诱导的 (L,M)-模糊凸结构 ············ 63

习题 3 ·· 68

第 4 章 L-模糊子群 ··· 72
4.1 群上的 L-模糊同余关系 ····································· 72
4.2 L-模糊子半群度 ·· 74
4.3 由 L-模糊子半群度确定的 L-模糊凸结构 ······················ 81
4.4 L-模糊子群度 ·· 83
4.5 L-模糊子群的运算 ··· 90
4.6 L-模糊正规子群度 ··· 93
4.7 由 L-模糊子群度确定的 L-模糊凸结构 ······················· 97
4.8 L-模糊商群 ··· 99
习题 4 ·· 102

第 5 章 L-模糊子环与理想 ······································· 106
5.1 L-模糊子环度 ··· 106
5.2 L-模糊子环的运算 ·· 113
5.3 由 L-模糊子环度确定的 L-模糊凸结构 ······················ 120
5.4 环上的 L-模糊同余关系 ···································· 123
5.5 L-模糊理想度 ··· 125
5.6 由 L-模糊理想度确定的 L-模糊凸结构 ······················ 131
5.7 L-模糊素理想度 ··· 135
5.8 L-模糊子域 ··· 139
5.9 由 L-模糊子域度确定的 L-模糊凸结构 ······················ 147
习题 5 ·· 152

第 6 章 L-模糊向量子空间 ······································· 154
6.1 模糊数与模糊自然数 ·· 154
6.2 模糊向量空间的模糊基和模糊维数 ···························· 157
6.3 L-模糊向量子空间度 ······································· 160
6.4 由 L-模糊子空间度确定的 L-模糊凸结构 ····················· 163
习题 6 ·· 165

第 7 章 L-模糊子格 ·· 166
7.1 L-模糊子格度 ··· 166
7.2 L-模糊凸子格度 ··· 174

7.3 由 L-模糊子格度确定的 L-模糊凸结构 ·············· 180
习题 7 ··· 181

第 8 章 L-模糊效应代数 ·· 182
8.1 效应代数基础 ··· 182
8.2 效应代数上的 L-模糊子代数度 ·························· 183
8.3 由模糊效应子代数确定的 L-模糊凸结构 ············· 189
8.4 效应代数中的 L-模糊理想度 ···························· 192
8.5 由 L-模糊理想度确定的 L-模糊凸结构 ·············· 200
习题 8 ··· 203

参考文献 ··· 204
《模糊数学与系统及其应用丛书》已出版书目

第 1 章 预 备 知 识

1.1 集合与映射

在这一节中, 我们先来介绍一些集合与映射的基本知识.

定义 1.1.1 假设 X, Y 是两个集合, 那么集合 $X \times Y$ 定义为 $\{(x,y) \mid x \in X, y \in Y\}$, 称为 X 与 Y 的笛卡儿积, 或者简称积.

定义 1.1.2 假设 X, Y 是两个集合. $X \times Y$ 的一个子集 R 被称为从 X 到 Y 的关系. $\forall (x,y) \in X \times Y$, 当 $(x,y) \in R$ 时, 我们就称 x 与 y 有关系 R, 记为 xRy, 否则就称 x 与 y 没有关系 R. 一个从 X 到 X 的关系称为 X 上的二元关系. 若 $A \subseteq X$, 则称集合 $\{y \in Y \mid \exists x \in A \text{ 使得 } xRy\}$ 为集合 A 关于 R 来说的像集, 并记为 $R(A)$. 关系 R^{-1} 就是 $Y \times X$ 的子集 $\{(y,x) \in Y \times X \mid (x,y) \in R\}$.

定义 1.1.3 设 R 是从 X 到 Y 的关系, S 是从 Y 到 Z 的关系. 则集合 $\{(x,z) \in X \times Z \mid \exists y \in Y \text{ 使得 } xRy \text{ 且 } ySz\}$ 是从 X 到 Z 的一个关系, 称为关系 R 与 S 的合成, 或者复合, 记为 $S \circ R$.

定义 1.1.4 集合 X 上的一个二元关系 R 叫做一个等价关系, 如果它满足下面三个条件:

(1) 自反性, 即 $\forall x \in X$, xRx;

(2) 对称性, 即若 xRy, 则 yRx;

(3) 传递性, 即若 xRy 且 yRz, 则 xRz.

定义 1.1.5 设 R 是集合 X 上的一个等价关系, 集合 $\{y \in X \mid xRy\}$ 称为 x 的 R-等价类, 记为 $[x]_R$. 集族 $\{[x]_R \mid x \in X\}$ 称为 X 关于 R 的商集, 记为 X/R.

定义 1.1.6 一个从 X 到 Y 的关系 f 称为从 X 到 Y 的一个映射, 如果对任意的 $x \in X$, 都存在唯一的一个 $y \in Y$ 使得 $(x,y) \in f$, 那就是 xfy, 此时它也可以记为 $f(x) = y$.

一个映射 f 称为单射, 如果对任意的 $x_1, x_2 \in X$, 当 $x_1 \neq x_2$ 时, 均有 $f(x_1) \neq f(x_2)$.

一个映射 f 称为满射, 如果对任意的 $y \in Y$, 均有 $x \in X$ 使得 $f(x) = y$.

如果映射 f 既是单射又是满射, 那么就称它是双射, 或者一一对应.

如果映射 $f : X \to Y$ 是双射, 那么定义 $f^{-1} : Y \to X$ 使得对任意的 $y \in Y$, $f^{-1}(y) = x$, 这里 $f(x) = y$. f^{-1} 称为 f 的逆映射.

今后我们把从 X 到 Y 的所有映射的集合记为 Y^X.

假设 X 是一个非空集合, 那么 X 的所有子集的集合记为 $\mathcal{P}(X)$, 称为 X 的幂集. 事实上我们也常常用 $\mathbf{2}^X$ 表示幂集, 下面我们来分析一下这样做的动机.

假设 A 是 X 的一个子集, 定义一个映射 $\chi_A : X \to \{0,1\} \triangleq \mathbf{2}$ 使得

$$\chi_A(x) = \begin{cases} 1, & x \in A, \\ 0, & x \in X - A. \end{cases}$$

不难验证下面性质:

(1) $\chi_{A \cup B}(x) = \max\{\chi_A(x), \chi_B(x)\}$;

(2) $\chi_{A \cap B}(x) = \min\{\chi_A(x), \chi_B(x)\}$;

(3) $\chi_{A'}(x) = 1 - \chi_A(x)$, 这里 A' 表示 A 的补集, 即 $A' = X - A$.

反过来, 给定一个映射 $f : X \to \{0, 1\}$, 我们定义一个集合

$$A_f = \{x \in X \mid f(x) = 1\},$$

那么容易证明 $\chi_{A_f} = f$ 而且 $A_{\chi_A} = A$. 因此, X 的一个子集和它的特征函数是一一对应的, 于是从同构的意义上说, $\mathcal{P}(X)$ 和 $\mathbf{2}^X$ 可看作一样的.

定义 1.1.7 设 f 是从 X 到 Y 的映射, 对任意的 $A \in \mathbf{2}^X$ 和 $B \in \mathbf{2}^Y$, 定义

$$f^{\to}(A) = \{f(x) \mid x \in A\},$$

$$f^{\leftarrow}(B) = \{x \in X \mid f(x) \in B\},$$

那么我们可以得到两个映射 $f^{\to} : \mathbf{2}^X \to \mathbf{2}^Y$ 和 $f^{\leftarrow} : \mathbf{2}^Y \to \mathbf{2}^X$, 称它们为 f 的扩张.

注 1.1.8 在很多数学教材中, 上边的 $f^{\to}(A)$ 一般记为 $f(A)$, 而 $f^{\leftarrow}(B)$ 一般记为 $f^{-1}(B)$, 这样做严格来说是不正确的, 因为我们在谈论映射时要把它看成一个关系, 而关系是要指明它的定义域和值域的, 很明显 $f^{\to}$ 的定义域不同于 f, $f^{\leftarrow}$ 的定义域也不同于 f^{-1}.

容易验证下面结论.

命题 1.1.9 设 $f : X \to Y$ 是一个映射, A 和 B 分别是 X 和 Y 的子集. 则

(1) $f^{\to}(f^{\leftarrow}(B)) \subseteq B$; 若 f 是满射, 则 $f^{\to}(f^{\leftarrow}(B)) = B$.

(2) $f^{\leftarrow}(f^{\to}(A)) \supseteq A$; 若 f 是单射, 则 $f^{\leftarrow}(f^{\to}(A)) = A$.

命题 1.1.10 设 f 是从 X 到 Y 的映射, $\{A_i \mid i \in \Omega\} \subseteq \mathbf{2}^X$ 和 $\{B_j \mid j \in \Lambda\} \subseteq \mathbf{2}^Y$, 那么

$$f^{\to}\left(\bigcup_{i \in \Omega} A_i\right) = \bigcup_{i \in \Omega} f^{\to}(A_i),$$

$$f^{\rightarrow}\left(\bigcap_{i\in\Omega}A_i\right)\subseteq\bigcap_{i\in\Omega}f^{\rightarrow}(A_i),$$

$$f^{\leftarrow}\left(\bigcup_{j\in\Lambda}B_j\right)=\bigcup_{j\in\Lambda}f^{\leftarrow}(B_j),$$

$$f^{\leftarrow}\left(\bigcap_{j\in\Lambda}B_j\right)=\bigcap_{j\in\Lambda}f^{\leftarrow}(B_j),$$

$$f^{\leftarrow}(B')=(f^{\leftarrow}(B))'.$$

该命题的证明是比较简单的, 故略去.

注 1.1.11 一般来说, $f^{\rightarrow}\left(\bigcap_{i\in\Omega}A_i\right)\neq\bigcap_{i\in\Omega}f^{\rightarrow}(A_i), f^{\rightarrow}(A')\neq(f^{\rightarrow}(A))'$.

定义 1.1.12 假设 X 是一个集合, 那么 X 中元素的个数称为 X 的势. 如果 X 的势 (可表示为 $|X|$) 是有限的, 就称它是一个有限集; 如果 X 与自然数集是一一对应的, 则称它是可数无限的, 此时它的势记为 $\aleph_0$. 如果 $|X|=m$ 且 $|Y|=n$, 那么 $|Y^X|=n^m$. 例如, $|\mathbf{2}^X|=2^m$.

定义 1.1.13 设 $\{X_i\}_{i\in\Omega}$ 是一族集合, 定义 $\{X_i\}_{i\in\Omega}$ 的乘积 $\prod_{i\in\Omega}X_i$ 为

$$\prod_{i\in\Omega}X_i=\left\{x\left|x:\Omega\rightarrow\bigcup_{i\in\Omega}X_i\text{ 使得 }\forall i\in\Omega, x_i=x(i)\in X_i\right.\right\}.$$

映射 $p_i:\prod_{i\in\Omega}X_i\rightarrow X_i$ 被定义为 $\forall x\in\prod_{i\in\Omega}X_i, p_i(x)=x_i\in X_i$, 也称它为投影映射.

当 Ω 仅有两个元素时, 定义 1.1.13 就退化为定义 1.1.1.

定义 1.1.14 设 X/R 是 X 关于等价关系 R 的商集, 称满足 $p(x)=[x]_R$ 的映射 $p:X\rightarrow X/R$ 为自然投射.

1.2 完全分配格

定义 1.2.1 设 X 是非空集, $\leqslant$ 是 X 上的一个二元关系, 如果
(1) $\leqslant$ 是自反的, 即 $\forall x\in X, x\leqslant x$;
(2) $\leqslant$ 是传递的, 即 $x\leqslant y, y\leqslant z\Rightarrow x\leqslant z$;
(3) $\leqslant$ 是反对称的, 即 $x\leqslant y, y\leqslant x\Rightarrow x=y$,

则称 $\leqslant$ 为 X 上的偏序关系, 称 $(X,\leqslant)$ 为偏序集, 简写为 X.

定义 1.2.2　一个偏序集 $(X, \leqslant)$ 称为全序集, 如果对 X 中的任二元素 a 与 b, 都有 $a \leqslant b$ 或者 $b \leqslant a$.

定义 1.2.3　一个偏序集 D 称为定向集, 如果对 D 中的任二元素 a 与 b, 都存在 $c \in D$ 使得 $a \leqslant c, b \leqslant c$. 称 D 为一个下定向集或者余定向集, 如果对 D 中的任二元素 a 与 b, 都存在 $c \in D$ 使得 $c \leqslant a, c \leqslant b$.

一个全序集一定是一个定向集.

定义 1.2.4　设 X 是偏序集, $A \subseteq X, a \in X$, 若 $\forall x \in A$, 均有 $x \leqslant a$, 就称 a 是 A 的一个上界. 若 $\forall x \in A$, 均有 $x \geqslant a$, 就称 a 是 A 的一个下界. 如果 A 有最小上界 a, 那么 a 就称为 A 的上确界, 记为 $a = \bigvee A$ 或者 $a = \sup A$. 如果 A 有最大下界 a, 那么 a 就称为 A 的下确界, 记为 $a = \bigwedge A$ 或者 $a = \inf A$. 当 $A = \{a_i \mid i \in \Omega\}$ 时, $\bigvee A$ 也记为 $\bigvee_{i \in \Omega} a_i$, $\bigwedge A$ 也记为 $\bigwedge_{i \in \Omega} a_i$. 如果 $A = \{x, y\}$, 那么 $\bigvee A$ 记为 $x \vee y$, $\bigwedge A$ 记为 $x \wedge y$.

定义 1.2.5　设 X 为偏序集, 若对 X 中的任二元素 a 与 b, $\sup\{a, b\}$ 与 $\inf\{a, b\}$ 恒存在, 则称 X 为格.

定义 1.2.6　设 X 是偏序集, 若 X 的每个子集都有上确界以及下确界, 即 $\forall A \subseteq X$, $\sup A$ 与 $\inf A$ 恒存在, 则称 X 为完备格.

完备格 X 中的最大元和最小元分别记为 1 和 0. 有时我们也称最小元为零元, 最大元为单位元. 以下我们常用 L 来表示一个格.

定义 1.2.7　一个完备格 L 称为一个完备的 Heyting 代数, 如果它满足下面无限分配律:

$$a \wedge \left(\bigvee_{i \in \Omega} b_i \right) = \bigvee_{i \in \Omega} (a \wedge b_i).$$

定义 1.2.8　设 L 是完备格, 如果下列两个等式成立:

$$\bigvee_{i \in I} \left(\bigwedge_{j \in J_i} a_{ij} \right) = \bigwedge_{f \in \prod_{i \in I} J_i} \left(\bigvee_{i \in I} a_{if(i)} \right),$$

$$\bigwedge_{i \in I} \left(\bigvee_{j \in J_i} a_{ij} \right) = \bigvee_{f \in \prod_{i \in I} J_i} \left(\bigwedge_{i \in I} a_{if(i)} \right),$$

则称 L 为完全分配格.

当 I 只有两个元素时, 上面两个等式变成了如下两种形式:

$$\left(\bigvee_{i \in I} a_i \right) \wedge \left(\bigvee_{j \in J} b_j \right) = \bigvee_{(i,j) \in I \times J} (a_i \wedge b_j),$$

1.2 完全分配格

$$\left(\bigwedge_{i\in I} a_i\right) \vee \left(\bigwedge_{j\in J} b_j\right) = \bigwedge_{(i,j)\in I\times J} (a_i \vee b_j).$$

当 I 中仅有一个元素时, 两个分配律又变成下列特殊形式:

$$a \wedge \left(\bigvee_{j\in J} b_j\right) = \bigvee_{j\in J} (a \wedge b_j),$$

$$a \vee \left(\bigwedge_{j\in J} b_j\right) = \bigwedge_{j\in J} (a \vee b_j).$$

因此一个完全分配格一定是一个完备的 Heyting 代数.

下面定理的证明留给读者.

定理 1.2.9 一族完全分配格的直积是完全分配的.

定义 1.2.10 设 L 是格, $a \in L$.

(1) 若对 L 中满足 $x \wedge y \leqslant a$ 的元素 x, y, 必有 $x \leqslant a$ 或 $y \leqslant a$ 成立, 就称 a 是 L 中的素元;

(2) 若对 L 中满足 $x \vee y \geqslant a$ 的元素 x, y, 必有 $a \leqslant x$, $a \leqslant y$ 成立, 就称 a 是 L 中的余素元.

我们用 $P(L)$ 表示 L 中所有非单位素元的集合, $J(L)$ 表示 L 中所有非零余素元的集合.

定义 1.2.11 L 中的一个二元关系 $\prec$ 被定义如下: 对于 $a, b \in L$, $a \prec b$ 当且仅当对每个子集 $D \subseteq L$, 当 $b \leqslant \sup D$ 时, 存在 $d \in D$ 使得 $a \leqslant d$. $\{a \in L \mid a \prec b\}$ 称为 b 的最大极小集, 用 $\beta(b)$ 表示.

此外, 定义 L 中的另一个二元关系 $\triangleleft$ 如下: 对于 $a, b \in L$, $b \triangleleft a$ 当且仅当对每个子集 $E \subseteq L$, 当 $b \geqslant \inf E$ 时, 存在 $e \in E$ 使得 $a \geqslant e$. $\{a \in L \mid b \triangleleft a\}$ 称为 b 的最大极大集, 用 $\alpha(b)$ 表示.

容易证明, 在完全分配格 L 中, α 是一个 $\bigwedge$-$\bigcup$ 映射, β 是保并映射, 那就是

$$\alpha\left(\bigwedge_{i\in\Omega} a_i\right) = \bigcup_{i\in\Omega} \alpha(a_i), \quad \beta\left(\bigvee_{i\in\Omega} a_i\right) = \bigcup_{i\in\Omega} \beta(a_i), \tag{1.2.1}$$

且对于每个 $b \in L$, 都存在 $\alpha(b)$ 和 $\beta(b)$ 使得 $b = \bigvee \beta(b) = \bigwedge \alpha(b)$.

在一个完全分配格 L 中, 容易验证

$$b = \bigvee \beta^*(b), \quad \text{这里 } \beta^*(b) \triangleq \beta(b) \cap J(L),$$

而且
$$b = \bigwedge \alpha^*(b), \quad \text{这里 } \alpha^*(b) \triangleq \alpha(b) \cap P(L).$$

因为完全分配格是完备的 Heyting 代数, 所以在 L 中存在一个二元运算 $\mapsto$, 它能够按下面公式给出:
$$a \mapsto b = \bigvee \{c \in L \mid a \wedge c \leqslant b\}.$$

称 $\mapsto$ 为蕴含. 下面我们列出了蕴含运算的一些性质.

定理 1.2.12 设 L 是一个完全分配格, $\mapsto$ 是对应于 $\wedge$ 的蕴含运算. 则对于所有 $a,b,c \in L, \{a_i\}_{i \in I} \subseteq L$, 下列条件成立:

(1) $1 \mapsto a = a$;

(2) $c \leqslant a \mapsto b \Leftrightarrow a \wedge c \leqslant b$;

(3) $a \mapsto b = 1 \Leftrightarrow a \leqslant b$;

(4) $a \mapsto \left(\bigwedge_{i \in I} a_i \right) = \bigwedge_{i \in I} (a \mapsto a_i)$, 特别地, 当 $b \leqslant c$ 时, 有 $a \mapsto b \leqslant a \mapsto c$;

(5) $\left(\bigvee_{i \in I} a_i \right) \mapsto b = \bigwedge_{i \in I} (a_i \mapsto b)$, 特别地, 当 $a \leqslant b$ 时, 有 $b \mapsto c \leqslant a \mapsto c$;

(6) $(a \mapsto c) \wedge (c \mapsto b) \leqslant a \mapsto b$;

(7) $(a \mapsto b) \wedge (c \mapsto d) \leqslant a \wedge c \mapsto b \wedge d$.

该定理的证明是容易的, 故略去.

定义 1.2.13 设 L 是完备格, $': L \to L$ 是 L 到自身的映射, 如果

(1) $'$ 是对合对应, 即 $\forall a \in L, (a')' = a$;

(2) $'$ 是逆序对应, 即 $a \leqslant b$ 蕴含 $b' \leqslant a'$,

则称 $'$ 为 L 上的逆序对合对应, 或简称为逆合对应.

1.3 代数系统

定义 1.3.1 设 G 是一个非空集合. 一个映射 $\circ : G \times G \to G$ 称为 G 上一个代数运算. $(G, \circ)$ 称为一个代数系统. $\forall a, b \in G, a \circ b$ 也简单记为 ab.

定义 1.3.2 一个代数系统 $(G, \circ)$ 称为一个半群, 如果它满足下面结合律:

(1) $\forall a, b, c \in G$, 有 $a \circ (b \circ c) = (a \circ b) \circ c$.

一个半群 $(G, \circ)$ 叫做一个幺半群, 如果它再满足下面条件:

(2) 在 G 中有一个元素 1(它叫做单位元或者幺元), 使得对 G 中任意元素 a, 都有 $1 \circ a = a \circ 1 = a$.

一个幺半群 $(G, \circ)$ 称作群, 如果它再满足下面条件:

(3) $\forall a \in G$, 都存在 G 中一个元素 b(叫做 a 的逆元, 记为 a^{-1}) 使得 $b \circ a = 1$.

定义 1.3.3 如果 (半) 群 G 的非空子集合 H 对于 G 的运算。也构成一个 (半) 群, 那么就称 H 为 G 的子 (半) 群.

定理 1.3.4 一个群 G 的非空子集合 H 对于 G 的运算构成子群当且仅当下列条件之一成立:

(1) $\forall a, b \in H$, $ab \in H$ 且 $b^{-1} \in H$;

(2) $\forall a, b \in H$, $ab^{-1} \in H$.

定义 1.3.5 设 H, K 是群 G 的两个子集, 则 G 的子集 $\{hk \mid h \in H, k \in K\}$ 叫做 H 与 K 的乘积, 记为 HK. 当 $H = \{a\}$ 时, HK 也记为 aK. 而当 $K = \{b\}$ 时, HK 也记为 Hb.

定理 1.3.6 设 H, K 是群 G 的两个子群, 则 HK 是 G 的子群当且仅当 $HK = KH$.

定义 1.3.7 设 N 是群 G 的子群, 如果对于任意的元素 $a \in G$, 都有 $aN = Na$, 那么就称 N 为 G 的正规子群.

定义 1.3.8 设 G_1, G_2 是两个 (半) 群, 映射 $f: G_1 \to G_2$ 称为一个 (半) 群同态, 如果 $\forall a, b \in G_1$, 都有 $f(ab) = f(a)f(b)$.

定理 1.3.9 设 G_1, G_2 是两个半群, 映射 $f: G_1 \to G_2$ 是一个半群同态, 则

(1) 如果 H_1 是 G_1 的子 (半) 群, 那么 $f^{\to}(H_1)$ 是 G_2 的子 (半) 群;

(2) 如果 H_2 是 G_2 的子 (半) 群, 那么 $f^{\leftarrow}(H_2)$ 是 G_1 的子 (半) 群.

定理 1.3.10 设 G_1, G_2 是两个群, 映射 $f: G_1 \to G_2$ 是一个群同态, 则

(1) 如果 N_2 是 G_2 的正规子群, 那么 $f^{\leftarrow}(N_2)$ 是 G_1 的正规子群;

(2) 如果 N_1 是 G_1 的正规子群, 那么当 f 是满射时, $f^{\to}(N_1)$ 是 G_2 的正规子群.

定理 1.3.11 设 N 是群 G 的正规子群, 则集族 $\{aN \mid a \in G\}$ 是一个群, 叫做 G 关于 N 的商群, 记为 G/N. 此时投射 $p: G \to G/N$ 是一个群同态.

定义 1.3.12 设 G_1, G_2 是两个 (半) 群, 在 $G_1 \times G_2$ 中定义乘法 "·" 运算如下:

$$\forall (a_1, b_1), (a_2, b_2) \in G_1 \times G_2, \quad (a_1, b_1) \cdot (a_2, b_2) = (a_1 a_2, b_1 b_2),$$

可以验证 $G_1 \times G_2$ 是一个 (半) 群, 称为 G_1 与 G_2 的直积.

定义 1.3.13 设 R 是一非空集合, 在 R 上定义了两个代数运算: 一个叫加法, 记为 $a + b$; 一个叫乘法, 记为 ab. 如果具有性质:

(1) R 对于加法成一个交换群;

(2) 乘法的结合律: 对所有的 $a, b, c \in R$,

$$a(bc) = (ab)c;$$

(3) 乘法对加法的分配律: 对所有的 $a, b, c \in R$,
$$a(b+c) = ab + ac,$$
$$(b+c)a = ba + ca,$$

那么 R 就称为一个环.

定义 1.3.14 设 S 是环 R 的一个非空子集合, 如果 S 对于 R 的两个运算也成一个环, 那么 S 称为 R 的一个子环.

定理 1.3.15 一个环 R 的非空子集合 S 对于 R 的运算构成子环当且仅当 $\forall a, b \in S, a - b, ab \in S$.

定义 1.3.16 设 R_1, R_2 是两个环, 映射 $f: R_1 \to R_2$ 称为一个环同态, 如果 $\forall a, b \in R_1$, 都有 $f(a-b) = f(a) - f(b)$ 和 $f(ab) = f(a)f(b)$ 成立.

定理 1.3.17 设 R_1, R_2 是两个环, 映射 $f: R_1 \to R_2$ 是一个环同态, 则

(1) 如果 S_1 是 R_1 的子环, 那么 $f^{\to}(S_1)$ 是 R_2 的子环;

(2) 如果 S_2 是 R_2 的子环, 那么 $f^{\leftarrow}(S_2)$ 是 R_1 的子环.

定义 1.3.18 设 R 是一环, I 是 R 的一个子集. 如果对于任意 $r \in R, a, b \in I$, 都有
$$a - b \in I, \quad ra \in I, \quad ar \in I,$$

那么称 I 为 R 的一个理想 (或称双边理想).

定理 1.3.19 设 R_1, R_2 是两个环, 映射 $f: R_1 \to R_2$ 是一个环同态, 则

(1) 如果 I_2 是 R_2 的理想, 那么 $f^{\leftarrow}(I_2)$ 是 R_1 的理想;

(2) 如果 I_1 是 R_1 的理想且 f 是满射, 那么 $f^{\to}(I_1)$ 是 R_2 的理想.

定义 1.3.20 设 R_1, R_2 是两个环, 在 $R_1 \times R_2$ 中定义加法 "+" 和乘法 "·" 运算如下: $\forall (a_1, b_1), (a_2, b_2) \in R_1 \times R_2$,
$$(a_1, b_1) + (a_2, b_2) = (a_1 + a_2, b_1 + b_2), \quad (a_1, b_1) \cdot (a_2, b_2) = (a_1 a_2, b_1 b_2),$$

可以验证 $R_1 \times R_2$ 是一个环, 称为 R_1 与 R_2 的直积.

定义 1.3.21 设 A 和 B 是环 R 的两个子集, 定义 R 的子集 $A \bullet B$ 使得它满足
$$A \bullet B = \left\{ x \,\middle|\, x = \sum_{i \in \Lambda} y_i z_i, \forall i \in \Lambda, y_i \in A, z_i \in B \right\},$$

在上述的表达式中, Λ 表示 Ω 的有限子集, 称 $A \bullet B$ 为 A 和 B 在环 R 中的内积.

定义 1.3.22 设 $\{A_i \mid i \in \Omega\}$ 是环 R 的一族子集, 定义 R 的子集 $\sum_{i \in \Omega} A_i$

使得
$$\sum_{i\in\Omega}A_i=\left\{x\,\bigg|\,x=\sum_{i\in\Lambda}x_i,\ x_i\in A_i\right\},$$

在上述的表达式中，Λ 表示 Ω 的有限子集，称 $\sum_{i\in\Omega}A_i$ 为 $\{A_i\mid i\in\Omega\}$ 在环 R 中的和.

定理 1.3.23 设 A 和 B 是交换环 R 的两个子环，则 $A\bullet B$ 是 R 的子环.

定理 1.3.24 设 $\{A_i\mid i\in\Omega\}$ 是环 R 的一族子环，则 $\sum_{i\in\Omega}A_i$ 是 R 的子环.

定义 1.3.25 设 R 为一个交换幺环，若 R 的一个理想 $P\neq R$，而且由 $ab\in P$ 恒有 $a\in P$ 或 $b\in P$，则 P 叫做 R 的一个素理想.

定义 1.3.26 如果环 F 是交换幺环，且至少含两个元素，且全体非零元素对乘法构成一个群，那么称环 F 为一个域.

如果域 F 的一个子环 S 也是一个域，则称 S 是 F 的一个子域.

定理 1.3.27 设 F_1,F_2 是两个环，映射 $f:F_1\to F_2$ 是一个域同态，则
(1) 如果 E_1 是 F_1 的子域，那么 $f^{\to}(E_1)$ 是 F_2 的子域；
(2) 如果 E_2 是 F_2 的子域，那么 $f^{\leftarrow}(E_2)$ 是 F_1 的子域.

定义 1.3.28 设 V 是一个带有加法 (记作 "+") 运算的非空集合，F 是一个域. 如果 V 关于加法运算构成一个交换群，并且对每个 $k\in F,v\in V$，在 V 中可唯一地确定一个元素 kv(称为 k 与 v 的标量乘法)，使得对所有的 $k,l\in F$，$u,v\in V$，满足

(M1) $(kl)v=k(lv)$;

(M2) $(k+l)v=kv+lv$;

(M3) $k(u+v)=ku+kv$;

(M4) $1v=v$,

则称 V 为域 F 上的一个向量空间或线性空间.

定义 1.3.29 设 V 是域 F 上的向量空间，U 是 V 的非空子集. 如果 U 关于 V 的运算也构成 F 上的向量空间，则称 U 为 V 的子空间.

定义 1.3.30 设 V,V' 是域 F 上的两个向量空间，若存在映射 $f:V\to V'$ 满足下列条件：
(1) 对任意 $\alpha,\beta\in V$，有 $f(\alpha+\beta)=f(\alpha)+f(\beta)$;
(2) 对任意 $k\in F,\alpha\in V$，有 $f(k\alpha)=kf(\alpha)$,
则称 f 是从 V 到 V' 的线性映射.

定理 1.3.31 设 V,V' 是域 F 上的两个向量空间，映射 $f:V\to V'$ 是一个线性映射，则

(1) 如果 W 是 V 的子空间,那么 $f^{\rightarrow}(W)$ 是 V' 的子空间;

(2) 如果 W' 是 V' 的子空间,那么 $f^{\leftarrow}(W')$ 是 V 的子空间.

习 题 1

1. 证明命题 1.1.9.

2. 证明命题 1.1.10.

3. 试举例说明一般 $f^{\rightarrow}\left(\bigcap\limits_{i\in\Omega} A_i\right) \neq \bigcap\limits_{i\in\Omega} f^{\rightarrow}(A_i)$.

4. 试举例说明一般 $f^{\rightarrow}(A') \neq (f^{\rightarrow}(A))'$.

5. 设 $f: X \to Y$ 是映射,$A \in \mathbf{2}^X$,$B \in \mathbf{2}^Y$. 试证明

(1) $f(A \cap f^{\leftarrow}(B)) = f(A) \cap B$;

(2) $f(A - f^{\leftarrow}(B)) = f(A) - B$.

6. 证明一个完备格是完全分配格当且仅当其每个元的极小集存在. (详细证明可见文献 [20].)

7. 证明一个完备格是完全分配格当且仅当其每个元的极大集存在. (详细证明可见文献 [20].)

8. 证明在一个完全分配格 L 中,极小映射 β 是一个保并映射,也就是说,$\forall \{a_i\}_{i\in\Omega} \subseteq L$,$\beta\left(\bigvee\limits_{i\in\Omega} a_i\right) = \bigcup\limits_{i\in\Omega} \beta(a_i)$.

9. 证明在一个完全分配格 L 中,极大映射 α 是一个 $\bigwedge$-$\bigcup$ 映射,也就是说,$\forall \{a_i\}_{i\in\Omega} \subseteq L$,$\alpha\left(\bigwedge\limits_{i\in\Omega} a_i\right) = \bigcup\limits_{i\in\Omega} \alpha(a_i)$.

10. 在一个完全分配格 L 中,极小映射 β 满足下面插入性质:对任意的 $a, b \in L$,当 $a \in \beta(b)$ 时,存在 $c \in \beta(b)$ 使得 $a \in \beta(c)$.

11. 在一个完全分配格 L 中,极大映射 α 满足下面插入性质:对任意的 $a, b \in L$,当 $a \in \alpha(b)$ 时,存在 $c \in \alpha(b)$ 使得 $a \in \alpha(c)$.

12. 证明定理 1.2.12.

13. 在一个完全分配格 L 中,验证 $b = \bigvee \beta^*(b)$.

14. 在一个完全分配格 L 中,验证 $b = \bigwedge \alpha^*(b)$.

15. 证明定理 1.3.9.

16. 证明定理 1.3.10.

17. 证明定理 1.3.17.

18. 证明定理 1.3.19.

19. 证明定理 1.3.23.

20. 证明定理 1.3.24.

21. 证明定理 1.3.27.

22. 证明定理 1.3.31.

23. 试举例说明定义 1.3.5 与定义 1.3.21 是不同的.

第 2 章 L-模糊集的表现定理

模糊集是由 Zadeh L A 于 1965 年首次在文献 [166] 中提出的. 随后被 Goguen J A 在 [77] 中推广到值为格的情形, 并称为格值模糊集, 简称 L-模糊集. 从模糊集理论诞生之日起, 其成果便得到爆炸式高速发展, 不但被广大数学工作者推广到数学的各个分支中, 而且也被应用到许许多多的应用科学中. 作者在 1986 年读硕士研究生二年级时, 恰好学院聘请北京师范大学的罗承忠老师为我们讲授 "模糊数学" 课程, 罗老师的言谈举止至今令作者记忆犹新, 也让作者终身受益. 在罗老师的课程中, 集合套思想贯穿始终, 其作用及重要性是众所周知的, 它使许多模糊概念变得直观, 同时也使模糊集理论更严密且更系统化. 后来罗老师把他的讲义整理出版, 见文献 [36]. 集合套理论是研究模糊集各种性质的有力工具, 因此进一步推广到格值模糊集上也是非常必要的. 不管怎么样, 作者觉得 L 集合套的表现略有不足, 其原因是 L 集合套定义形式上多了个条件且仅适用于稠密的完备格. 正因如此, 作者在 L 是完全分配格时, 借助极小集与极大集, 在文献 [25] 中引入了 L_β 集合套与 L_α 集合套两种概念. 本章将首先介绍这两种集合套的基本概念和表现定理, 后续几章的结果始终把它们作为主要工具.

2.1—2.3 节和 2.5 节的主要结果均来自于文献 [25-27]. 2.4 节来源于文献 [131].

2.1 L-模糊集的分解定理

定义 2.1.1 设 X 是一个非空集合, L 是一个完备格, 称映射 $A: X \to L$ 为 X 的一个格值模糊子集, 简称 L-模糊集. 当 $L = [0,1]$ 时, 就称 A 为一个模糊集. X 上的 L-模糊集的全体记为 L^X.

对于两个 L-模糊集 $A, B \in L^X$, 规定

$$A \leqslant B \text{ 当且仅当 } \forall x \in X, \text{ 均有} A(x) \leqslant B(x).$$

另外, 对于一族 L-模糊集 $\{A_i \mid i \in \Omega\} \subseteq L^X$, 我们能够点态地定义它的两种运算如下: $\forall x \in X$,

$$\left(\bigvee_{i \in \Omega} A_i\right)(x) = \bigvee_{i \in \Omega} A_i(x), \quad \left(\bigwedge_{i \in \Omega} A_i\right)(x) = \bigwedge_{i \in \Omega} A_i(x).$$

于是按照上面的运算, L^X 也构成一个完备格. 如果 L 是完全分配的, 那么 L^X 也是完全分配的.

下面我们提到的 L-模糊集中的 L 均为完全分配格. 借助 L 中的极小集和极大集概念, 我们先给出下列四种截集的定义.

定义 2.1.2 设 $A \in L^X, a \in L$. 定义

$$A_{[a]} = \{x \in X \mid a \leqslant A(x)\}, \quad A^{(a)} = \{x \in X \mid A(x) \nleqslant a\},$$

$$A_{(a)} = \{x \in X \mid a \in \beta(A(x))\}, \quad A^{[a]} = \{x \in X \mid a \notin \alpha(A(x))\}.$$

从极小集和极大集的性质可以很容易地验证下面两条件成立:

$$a \in \beta(b) \Rightarrow A_{[b]} \subseteq A_{(a)} \subseteq A_{[a]}, \quad a \in \alpha(b) \Rightarrow A^{[a]} \subseteq A^{(b)} \subseteq A^{[b]}.$$

读者应该知道这样的事实: 上述四种截集的前两个是早有的, 后两个才是我们在文献 [25–27] 中首次提出的.

下面定理给出了几种截集的性质.

定理 2.1.3 对于一族 L-模糊集 $\{A_i \mid i \in \Omega\} \subseteq L^X$ 和任意的 $a \in L$, 下面条件成立:

(1) $\left(\bigwedge_{i \in \Omega} A_i\right)_{[a]} = \bigcap_{i \in \Omega} (A_i)_{[a]}$;

(2) $\left(\bigvee_{i \in \Omega} A_i\right)_{(a)} = \bigcup_{i \in \Omega} (A_i)_{(a)}$;

(3) $\left(\bigwedge_{i \in \Omega} A_i\right)^{[a]} = \bigcap_{i \in \Omega} (A_i)^{[a]}$;

(4) $\left(\bigvee_{i \in \Omega} A_i\right)^{(a)} = \bigcup_{i \in \Omega} (A_i)^{(a)}$.

证明 (1) 和 (4) 的证明在其他专著中均有结论, 这里我们就略去证明, 仅证明 (2) 和 (3).

如果 $A, B \in L^X$ 且 $A \leqslant B$, 那么容易看到 $A_{(a)} \subseteq B_{(a)}$. 因此, 我们有

$$\left(\bigvee_{i \in \Omega} A_i\right)_{(a)} \supseteq \bigcup_{i \in \Omega} (A_i)_{(a)}.$$

为了证明 (2), 我们只需要证明

$$\left(\bigvee_{i \in \Omega} A_i\right)_{(a)} \subseteq \bigcup_{i \in \Omega} (A_i)_{(a)}.$$

为此，假设 $x \in \left(\bigvee_{i \in \Omega} A_i\right)_{(a)}$，那么

$$a \in \beta\left(\bigvee_{i \in \Omega} A_i(x)\right) = \bigcup_{i \in \Omega} \beta(A_i(x)),$$

从而存在 $i \in \Omega$ 使得 $a \in \beta(A_i(x))$，也就是 $x \in (A_i)_{(a)} \subseteq \bigcup_{i \in \Omega} (A_i)_{(a)}$. 这样就完成了 (2) 的证明.

为了证明 (3)，我们先证明下面结论.

如果 $A, B \in L^X$ 且 $A \leqslant B$，那么 $A^{[a]} \subseteq B^{[a]}$.

假如 $x \notin B^{[a]}$，那么 $a \in \alpha(B(x))$. 由 $A \leqslant B$ 我们知道 $\alpha(B(x)) \subseteq \alpha(A(x))$. 因此 $a \in \alpha(A(x))$，这意味着 $x \notin A^{[a]}$. 所以我们有 $A^{[a]} \subseteq B^{[a]}$. 由此结论我们可以知道

$$\left(\bigwedge_{i \in \Omega} A_i\right)^{[a]} \subseteq \bigcap_{i \in \Omega} (A_i)^{[a]}.$$

这样，为了证明 (3)，我们只需要证明

$$\left(\bigwedge_{i \in \Omega} A_i\right)^{[a]} \supseteq \bigcap_{i \in \Omega} (A_i)^{[a]}.$$

为此，我们设 $x \notin \left(\bigwedge_{i \in \Omega} A_i\right)^{[a]}$，则

$$a \in \alpha\left(\left(\bigwedge_{i \in \Omega} A_i\right)(x)\right) = \bigcup_{i \in \Omega} \alpha(A_i(x)),$$

这意味着 $\exists i \in \Omega$ 使得 $a \in \alpha(A_i(x))$，也就是 $x \notin (A_i)^{[a]}$，从而 $x \notin \bigcap_{i \in \Omega} (A_i)^{[a]}$. 这表明

$$\left(\bigwedge_{i \in \Omega} A_i\right)^{[a]} \supseteq \bigcap_{i \in \Omega} (A_i)^{[a]}. \qquad \square$$

对于 $a \in L$ 和 $D \subseteq X$，我们定义两个 L-模糊集 $a \wedge D$ 和 $a \vee D$ 如下:

$$(a \wedge D)(x) = \begin{cases} a, & x \in D, \\ 0, & x \notin D, \end{cases} \quad (a \vee D)(x) = \begin{cases} 1, & x \in D, \\ a, & x \notin D. \end{cases}$$

我们将不区别一个分明集合 A 和它的特征函数 χ_A. 于是可以得到下面的分解定理.

定理 2.1.4 (分解定理) $\forall A \in L^X$, 下面条件成立.

(1) $A = \bigvee_{a \in L} (a \wedge A_{[a]}) = \bigvee_{a \in J(L)} (a \wedge A_{[a]});$

(2) $A = \bigvee_{a \in L} (a \wedge A_{(a)}) = \bigvee_{a \in J(L)} (a \wedge A_{(a)});$

(3) $A = \bigwedge_{a \in L} (a \vee A^{[a]}) = \bigwedge_{a \in P(L)} (a \vee A^{[a]});$

(4) $A = \bigwedge_{a \in L} (a \vee A^{(a)}) = \bigwedge_{a \in P(L)} (a \vee A^{(a)});$

(5) $\forall a \in L, A_{[a]} = \bigcap_{b \in \beta(a)} A_{[b]} = \bigcap_{b \in \beta(a)} A_{(b)} = \bigcap_{b \in \beta^*(a)} A_{[b]} = \bigcap_{b \in \beta^*(a)} A_{(b)};$

(6) $\forall a \in L, A_{(a)} = \bigcup_{a \in \beta(b)} A_{[b]} = \bigcup_{a \in \beta(b)} A_{(b)} = \bigcup_{\substack{a \in \beta(b) \\ b \in J(L)}} A_{[b]} = \bigcup_{\substack{a \in \beta(b) \\ b \in J(L)}} A_{(b)};$

(7) $\forall a \in L, A^{[a]} = \bigcap_{a \in \alpha(b)} A^{[b]} = \bigcap_{a \in \alpha(b)} A^{(b)} = \bigcap_{\substack{a \in \alpha(b) \\ b \in P(L)}} A^{[b]} = \bigcap_{\substack{a \in \alpha(b) \\ b \in P(L)}} A^{(b)};$

(8) $\forall a \in L, A^{(a)} = \bigcup_{b \in \alpha(a)} A^{[b]} = \bigcup_{b \in \alpha(a)} A^{(b)} = \bigcup_{b \in \alpha^*(a)} A^{[b]} = \bigcup_{b \in \alpha^*(a)} A^{(b)};$

(9) $A_{[a]} = \bigcap_{b \not\geq a} A^{(b)} = \bigcap_{b \not\geq a} A^{[b]} = \bigcap_{\substack{b \in P(L) \\ b \not\geq a}} A^{(b)} = \bigcap_{\substack{b \in P(L) \\ b \not\geq a}} A^{[b]};$

(10) $A^{(a)} = \bigcup_{b \not\leq a} A_{(b)} = \bigcup_{b \not\leq a} A_{[b]}.$

证明 (1) 的证明由下面等式可得.

$$A(x) = \bigvee \{a \in J(L) \mid a \leqslant A(x)\} = \bigvee \{a \in J(L) \mid A_{[a]}(x) = 1\}$$
$$= \bigvee_{a \in J(L)} (a \wedge A_{[a]}(x)) \quad (\text{这里的 } J(L) \text{ 能够被 } L \text{ 取代}).$$

(2) 成立由下式可证.

$$A(x) = \bigvee \{a \in J(L) \mid a \in \beta^*(A(x))\} = \bigvee \{a \in J(L) \mid A_{(a)}(x) = 1\}$$
$$= \bigvee_{a \in J(L)} (a \wedge A_{(a)}(x)) \quad (\text{这里的 } J(L) \text{ 能够被 } L \text{ 取代}).$$

(3) 的证明由下式可得.

$$A(x) = \bigwedge \{a \in L \mid a \in \alpha(A(x))\} = \bigwedge \{a \in L \mid A^{[a]}(x) = 0\}$$
$$= \bigwedge_{a \in L} (a \vee A^{[a]}(x)) \quad (\text{这里的 } L \text{ 能够被 } P(L) \text{ 取代}).$$

2.1 L-模糊集的分解定理

(4) 的证明从下式可见.

$$A(x) = \bigwedge\{a \in L \mid A(x) \leqslant a\} = \bigwedge\{a \in L \mid A^{(a)}(x) = 0\}$$
$$= \bigwedge_{a \in L}(a \vee A^{(a)}(x)) \quad \text{(这里的 } L \text{ 能够被 } P(L) \text{ 取代)}.$$

(5) $\forall b \in \beta(a)$ ($b \in \beta^*(a)$ 情形的证明相同), 由 $A_{[a]} \subseteq A_{(b)} \subseteq A_{[b]}$ 可知

$$A_{[a]} \subseteq \bigcap_{b \in \beta(a)} A_{(b)} \subseteq \bigcap_{b \in \beta(a)} A_{[b]}.$$

而由

$$x \in \bigcap_{b \in \beta(a)} A_{[b]} \Rightarrow A(x) \geqslant \bigvee\{b \in L \mid b \in \beta(a)\} = a \Rightarrow x \in A_{[a]},$$

我们知道 $\bigcap_{b \in \beta(a)} A_{[b]} \subseteq A_{[a]}$, 于是完成了 (5) 的证明.

(6) 对于满足条件 $a \in \beta(b)$ 的任意 $b \in L$ ($b \in J(L)$ 时的证明相同), 由 $A_{(b)} \subseteq A_{[b]} \subseteq A_{(a)}$ 可知

$$\bigcup_{a \in \beta(b)} A_{(b)} \subseteq \bigcup_{a \in \beta(b)} A_{[b]} \subseteq A_{(a)}.$$

为了证明 $\bigcup_{a \in \beta(b)} A_{(b)} \supseteq A_{(a)}$, 取 $x \in A_{(a)}$, 则 $a \in \beta(A(x))$. 因为

$$\beta(A(x)) = \beta\left(\bigvee\{c \in L \mid c \in \beta(A(x))\}\right) = \bigcup\{\beta(c) \mid c \in \beta(A(x))\},$$

所以存在 $c \in \beta(A(x))$ 使得 $a \in \beta(c)$, 也就是说存在 $c \in L$ 使得 $a \in \beta(c)$ 且 $x \in A_{(c)}$, 进一步得到 $x \in \bigcup_{a \in \beta(b)} A_{(b)}$, 这说明 $\bigcup_{a \in \beta(b)} A_{(b)} \supseteq A_{(a)}$ 是成立的, 完成了 (6) 的证明.

(7) 对于满足条件 $a \in \alpha(b)$ 的任意 $b \in L$ ($b \in P(L)$ 时的证明相同), 由 $A^{[a]} \subseteq A^{(b)} \subseteq A^{[b]}$ 可知

$$A^{[a]} \subseteq \bigcap_{a \in \alpha(b)} A^{(b)} \subseteq \bigcap_{a \in \alpha(b)} A^{[b]}.$$

为了证明 (7), 我们只需要证明 $A^{[a]} \supseteq \bigcap_{a \in \alpha(b)} A^{[b]}$ 即可. 取 $x \notin A^{[a]}$, 则 $a \in \alpha(A(x))$. 因为

$$\alpha(A(x)) = \alpha\left(\bigwedge\{c \in L \mid c \in \alpha(A(x))\}\right) = \bigcup\{\alpha(c) \mid c \in \alpha(A(x))\},$$

所以存在 $c \in \alpha(A(x))$ 使得 $a \in \alpha(c)$, 也就是说存在 $c \in L$ 使得 $a \in \alpha(c)$ 且 $x \notin A^{[c]}$, 进一步得到 $x \notin \bigcap_{a \in \alpha(b)} A^{[b]}$, 因此 $A^{[a]} \supseteq \bigcap_{a \in \alpha(b)} A^{[b]}$, 这样就证明了 (7).

(8) $\forall b \in \alpha(a)$ ($b \in \alpha^*(a)$ 情形的证明相同), 由 $A^{(b)} \subseteq A^{[b]} \subseteq A^{(a)}$ 可知
$$\bigcup_{b \in \alpha(a)} A^{(b)} \subseteq \bigcup_{b \in \alpha(a)} A^{[b]} \subseteq A^{(a)}.$$

现在我们只需证明 $\bigcup_{b \in \alpha(a)} A^{(b)} \supseteq A^{(a)}$ 即可. 设 $x \in A^{(a)}$, 则
$$A(x) \not\leqslant a = \bigwedge \{b \in L \mid b \in \alpha(a)\},$$

这意味着存在 $b \in \alpha(a)$ 使得 $A(x) \not\leqslant b$, 也就是 $x \in A^{(b)}$. 因此 $x \in \bigcup_{b \in \alpha(a)} A^{(b)}$. 这证明了 $\bigcup_{b \in \alpha(a)} A^{(b)} \supseteq A^{(a)}$.

(9) 设 $x \in A_{[a]}$, 则由 (4) 知 $\forall b \in L$(或者 $b \in P(L)$), 都有 $b \vee A^{(b)}(x) \geqslant a$. 于是 $\forall b \not\geqslant a, x \in A^{(b)}$, 从而 $A_{[a]} \subseteq \bigcap_{b \not\geqslant a} A^{(b)} \subseteq \bigcap_{b \not\geqslant a} A^{[b]}$.

反之, 为了证明 $\bigcap_{b \not\geqslant a} A^{[b]} \subseteq A_{[a]}$, 我们假设 $x \notin A_{[a]}$, 那么 $A(x) \not\geqslant a$, 也就是
$$A(x) = \bigwedge \{b \in L \mid b \in \alpha^*(A(x))\} = \bigwedge \{b \in L \mid b \in \alpha(A(x))\} \not\geqslant a.$$

这意味着存在 $b \in \alpha(A(x))$ (或者 $b \in \alpha^*(A(x))$) 使得 $b \not\geqslant a$, 这等价于说存在 $b \not\geqslant a$ 使得 $x \notin A^{[b]}$, 即 $x \notin \bigcap_{b \not\geqslant a} A^{[b]}$.

(10) 设 $x \in A^{(a)}$, 则由 (2) 知存在 $b \in L$ 使 $b \wedge A_{(b)}(x) \not\leqslant a$. 于是 $b \not\leqslant a$ 且 $x \in A_{(b)}$. 从而 $A^{(a)} \subseteq \bigcup_{b \not\leqslant a} A_{(b)} \subseteq \bigcup_{b \not\leqslant a} A_{[b]}$. 反之, 为了证明 $\bigcup_{b \not\leqslant a} A_{[b]} \subseteq A^{(a)}$, 我们假设 $x \in \bigcup_{b \not\leqslant a} A_{[b]}$, 那么存在 $b \not\leqslant a$ 使得 $x \in A_{[b]}$, 也就是 $A(x) \geqslant b$, 从而 $A(x) \not\leqslant a$, 于是 $x \in A^{(a)}$. □

由上述分解定理, 我们不难得到下面结论.

定理 2.1.5 设 $A \in L^X$, 且 $H: L \to 2^X$ 是映射.

(1) 若 $\forall a \in L, A_{(a)} \subseteq H(a) \subseteq A_{[a]}$, 则 $A = \bigvee_{a \in L} (a \wedge H(a))$;

(2) 若 $\forall a \in L, A^{(a)} \subseteq H(a) \subseteq A^{[a]}$, 则 $A = \bigwedge_{a \in L} (a \vee H(a))$.

2.2 表现定理 I

设 $\mathscr{P}(X)^L$ 表示映射 $H: L \to \mathscr{P}(X)$ 的全体, 则由 $\mathscr{P}(X)$ 是完全分配格可知, $\mathscr{P}(X)^L$ 从 $\mathscr{P}(X)$ 中点式地诱导出格运算 "$\leqslant, \vee, \wedge$" 使得 $(\mathscr{P}(X)^L, \leqslant, \vee, \wedge)$ 仍是完全分配格.

定义 2.2.1 设 $H \in \mathscr{P}(X)^L$,
(1) 若 $a \in \beta(b) \Rightarrow H(b) \subseteq H(a)$, 则称 H 为 X 上的 L_β 集合套;
(2) 若 $a \in \alpha(b) \Rightarrow H(a) \subseteq H(b)$, 则称 H 为 X 上的 L_α 集合套.

显然当 $L = [0,1]$ 时, L_β 集合套与 L_α 集合套就是 [36] 中罗承忠老师的集合套.

定理 2.2.2 对 $H \in \mathscr{P}(X)^L$,
(1) 若 $\forall a \in L, A_{(a)} \subseteq H(a) \subseteq A_{[a]}$, 则 H 是 L_β 集合套;
(2) 若 $\forall a \in L, A^{(a)} \subseteq H(a) \subseteq A^{[a]}$, 则 H 是 L_α 集合套.

下面我们分别以 $\mathscr{U}_{L_\beta}(X)$ 与 $\mathscr{U}_{L_\alpha}(X)$ 记 X 上的 L_β 集合套全体与 L_α 集合套全体.

定理 2.2.3 $\mathscr{U}_{L_\beta}(X)$ 与 $\mathscr{U}_{L_\alpha}(X)$ 皆是 $\mathscr{P}(X)^L$ 的完备子格, 从而皆是完全分配格.

证明 只需证 $\mathscr{U}_{L_\beta}(X)$ (或 $\mathscr{U}_{L_\alpha}(X)$) 对任意交与并封闭. 设 $\{H_t\}_{t \in T} \subseteq \mathscr{U}_{L_\beta}(X)$, 则 $\forall a, b \in L$, 当 $a \in \beta(b)$ 时, 有

$$\left(\bigwedge_{t \in T} H_t\right)(b) = \bigcap_{t \in T} H_t(b) \subseteq \bigcap_{t \in T} H_t(a) = \left(\bigwedge_{t \in T} H_t\right)(a),$$

即 $\bigwedge_{t \in T} H_t \in \mathscr{U}_{L_\beta}(X)$. 类似可证 $\bigvee_{t \in T} H_t \in \mathscr{U}_{L_\beta}(X)$. 于是 $\mathscr{U}_{L_\beta}(X)$ 是 $\mathscr{P}(X)^L$ 的完备子格, 从而是完全分配格. 同样可证 $\mathscr{U}_{L_\alpha(X)}$ 亦然. □

注 2.2.4 [36] 与 [17] 中的 L 集合套全体一般不是 $\mathscr{P}(X)^L$ 的完备子格, 因为 $\mathscr{U}_L(X)$ 对并运算一般不封闭.

定理 2.2.5 对 $H \in \mathscr{U}_{L_\beta}(X)$, 令 $f(H) = \bigvee_{a \in L}(a \wedge H(a))$, 则

(1) $f(H)_{(a)} \subseteq H(a) \subseteq f(H)_{[a]}$;
(2) $f(H)_{[a]} = \bigcap_{b \in \beta(a)} H(b)$;
(3) $f(H)_{(a)} = \bigcup_{a \in \beta(b)} H(b)$;
(4) f 是 $\mathscr{U}_{L_\beta}(X)$ 到 L^X 的同态满射.

证明 (1) 下面蕴含式表明 $f(H)_{(a)} \subseteq H(a)$.

$$x \in f(H)_{(a)} \Rightarrow a \in \beta(f(H)(x))$$
$$\Rightarrow \exists b \in L \text{ 使 } a \in \beta(b \wedge H(b)(x))$$
$$\Rightarrow a \in \beta(b) \text{ 且 } x \in H(b)$$
$$\Rightarrow x \in H(a).$$

而 $H(a) \subseteq f(H)_{[a]}$ 是显然的.

(2) 与 (3) 由定理 2.1.4 中 (5) 与 (6) 可知.

(4) $\forall A \in L^X, \forall a \in L$, 令 $H(a) = A_{(a)}$, 则 $H \in \mathscr{U}_{L_\beta}(X)$. 由定理 2.1.4 知 $A = f(H)$, 故 f 是满射. 下证 f 保交与并运算.

设 $\{H_t\}_{t \in T} \subseteq \mathscr{U}_{L_\beta}(X)$, 则由 $\bigwedge\limits_{t \in T} H_t \in \mathscr{U}_{L_\beta}(X)$ 与 (2) 知, 对于任意的 $a \in L$,

$$\left(f\left(\bigwedge_{t \in T} H_t\right)\right)_{[a]} = \bigcap_{b \in \beta(a)} \left(\bigwedge_{t \in T} H_t\right)(b) = \bigcap_{b \in \beta(a)} \bigcap_{t \in T} H_t(b)$$
$$= \bigcap_{t \in T} \bigcap_{b \in \beta(a)} H_t(b) = \bigcap_{t \in T} (f(H_t))_{[a]}$$
$$= \left(\bigwedge_{t \in T} f(H_t)\right)_{[a]}.$$

故由定理 2.1.4 知, $f\left(\bigwedge\limits_{t \in T} H_t\right) = \bigwedge\limits_{t \in T} f(H_t)$, 即 f 保交. 类似地, 借助 (3) 可证 f 保并. □

借助极大集性质与定理 2.1.4 不难得到下面的结论.

定理 2.2.6 对 $H \in \mathscr{U}_{L_\alpha}(X)$, 令 $g(H) = \bigwedge\limits_{a \in L}(a \vee H(a))$, 则

(1) $g(H)^{(a)} \subseteq H(a) \subseteq g(H)^{[a]}$;

(2) $g(H)^{[a]} = \bigcap\limits_{a \in \alpha(b)} H(b)$;

(3) $g(H)^{(a)} = \bigcup\limits_{b \in \alpha(a)} H(b)$;

(4) g 是从 $\mathscr{U}_{L_\alpha}(X)$ 到 L^X 的同态满射.

定义 2.2.7 设 $H_1, H_2 \in \mathscr{U}_{L_\beta}(X)$, 若 $\forall a \in L, \bigcap\limits_{b \in \beta(a)} H_1(b) = \bigcap\limits_{b \in \beta(a)} H_2(b)$, 则称 H_1 与 H_2 在 $\mathscr{U}_{L_\beta}(X)$ 中等价.

定义 2.2.8 设 $H_3, H_4 \in \mathscr{U}_{L_\alpha}(X)$, 若 $\forall a \in L, \bigcap\limits_{a \in \alpha(b)} H_3(b) = \bigcap\limits_{a \in \alpha(b)} H_4(b)$, 则称 H_3 与 H_4 在 $\mathscr{U}_{L_\alpha}(X)$ 中等价.

2.2 表现定理 I

上述定义的两种等价均是等价关系.

例 2.2.9 对一个 L-模糊集 $A \in L^X$, 令 $H_1(a) = A_{(a)}, H_2(a) = A_{[a]}, H_3(a) = A^{(a)}, H_4(a) = A^{[a]}$, 那么 H_1 与 H_2 是两个等价的 L_β 集合套, H_3 与 H_4 是两个等价的 L_α 集合套.

定理 2.2.10 设 $H_1 \in \mathscr{U}_{L_\beta}(X), H_2 \in \mathscr{U}_{L_\alpha}(X)$, 则 $\forall a \in L$,

(1) $\bigcup\limits_{a \in \beta(b)} H_1(b) = \bigcup\limits_{a \in \beta(b)} \bigcap\limits_{c \in \beta(b)} H_1(c)$;

(2) $\bigcap\limits_{b \in \beta(a)} H_1(b) = \bigcap\limits_{b \in \beta(a)} \bigcup\limits_{b \in \beta(c)} H_1(c)$;

(3) $\bigcup\limits_{b \in \alpha(a)} H_2(b) = \bigcup\limits_{b \in \alpha(a)} \bigcap\limits_{b \in \alpha(c)} H_2(c)$;

(4) $\bigcap\limits_{a \in \alpha(b)} H_2(b) = \bigcap\limits_{a \in \alpha(b)} \bigcup\limits_{c \in \alpha(b)} H_2(c)$.

证明 (1) 从下面蕴含式

$$c \in \beta(b) \Rightarrow H_1(b) \subseteq H_1(c) \Rightarrow H_1(b) \subseteq \bigcap\limits_{c \in \beta(b)} H_1(c),$$

我们可得

$$\bigcup\limits_{a \in \beta(b)} H_1(b) \subseteq \bigcup\limits_{a \in \beta(b)} \bigcap\limits_{c \in \beta(b)} H_1(c).$$

反之, 由

$$a \in \beta(b) = \beta\left(\bigvee \beta(b)\right) = \bigcup\limits_{r \in \beta(b)} \beta(r)$$

可知, 存在 $r \in \beta(b)$ 使 $a \in \beta(r)$. 于是

$$\bigcap\limits_{c \in \beta(b)} H_1(c) \subseteq H_1(r) \subseteq \bigcup\limits_{a \in \beta(b)} H_1(b).$$

从而

$$\bigcup\limits_{a \in \beta(b)} \bigcap\limits_{c \in \beta(b)} H_1(c) \subseteq \bigcup\limits_{a \in \beta(b)} H_1(b).$$

得证 (1). 由极小集与极大集性质同样可证 (2), (3) 与 (4). □

推论 2.2.11 设 $H_1, H_2 \in \mathscr{U}_{L_\beta}(X), H_3, H_4 \in \mathscr{U}_{L_\alpha}(X)$, 则

(1) H_1 与 H_2 等价 $\Leftrightarrow \forall a \in L, \bigcup\limits_{a \in \beta(b)} H_1(b) = \bigcup\limits_{a \in \beta(b)} H_2(b)$;

(2) H_3 与 H_4 等价 $\Leftrightarrow \forall a \in L, \bigcup\limits_{b \in \alpha(a)} H_3(b) = \bigcup\limits_{b \in \alpha(a)} H_4(b)$.

由于 $\mathscr{U}_{L_\beta}(X)$ 与 $\mathscr{U}_{L_\alpha}(X)$ 中的等价是等价关系，所以它们分别把 $\mathscr{U}_{L_\beta}(X)$ 与 $\mathscr{U}_{L_\alpha}(X)$ 分类，即分成若干等价类，分别以 $\widetilde{\mathscr{U}_{L_\beta}}(X)$ 与 $\widetilde{\mathscr{U}_{L_\alpha}}(X)$ 记它们的等价类全体. 对 $H \in \widetilde{\mathscr{U}_{L_\beta}}(X)$(或 $\widetilde{\mathscr{U}_{L_\alpha}}(X)$)，以 $[H]$ 表示 H 所在的等价类.

设 $[H_t] \in \widetilde{\mathscr{U}_{L_\beta}}(X)$(或 $\widetilde{\mathscr{U}_{L_\alpha}}(X)$)$(t \in T)$，令

$$\bigwedge_{t \in T}[H_t] = \left[\bigwedge_{t \in T} H_t\right], \quad \bigvee_{t \in T}[H_t] = \left[\bigvee_{t \in T} H_t\right].$$

由下面定理知这样的规定是合理的.

定理 2.2.12 设 $E_t, H_t \in \mathscr{U}_{L_\beta}(X)$(或 $\mathscr{U}_{L_\alpha}(X)$) 且 E_t 与 H_t 等价，这里 $t \in T$，则 $\bigwedge_{t \in T} E_t$ 与 $\bigwedge_{t \in T} H_t$ 等价，$\bigvee_{t \in T} E_t$ 与 $\bigvee_{t \in T} H_t$ 等价.

证明 由定义 2.2.7、定义 2.2.8 及推论 2.2.11 易得. □

定理 2.2.13 $(\widetilde{\mathscr{U}_{L_\beta}}(X), \vee, \wedge) \cong (\widetilde{\mathscr{U}_{L_\alpha}}(X), \vee, \wedge) \cong (L^X, \vee, \wedge)$，这里 $\cong$ 表示格同构.

证明 $\forall [H_1] \in \widetilde{\mathscr{U}_{L_\beta}}(X), \forall [H_2] \in \widetilde{\mathscr{U}_{L_\alpha}}(X)$，令

$$f([H_1]) = \bigvee_{a \in L}(a \wedge H_1(a)),$$

$$g([H_2]) = \bigwedge_{a \in L}(a \vee H_2(a)).$$

由定理 2.2.5 与定理 2.2.6 易见 f, g 皆是同构. □

2.3 表现定理 II

定义 2.3.1 设 $F, G \in \mathscr{P}(X)^L$，称
(1) F 为 L_β 集轮，如果 $\forall a \in L$，$F(a) = \bigcap_{b \in \beta(a)} F(b)$;
(2) F 为 L_α 集轮，如果 $\forall a \in L$，$F(a) = \bigcap_{a \in \alpha(b)} F(b)$;
(3) G 为 L_β 开集轮，如果 $\forall a \in L$，$G(a) = \bigcup_{a \in \beta(b)} G(b)$;
(4) G 为 L_α 开集轮，如果 $\forall a \in L$，$G(a) = \bigcup_{b \in \alpha(a)} G(b)$.

显然，L_β(开) 集轮是 L_β 集合套，L_α(开) 集轮是 L_α 集合套.

2.3 表现定理 II

例 2.3.2 对一个 L-模糊集 $A \in L^X$, 令 $H_1(a) = A_{(a)}, H_2(a) = A_{[a]}, H_3(a) = A^{(a)}, H_4(a) = A^{[a]}$, 那么 H_1 是 L_β 开集轮, H_2 是 L_α 集轮, H_3 是 L_α(开) 集轮, H_4 是 L_α 开集轮.

定理 2.3.3 设 $H_1 \in \mathscr{U}_{L_\beta}(X), H_2 \in \mathscr{U}_{L_\alpha}(X), \forall a \in L$, 令

$$F_{H_1}(a) = \bigcap_{b \in \beta(a)} H_1(b), \qquad F_{H_2}(a) = \bigcap_{a \in \alpha(b)} H_2(b),$$

$$G_{H_1}(a) = \bigcup_{a \in \beta(b)} H_1(b), \qquad G_{H_2}(a) = \bigcup_{b \in \alpha(a)} H_2(b),$$

则 F_{H_1}, F_{H_2} 分别是 L_β 集轮与 L_α 集轮, G_{H_1}, G_{H_2} 分别是 L_β 开集轮与 L_α 开集轮.

证明 由

$$\bigcap_{b \in \beta(a)} F_{H_1}(b) = \bigcap_{b \in \beta(a)} \bigcap_{c \in \beta(b)} H_1(c) = \bigcap_{c \in \beta(a)} H_1(c) = F_{H_1}(a)$$

可知 F_{H_1} 是 L_β 集轮, 其余类似可证. □

记全体 L_β 集轮 (L_α 集轮) 为 $\Phi_{L_\beta}(X)(\Phi_{L_\alpha}(X))$, 全体 L_β 开集轮 (L_α 开集轮) 为 $\Psi_{L_\beta}(X)(\Psi_{L_\alpha}(X))$. 易证 $\Phi_{L_\beta}(X)$(或 $\Phi_{L_\alpha}(X)$) 对 $\mathscr{P}(X)^L$ 的交封闭且它们皆有最大元 F(这里 $F(a) = X, \forall a \in L$), $\Psi_{L_\beta}(X)$(或 $\Psi_{L_\alpha}(X)$) 对 $\mathscr{P}(X)^L$ 的并封闭且它们皆有最小元 G (这里 $G(a) = \varnothing, \forall a \in L$). 故 $\Phi_{L_\beta}(X), \Phi_{L_\alpha}(X), \Psi_{L_\beta}(X), \Psi_{L_\alpha}(X)$ 皆是完备格且 $\Phi_{L_\beta}(X), \Phi_{L_\alpha}(X)$ 是 $\mathscr{P}(X)^L$ 的完备交子半格, $\Psi_{L_\beta}(X), \Psi_{L_\alpha}(X)$ 是 $\mathscr{P}(X)^L$ 的完备并子半格.

下面定理 2.3.4—定理 2.3.7 的证明是简单的, 证明留给读者.

定理 2.3.4 设 $\{F_t\}_{t \in T} \subseteq \Phi_{L_\beta}(X)$, 令

$$F(a) = \bigcap_{b \in \beta(a)} \left(\bigcup_{t \in T} F_t(b) \right) \quad (\forall a \in L),$$

则 F 是 $\{F_t\}_{t \in T}$ 在 $\Phi_{L_\beta}(X)$ 中的上确界.

定理 2.3.5 设 $\{F_t\}_{t \in T} \subseteq \Phi_{L_\alpha}(X)$, 令

$$F(a) = \bigcap_{a \in \alpha(b)} \left(\bigcup_{t \in T} F_t(b) \right) \quad (\forall a \in L),$$

则 F 是 $\{F_t\}_{t \in T}$ 在 $\Phi_{L_\alpha}(X)$ 中的上确界.

定理 2.3.6 设 $\{G_t\}_{t\in T} \subseteq \Psi_{L_\beta}(X)$, 令

$$G(a) = \bigcup_{a\in\beta(b)}\left(\bigcap_{t\in T} G_t(b)\right) \quad (\forall a \in L),$$

则 G 是 $\{G_t\}_{t\in T}$ 在 $\Psi_{L_\beta}(X)$ 中的下确界.

定理 2.3.7 设 $\{G_t\}_{t\in T} \subseteq \Psi_{L_\alpha}(X)$, 令

$$G(a) = \bigcup_{b\in\alpha(a)}\left(\bigcap_{t\in T} G_t(b)\right) \quad (\forall a \in L),$$

则 G 是 $\{G_t\}_{t\in T}$ 在 $\Psi_{L_\alpha}(X)$ 中的下确界.

定理 2.3.8 作为完备格, $\Phi_{L_\beta}(X), \Phi_{L_\alpha}(X), \Psi_{L_\beta}(X), \Psi_{L_\alpha}(X)$ 皆与 L^X 同构, 从而皆是完全分配格.

证明 这从定理 2.2.5 与定理 2.2.6 容易得出. □

2.4 L-模糊关系

在这一节中, 作为两种集合套理论的应用, 我们将给出模糊关系的刻画, 首先给出两个 L-模糊集乘积的刻画.

定义 2.4.1 设 $A \in L^X$ 且 $B \in L^Y$. 定义 $X \times Y$ 上的一个 L-模糊集合 $A \times B$ 如下:

$$(A \times B)(x, y) = A(x) \wedge B(y), \quad \forall (x, y) \in X \times Y.$$

$A \times B$ 叫做 A 和 B 的乘积.

上述乘积定义也可以推广到如下任意多个 L-模糊集的乘积.

定义 2.4.2 令 $\{X_i \mid i \in \Omega\}$ 是一族集合, A_i 是 X_i 的 L-模糊子集, $\forall i \in \Omega$, $x_i \in X_i$, 定义 $\{A_i \mid i \in \Omega\}$ 的笛卡儿积 $\prod_{i\in\Omega} A_i$ 为

$$\left(\prod_{i\in\Omega} A_i\right)(x) = \bigwedge_{i\in\Omega} A_i(x_i), \quad \text{这里 } x \in \prod_{i\in\Omega} X_i.$$

下面我们仅给出两个 L-模糊集乘积的表示定理. 任意多个的 L-模糊集乘积表示留给读者.

2.4 L-模糊关系

定理 2.4.3 设 $A \in L^X, B \in L^Y$. 那么 $\forall a \in L$, 下面条件成立:
(1) $(A \times B)_{(a)} \subseteq A_{(a)} \times B_{(a)} \subseteq A_{[a]} \times B_{[a]} = (A \times B)_{[a]}$;
(2) $(A \times B)^{(a)} \subseteq A^{(a)} \times B^{(a)} \subseteq A^{[a]} \times B^{[a]} = (A \times B)^{[a]}$;
(3) $A \times B = \bigwedge\limits_{a \in L} \{a \vee (A^{[a]} \times B^{[a]})\} = \bigwedge\limits_{a \in P(L)} \{a \vee (A^{[a]} \times B^{[a]})\}$;
(4) $A \times B = \bigwedge\limits_{a \in L} \{a \vee (A^{(a)} \times B^{(a)})\} = \bigwedge\limits_{a \in P(L)} \{a \vee (A^{(a)} \times B^{(a)})\}$;
(5) $A \times B = \bigvee\limits_{a \in L} \{a \wedge (A_{[a]} \times B_{[a]})\} = \bigvee\limits_{a \in J(L)} \{a \wedge (A_{[a]} \times B_{[a]})\}$;
(6) $A \times B = \bigvee\limits_{a \in L} \{a \wedge (A_{(a)} \times B_{(a)})\} = \bigvee\limits_{a \in J(L)} \{a \wedge (A_{(a)} \times B_{(a)})\}$.

证明 (1) $\forall a \in L, (A \times B)_{(a)} \subseteq A_{(a)} \times B_{(a)}$ 由下面蕴含式可证.

$$(x, y) \in (A \times B)_{(a)}$$
$$\Rightarrow a \in \beta((A \times B)(x, y)) = \beta(A(x) \wedge B(y))$$
$$\Rightarrow a \in \beta(A(x)) \text{ 和 } a \in \beta(B(y))$$
$$\Rightarrow x \in A_{(a)} \text{ 和 } y \in B_{(a)}$$
$$\Rightarrow (x, y) \in A_{(a)} \times B_{(a)}.$$

显然 $\forall a \in L, A_{(a)} \times B_{(a)} \subseteq A_{[a]} \times B_{[a]} = (A \times B)_{[a]}$.
(2) $\forall a \in L, (A \times B)^{(a)} \subseteq A^{(a)} \times B^{(a)}$ 由下面蕴含式可证.

$$(x, y) \in (A \times B)^{(a)}$$
$$\Rightarrow (A \times B)(x, y) = A(x) \wedge B(y) \not\leq a$$
$$\Rightarrow A(x) \not\leq a \text{ 和 } B(y) \not\leq a$$
$$\Rightarrow x \in A^{(a)} \text{ 和 } y \in B^{(a)}$$
$$\Rightarrow (x, y) \in A^{(a)} \times B^{(a)}.$$

显然 $\forall a \in L, A^{(a)} \times B^{(a)} \subseteq A^{[a]} \times B^{[a]}$.
$\forall a \in L, A^{[a]} \times B^{[a]} = (A \times B)^{[a]}$ 由下面蕴含式可证.

$$(x, y) \in A^{[a]} \times B^{[a]}$$
$$\Leftrightarrow x \in A^{[a]} \text{ 和 } y \in B^{[a]}$$
$$\Leftrightarrow a \notin \alpha(A(x)) \text{ 和 } a \notin \alpha(B(y))$$
$$\Leftrightarrow a \notin \alpha(A(x) \wedge B(y)) = \alpha(A(x)) \cup \alpha(B(y))$$
$$\Leftrightarrow a \notin \alpha((A \times B)(x, y))$$
$$\Leftrightarrow (x, y) \in (A \times B)^{[a]}.$$

(1), (2) 和定理 2.1.5 蕴含 (3), (4), (5) 和 (6). □

定义 2.4.4 设 $A \in L^X, B \in L^Y, R \in L^{X \times Y}$. R 叫做从 A 到 B 的一个 L-模糊关系, 如果 $R \leqslant A \times B$.

由定理 2.4.3 易得下面结论.

定理 2.4.5 设 $A \in L^X, B \in L^Y, R \in L^{X \times Y}$. 则下列条件等价:

(1) R 是从 A 到 B 的一个 L-模糊关系;

(2) $\forall a \in L, R_{[a]}$ 是从 $A_{[a]}$ 到 $B_{[a]}$ 的关系;

(3) $\forall a \in M(L), R_{[a]}$ 是从 $A_{[a]}$ 到 $B_{[a]}$ 的关系;

(4) $\forall a \in L, R_{(a)}$ 是从 $A_{(a)}$ 到 $B_{(a)}$ 的关系;

(5) $\forall a \in M(L), R_{(a)}$ 是从 $A_{(a)}$ 到 $B_{(a)}$ 的关系;

(6) $\forall a \in L, R^{[a]}$ 是从 $A^{[a]}$ 到 $B^{[a]}$ 的关系;

(7) $\forall a \in P(L), R^{[a]}$ 是从 $A^{[a]}$ 到 $B^{[a]}$ 的关系;

(8) $\forall a \in L, R^{(a)}$ 是从 $A^{(a)}$ 到 $B^{(a)}$ 的关系;

(9) $\forall a \in P(L), R^{(a)}$ 是从 $A^{(a)}$ 到 $B^{(a)}$ 的关系.

定义 2.4.6 设 $A \in L^X, B \in L^Y, C \in L^Z, R \leqslant A \times B$ 且 $Q \leqslant B \times C$. 定义一个从 A 到 C 的 L-模糊关系 $Q \circ R$, 使得 $\forall (x,z) \in X \times Z$,

$$(Q \circ R)(x,z) = \bigvee_{y \in Y} \{R(x,y) \wedge Q(y,z)\},$$

它叫做 R 与 Q 的合成.

定理 2.4.7 设 $A \in L^X, B \in L^Y, C \in L^Z, R \leqslant A \times B$ 且 $Q \leqslant B \times C$. 那么 $\forall a \in L$, 下面条件成立:

(1) $(Q \circ R)_{(a)} \subseteq Q_{(a)} \circ R_{(a)} \subseteq Q_{[a]} \circ R_{[a]} \subseteq (Q \circ R)_{[a]}$;

(2) $(Q \circ R)^{(a)} \subseteq Q^{(a)} \circ R^{(a)} \subseteq Q^{[a]} \circ R^{[a]} \subseteq (Q \circ R)^{[a]}$, 特别地, $\forall a \in P(L)$, $(Q \circ R)^{(a)} = Q^{(a)} \circ R^{(a)}$;

(3) $Q \circ R = \bigvee_{a \in L} \{a \wedge (Q_{[a]} \circ R_{[a]})\} = \bigvee_{a \in J(L)} \{a \wedge (Q_{[a]} \circ R_{[a]})\}$;

(4) $Q \circ R = \bigvee_{a \in L} \{a \wedge (Q_{(a)} \circ R_{(a)})\} = \bigvee_{a \in J(L)} \{a \wedge (Q_{(a)} \circ R_{(a)})\}$;

(5) $Q \circ R = \bigwedge_{a \in L} \{a \vee (Q^{[a]} \circ R^{[a]})\} = \bigwedge_{a \in P(L)} \{a \vee (Q^{[a]} \circ R^{[a]})\}$;

(6) $Q \circ R = \bigwedge_{a \in L} \{a \vee (Q^{(a)} \circ R^{(a)})\} = \bigwedge_{a \in P(L)} \{a \vee (Q^{(a)} \circ R^{(a)})\}$.

证明 (1) $\forall a \in L, (Q \circ R)_{(a)} \subseteq Q_{(a)} \circ R_{(a)}$ 从下面蕴含式可证.

$$(x,z) \in (Q \circ R)_{(a)}$$
$$\Rightarrow a \in \beta((Q \circ R)(x,z))$$

2.4 L-模糊关系

$\Rightarrow \exists y \in Y$ 使得 $a \in \beta(R(x,y) \wedge Q(y,z))$

$\Rightarrow \exists y \in Y$ 使得 $a \in \beta(R(x,y))$ 且 $a \in \beta(Q(y,z))$

$\Rightarrow \exists y \in Y$ 使得 $(x,y) \in R_{(a)}$ 且 $(y,z) \in Q_{(a)}$

$\Rightarrow (x,z) \in Q_{(a)} \circ R_{(a)}$.

容易验证 $\forall a \in L, Q_{(a)} \circ R_{(a)} \subseteq Q_{[a]} \circ R_{[a]} \subseteq (Q \circ R)_{[a]}$.

(2) $\forall a \in L, (Q \circ R)^{(a)} \subseteq Q^{(a)} \circ R^{(a)}$ 从下面蕴含式可证.

$(x,z) \in (Q \circ R)^{(a)}$

$\Rightarrow (Q \circ R)(x,z) \not\leqslant a$

$\Rightarrow \exists y \in Y$ 使得 $R(x,y) \wedge Q(y,z) \not\leqslant a$

$\Rightarrow \exists y \in Y$ 使得 $R(x,y) \not\leqslant a$ 且 $Q(y,z) \not\leqslant a$

$\Rightarrow \exists y \in Y$ 使得 $(x,y) \in R^{(a)}$ 且 $(y,z) \in Q^{(a)}$

$\Rightarrow (x,z) \in Q^{(a)} \circ R^{(a)}$.

$\forall a \in P(L)$, 上面蕴含式的逆也成立. 这样就有

$$(Q \circ R)^{(a)} = Q^{(a)} \circ R^{(a)}.$$

$\forall a \in L$, 显然 $Q^{(a)} \circ R^{(a)} \subseteq Q^{[a]} \circ R^{[a]}$.

$\forall a \in L, Q^{[a]} \circ R^{[a]} \subseteq (Q \circ R)^{[a]}$ 从下面蕴含式可证.

$(x,z) \notin (Q \circ R)^{[a]}$

$\Rightarrow a \in \alpha((Q \circ R)(x,z))$

$\Rightarrow \forall y \in Y, a \in \alpha(R(x,y) \wedge Q(y,z)) = \alpha(R(x,y)) \cup \alpha(Q(y,z))$

$\Rightarrow \forall y \in Y, a \in \alpha(R(x,y))$ 或 $a \in \alpha(Q(y,z))$

$\Rightarrow \forall y \in Y, (x,y) \notin R^{[a]}$ 或 $(y,z) \notin Q^{[a]}$

$\Rightarrow (x,z) \notin Q^{[a]} \circ R^{[a]}$.

(1), (2) 和定理 2.1.5 蕴含 (3), (4), (5) 和 (6). $\square$

等价关系是非常重要的概念, 下面给出它的模糊形式的推广.

定义 2.4.8 一个从 X 到 X 的 L-模糊关系 R 叫做 X 上的一个 L-模糊等价关系, 如果 R 满足

(1) $\forall x \in X, R(x,x) = 1$;

(2) $\forall x, y \in X, R(x,y) = R(y,x)$;

(3) $R \circ R \subseteq R$.

下面定理给出了模糊关系的截集式刻画, 其证明是简单的, 故略去.

定理 2.4.9 对 X 上的一个 L-模糊关系 R, 我们有 $(4) \Rightarrow (1)$ 且 $(1) \Leftrightarrow (2) \Leftrightarrow (3) \Leftrightarrow (5) \Leftrightarrow (6) \Leftrightarrow (7)$.

(1) R 是 X 上的一个 L-模糊等价关系;
(2) $\forall a \in L$, $R_{[a]}$ 是 X 上的一个等价关系;
(3) $\forall a \in M(L)$, $R_{[a]}$ 是 X 上的一个等价关系;
(4) $\forall a \in \beta(1)$, $R_{(a)}$ 是 X 上的一个等价关系;
(5) $\forall a \in L$, $R^{[a]}$ 是 X 上的一个等价关系;
(6) $\forall a \in P(L)$, $R^{[a]}$ 是 X 上的一个等价关系;
(7) $\forall a \in P(L)$, $R^{(a)}$ 是 X 上的一个等价关系.

一般地, 定理中的 $(1) \Rightarrow (4)$ 不成立.

例 2.4.10 设 $X = \{x, y, z\}, L = \{1, a, b, c\} \cup \left[0, \dfrac{1}{2}\right]$, 这里 x, y, z 是不同的 且 $\left[0, \dfrac{1}{2}\right]$ 是一个区间. 定义 L 中的序如下:

$\forall x \in \left[0, \dfrac{1}{2}\right], x \leqslant c = \dfrac{1}{2}, c < a, c < b, a \not\leqslant b, b \not\leqslant a, a < 1, b < 1;$

$\left[0, \dfrac{1}{2}\right]$ 中的序是通常的.

那么 L 是一个完全分配格. 取 $R \in L^{X \times X}$ 使得

$$R(x,x) = R(y,y) = R(z,z) = 1, \quad R(x,y) = R(y,x) = a,$$
$$R(y,z) = R(z,y) = b, \qquad\qquad R(x,z) = R(z,x) = \dfrac{1}{2}.$$

显然

$$(R \circ R)(x,x) = (R \circ R)(y,y) = (R \circ R)(z,z) = 1,$$
$$(R \circ R)(x,y) = (R \circ R)(y,x) = a,$$
$$(R \circ R)(y,z) = (R \circ R)(z,y) = b,$$
$$(R \circ R)(x,z) = (R \circ R)(z,x) = \dfrac{1}{2}.$$

所以 $R \circ R \subseteq R$. 从而 R 是一个 L-模糊等价关系. 容易验证

$$R_{(\frac{1}{2})} = \{(x,x), (y,y), (z,z), (x,y), (y,x), (y,z), (z,y)\},$$
$$R_{(\frac{1}{2})} \circ R_{(\frac{1}{2})} = X \times X \not\subseteq R_{(\frac{1}{2})}.$$

这表明 $R_{(\frac{1}{2})}$ 不是等价关系.

2.5 L 值 Zadeh 型函数

Zadeh 型函数是联系两个模糊集的重要桥梁, 它是普通集合间映射的扩张, 最初它的定义如下.

定义 2.5.1 设 $f: X \to Y$ 是一个映射, $A \in L^X$, $B \in L^Y$. 定义两个映射 $f_L^{\to}: L^X \to L^Y$ 和 $f_L^{\leftarrow}: L^Y \to L^X$ 分别为

$$f_L^{\to}(A)(y) = \bigvee \{A(x) \mid f(x) = y\}, \quad \forall y \in Y;$$

$$f_L^{\leftarrow}(B) = B \circ f.$$

那么 $f_L^{\to}: L^X \to L^Y$ 称为 L 值 Zadeh 型函数, $f_L^{\leftarrow}: L^Y \to L^X$ 称为 Zadeh 型函数的逆函数.

注 2.5.2 在很多文章中 $f_L^{\to}(A)$ 和 $f_L(A)$ 不加区别, 我们之所以区别它们无非就是强调它们的定义域和值域不同, 在不至于引起混淆的情况下, 可以不加区别.

下面定理的证明是容易的, 证明就留给读者吧.

定理 2.5.3 设 $f: X \to Y$ 是一个映射, A 和 B 分别是 X 和 Y 中的 L-模糊子集. 则

(1) $f_L^{\to}(f_L^{\leftarrow}(B)) \leqslant B$; 若 f 是满射, 则 $f_L^{\to}(f_L^{\leftarrow}(B)) = B$.

(2) $f_L^{\leftarrow}(f_L^{\to}(A)) \geqslant A$; 若 f 是单射, 则 $f_L^{\leftarrow}(f_L^{\to}(A)) = A$.

下面定理给出了 L 值 Zadeh 型函数及其逆函数的各种等价形式.

定理 2.5.4 设 $f: X \to Y$ 是映射, $A \in L^X, B \in L^Y$, 则

(1) $f_L^{\to}(A) = \bigvee_{a \in L} (a \wedge f^{\to}(A_{[a]})) = \bigvee_{a \in L}(a \wedge f^{\to}(A_{(a)}))$;

(2) $(f_L^{\to}(A))_{[a]} = \bigcap_{b \in \beta(a)} f^{\to}(A_{[b]}) = \bigcap_{b \in \beta(a)} f^{\to}(A_{(b)})$;

(3) $(f_L^{\to}(A))_{(a)} = f^{\to}(A_{(a)})$;

(4) $f_L^{\to}(A) = \bigwedge_{a \in L}(a \vee f^{\to}(A^{[a]})) = \bigwedge_{a \in L}(a \vee f^{\to}(A^{(a)}))$;

(5) $(f_L^{\to}(A))^{[a]} = \bigcap_{a \in \alpha(b)} f^{\to}(A^{[b]}) = \bigcap_{a \in \alpha(b)} f^{\to}(A^{(b)})$;

(6) $(f_L^{\to}(A))^{(a)} = f^{\to}(A^{(a)})$;

(7) $f_L^{\leftarrow}(B) = \bigvee_{a \in L}(a \wedge f^{\leftarrow}(B_{[a]})) = \bigvee_{a \in L}(a \wedge f^{\leftarrow}(B_{(a)}))$;

(8) $(f_L^{\leftarrow}(B))_{[a]} = f^{\leftarrow}(B_{[a]})$;

(9) $(f_L^{\leftarrow}(B))_{(a)} = f^{\leftarrow}(B_{(a)})$;

(10) $f_L^{\leftarrow}(B) = \bigwedge_{a \in L}(a \vee f^{\leftarrow}(B^{[a]})) = \bigwedge_{a \in L}(a \vee f^{\leftarrow}(B^{(a)}))$;

(11) $(f_L^{\leftarrow}(B))^{[a]} = f^{\leftarrow}(B^{[a]})$;

(12) $(f_L^{\leftarrow}(B))^{(a)} = f^{\leftarrow}(B^{(a)})$.

证明 设 $y \in Y, a \in L$, 则由下面蕴含式可得 (3) 的证明.

$$\begin{aligned} y \in (f_L^{\rightarrow}(A))_{(a)} &\Leftrightarrow a \in \beta\left(f_L^{\rightarrow}(A)(y)\right) \\ &\Leftrightarrow a \in \beta\left(\bigvee\{A(x) \mid f(x) = y\}\right) \\ &\Leftrightarrow a \in \bigcup\{\beta(A(x)) \mid f(x) = y\} \\ &\Leftrightarrow \exists x \in X \ \text{使得} \ f(x) = y \ \text{且} \ x \in A_{(a)} \\ &\Leftrightarrow y \in f^{\rightarrow}(A_{(a)}). \end{aligned}$$

进一步由定理 2.1.4 和 $f^{\rightarrow}$ 的保序性我们能够得到

$$f_L^{\rightarrow}(A) = \bigvee_{a \in L} \left(a \wedge f^{\rightarrow}(A_{(a)})\right) \leqslant \bigvee_{a \in L} \left(a \wedge f^{\rightarrow}(A_{[a]})\right).$$

为了证明 (1), 只需要证明 $\forall y \in Y$,

$$f_L^{\rightarrow}(A)(y) \geqslant \bigvee_{a \in L} \left(a \wedge f^{\rightarrow}(A_{[a]})\right)(y).$$

设 $b \in L$ 且 $b \prec \bigvee_{a \in L}\left(a \wedge f^{\rightarrow}(A_{[a]})\right)(y)$, 则存在 $a \in L$ 使得 $b \prec a \wedge \left(f^{\rightarrow}(A_{[a]})\right)(y)$, 那就是 $b \in \beta(a)$ 且 $\left(f^{\rightarrow}(A_{[a]})\right)(y) = 1$. 于是存在 $x \in A_{[a]}$ 使得 $f(x) = y$, 这意味着

$$f_L^{\rightarrow}(A)(y) = \bigvee\{A(x) \mid f(x) = y\} \geqslant a \geqslant b.$$

从而

$$\begin{aligned} f_L^{\rightarrow}(A)(y) &\geqslant \bigvee\left\{b \in L \mid b \prec \bigvee_{a \in L}\left(a \wedge f^{\rightarrow}(A_{[a]})\right)(y)\right\} \\ &= \bigvee_{a \in L}\left(a \wedge f^{\rightarrow}(A_{[a]})\right)(y). \end{aligned}$$

这样就完成了 (1) 的证明.

再由 (1), (3) 以及定理 2.1.4 不难证明 (2).

设 $y \in Y, a \in L$, 则由下面蕴含式可得 (6) 的证明.

$$\begin{aligned} y \in (f_L^{\rightarrow}(A))^{(a)} &\Leftrightarrow f_L^{\rightarrow}(A)(y) \not\leqslant a \\ &\Leftrightarrow \bigvee\{A(x) \mid f(x) = y\} \not\leqslant a \\ &\Leftrightarrow \exists x \in X \ \text{使得} \ f(x) = y \ \text{而且} \ A(x) \not\leqslant a \\ &\Leftrightarrow \exists x \in X \ \text{使得} \ f(x) = y \ \text{而且} \ x \in A^{(a)} \\ &\Leftrightarrow y \in f^{\rightarrow}(A^{(a)}). \end{aligned}$$

进一步, 由定理 2.1.4 和 $f^{\to}$ 的保序性我们能够得到

$$f_L^{\to}(A) = \bigwedge_{a \in L} \left(a \vee f^{\to}(A^{(a)})\right) \leqslant \bigwedge_{a \in L} \left(a \vee f^{\to}(A^{[a]})\right).$$

为了证明 (4), 只需要证明 $\forall y \in Y$,

$$f_L^{\to}(A)(y) \geqslant \bigwedge_{a \in L} \left(a \vee f^{\to}(A^{[a]})\right)(y).$$

设 $a \in \alpha(f_L^{\to}(A))$, 则对满足 $f(x) = y$ 的任何 $x \in X$, 都有

$$a \in \alpha\left(\bigvee_{a \in L}\{A(x) \mid f(x) = y\}\right) \subseteq \alpha(A(x)),$$

也就是说对满足 $f(x) = y$ 的任何 $x \in X$, 都有 $x \notin A^{[a]}$, 这意味着 $y \notin f^{\to}(A^{[a]})$. 从而

$$\bigwedge_{a \in L} \left(a \vee f^{\to}(A^{[a]})\right)(y) \leqslant a.$$

因此有

$$f_L^{\to}(A)(y) \geqslant \bigwedge_{a \in L} \left(a \vee f^{\to}(A^{[a]})\right)(y).$$

这样就完成了 (4) 的证明.

再由 (4), (6) 以及定理 2.1.4 不难证明 (5).

同理可证 (7)—(12). □

定理 2.5.5 设 f 是从 X 到 Y 的映射, $\{A_i \mid i \in \Omega\} \subseteq L^X$ 和 $\{B_j \mid j \in \Lambda\} \subseteq L^Y$, 那么

$$f_L^{\to}\left(\bigvee_{i \in \Omega} A_i\right) = \bigvee_{i \in \Omega} f_L^{\to}(A_i);$$

$$f_L^{\to}\left(\bigwedge_{i \in \Omega} A_i\right) \subseteq \bigwedge_{i \in \Omega} f_L^{\to}(A_i);$$

$$f_L^{\leftarrow}\left(\bigvee_{j \in \Lambda} B_j\right) = \bigvee_{j \in \Lambda} f_L^{\leftarrow}(B_j);$$

$$f_L^{\leftarrow}\left(\bigwedge_{j \in \Lambda} B_j\right) = \bigwedge_{j \in \Lambda} f_L^{\leftarrow}(B_j);$$

$$f_L^{\leftarrow}(B') = (f_L^{\leftarrow}(B))'.$$

证明 由分明映射的性质和定理 2.5.4 易证. □

2.6 L-模糊偏序

在这一节中, 作为集合套理论的应用, 我们将给出 L-模糊偏序的表达形式.

定义 2.6.1 让 X 是一个集, $A \in L^X$ 且 $A \neq \emptyset$. 一个从 A 到 A 的 L-模糊关系 R 叫做 A 上的一个 L-模糊偏序, 如果 R 满足下面条件:

(1) $\forall x \in A^{(0)}, R(x,x) = A(x)$;

(2) $R \circ R \leqslant R$;

(3) $\forall x, y \in A^{(0)}, R(x,y) \wedge R(y,x) \neq 0 \Rightarrow x = y$.

当 R 是 A 上的一个 L-模糊偏序时, 就称 (A, R) 是一个 L-模糊偏序集.

L-模糊偏序集可以借助分明偏序集刻画.

定理 2.6.2 对于 A 上的一个 L-模糊关系 R, 下面蕴含式 (4) $\Rightarrow$ (1) 和 (1) $\Leftrightarrow$ (2) $\Leftrightarrow$ (3) $\Leftrightarrow$ (5) $\Leftrightarrow$ (6) $\Leftrightarrow$ (7) 成立.

(1) R 是 A 上的一个 L-模糊偏序;

(2) $\forall a \in L$, 如果 $R_{[a]}$ 是非空集合, 那么它是 $A_{[a]}$ 上的一个偏序;

(3) $\forall a \in J(L)$, 如果 $R_{[a]}$ 是非空集合, 那么它是 $A_{[a]}$ 上的一个偏序;

(4) $\forall a \in \beta(1)$, 如果 $R_{(a)}$ 是非空集合, 那么它是 $A_{(a)}$ 上的一个偏序;

(5) $\forall a \in \alpha(0)$, 如果 $R^{[a]}$ 是非空集合, 那么它是 $A^{[a]}$ 上的一个偏序;

(6) $\forall a \in \alpha^*(0)$, 如果 $R^{[a]}$ 是非空集合, 那么它是 $A^{[a]}$ 上的一个偏序;

(7) $\forall a \in P(L)$, 如果 $R^{(a)}$ 是非空集合, 那么它是 $A^{(a)}$ 上的一个偏序.

证明是简单的, 留给读者自己完成.

注 2.6.3 一般地, 定理 2.6.2 中的 (1) $\Rightarrow$ (4) 不成立, 此由下例可见.

例 2.6.4 设 $X = \{x, y, z\}, L = \{1, a, b, c\} \cup \left[0, \dfrac{1}{2}\right]$, 这里 x, y, z 是不同的且 $\left[0, \dfrac{1}{2}\right]$ 是一个区间. 我们定义 L 中的序如下:

$$\forall e \in \left[0, \dfrac{1}{2}\right], e \leqslant c = \dfrac{1}{2}, c < a, c < b, a \not\leqslant b, b \not\leqslant a, a < 1, b < 1;$$

$\left[0, \dfrac{1}{2}\right]$ 中的序是通常序.

那么 L 是一个完全分配格. 取 $R \in L^{X \times X}$ 使得

$$R(x, x) = R(y, y) = R(z, z) = 1,$$
$$R(y, x) = R(z, y) = R(z, x) = 0,$$
$$R(x, y) = a, R(y, z) = b, R(x, z) = \dfrac{1}{2}.$$

显然
$$(R \circ R)(x,x) = (R \circ R)(y,y) = (R \circ R)(z,z) = 1,$$
$$(R \circ R)(z,y) = (R \circ R)(y,x) = (R \circ R)(z,x) = 0,$$
$$(R \circ R)(x,y) = a, (R \circ R)(y,z) = b, (R \circ R)(x,z) = \frac{1}{2}.$$

因此 R 是 X 上的一个 L-模糊偏序. 但是容易证明
$$R_{(\frac{1}{2})} = \{(x,x),(y,y),(z,z),(x,y),(y,z)\},$$
$$R_{(\frac{1}{2})} \circ R_{(\frac{1}{2})} = \{(x,x),(y,y),(z,z),(x,y),(y,z),(x,z)\} \nsubseteq R_{(\frac{1}{2})}.$$
这表明 $R_{(\frac{1}{2})}$ 不是 X 上的偏序.

2.7 模糊映射

从模糊关系角度出发, 去定义模糊映射是由李洪兴、罗承忠和汪培庄三位老师于 1993 年提出的, 详细可见文献 [1], 作者于 2000 年在文献 [30] 中将其推广到了格值模糊集之间, 引入了 L-模糊映射的概念.

定义 2.7.1 设 $A \in L^X, B \in L^Y$. 一个从 A 到 B 的 L-模糊关系 f 叫做从 A 到 B 的 L-模糊映射, 如果 $\forall a \in J(L)$, $f_{[a]}$ 是从 $A_{[a]}$ 到 $B_{[a]}$ 的映射. 如果 f 是从 A 到 B 的 L-模糊映射, 那么它记为 $f: A \to B$.

定理 2.7.2 设 $A \in L^X, B \in L^Y$ 且 $f \leqslant A \times B$, 则下列条件等价:
(1) 映射 f 是 A 到 B 的 L-模糊映射;
(2) 对任意的 $a \in J(L)$, $f_{(a)}$ 是 $A_{(a)}$ 到 $B_{(a)}$ 的映射;
(3) 对任意的 $a \in P(L)$, $f^{(a)}$ 是 $A^{(a)}$ 到 $B^{(a)}$ 的映射.

证明 (1) $\Rightarrow$ (2). 设 $a \in J(L)$ 且 $x \in A_{(a)}$. 那么 $a \in \beta^*(A(x))$. 从而存在 $b \in \beta^*(A(x))$ 使得 $a \in \beta^*(b)$. 显然 $x \in A_{[b]} \subseteq A_{(a)}$. 由 (1) 可知存在 $y \in B_{[b]} \subseteq B_{(a)}$ 使得 $(x,y) \in f_{[b]} \subseteq f_{(a)}$.

另外, 如果存在另一个 $z \in B_{(a)}$ 使得 $(x,z) \in f_{(a)}$, 那么我们能够得到
$$(x,y),(x,z) \in f_{(a)} \subseteq f_{[a]} \subseteq A_{[a]} \times B_{[a]}.$$
从而由 (1) 可知有 $y = z$ 成立. 因此 $f_{(a)}$ 是从 $A_{(a)}$ 到 $B_{(a)}$ 的映射.

(2) $\Rightarrow$ (3). 设 $a \in P(L)$ 且 $x \in A^{(a)}$. 则 $A(x) \nleqslant a$. 取 $b \in \beta^*(A(x))$ 使得 $b \nleqslant a$. 则 $x \in A_{(b)}$. 于是由 (2) 我们知道存在 $y \in B_{(b)} \subseteq B^{(a)}$ 使得 $(x,y) \in f_{(b)} \subseteq f^{(a)}$.

另外, 如果存在 $z \in B^{(a)}$ 使得 $(x,z) \in f^{(a)}$, 那么从 $f(x,y) \nleqslant a$ 和 $f(x,z) \nleqslant a$ 可以知道 $f(x,y) \wedge f(x,z) \nleqslant a$. 取 $c \in \beta^*(f(x,y) \wedge f(x,z))$ 使得 $c \nleqslant a$. 则

$(x,y) \in f_{(c)}$ 且 $(x,z) \in f_{(c)}$. 由 (2) 可得 $y = z$. 这表明 $f^{(a)}$ 是从 $A^{(a)}$ 到 $B^{(a)}$ 的映射.

(3) $\Rightarrow$ (1). 设 $a \in J(L)$ 且 $x \in A_{[a]}$. 则对满足 $b \not\geqslant a$ 的任意 $b \in P(L)$, 都有 $x \in A^{(b)}$. 由 (3) 我们知道存在 $y_b \in B^{(b)}$ 使得 $(x, y_b) \in f^{(b)}$. 取 $c \in P(L)$ 使得 $c \not\geqslant a$. 下面我们来证明 $y_b = y_c$.

显然 $a \not\leqslant b \vee c$. 取 $e \in P(L)$ 使得 $e \geqslant b \vee c$ 且 $e \not\geqslant a$. 则 $(x, y_e) \in f^{(e)} \subseteq f^{(b)}$ 且 $(x, y_e) \in f^{(e)} \subseteq f^{(c)}$. 由 (3) 我们可以得到 $y_b = y_e = y_c$. 令 $y = y_c$. 则

$$(x,y) \in \bigcap \{f^{(b)} \mid b \in P(L),\ a \not\leqslant b\} = f_{[a]} \subseteq A_{[a]} \times B_{[a]}.$$

另外, 如果存在 $z \in B_{[a]}$ 使得 $(x, z) \in f_{[a]}$, 那么 $f(x,y) \geqslant a$ 且 $f(x, z) \geqslant a$. 取 $b \in P(L)$ 使得 $b \not\geqslant a$. 则 $(x,y), (x,z) \in f_{[a]} \subseteq f^{(b)}$. 于是由 (3) 我们知道 $y = z$. 这样我们就证明了 $f_{[a]}$ 是从 $A_{[a]}$ 到 $B_{[a]}$ 的映射. □

定义 2.7.3 设 $A \in L^X$, $B \in L^Y$. 一个 L-模糊映射 $f: A \to B$ 叫做一个 L-模糊单射, 如果对每个 $a \in J(L)$, $f_{[a]}: A_{[a]} \to B_{[a]}$ 是单射. $f: A \to B$ 叫做一个 L-模糊满射, 如果对任意的 $a \in J(L)$, $B_{(a)} \subseteq f_{[a]}(A_{[a]}) \subseteq B_{[a]}$. $f: A \to B$ 叫做一个 L-模糊双射, 如果 f 既是单的又是满的.

相似于定理 2.7.2 的证明我们能够得到下面定理.

定理 2.7.4 设 $A \in L^X$, $B \in L^Y$ 且 $f: A \to B$ 是 L-模糊映射. 则下列条件等价:

(1) 映射 f 是一个从 A 到 B 的 L-模糊单射;

(2) 任意的 $a \in J(L)$, $f_{(a)}$ 是一个从 $A_{(a)}$ 到 $B_{(a)}$ 的单射;

(3) 任意的 $a \in P(L)$, $f^{(a)}$ 是一个从 $A^{(a)}$ 到 $B^{(a)}$ 的单射.

定理 2.7.5 设 $A \in L^X$, $B \in L^Y$ 且 $f: A \to B$ 是一个 L-模糊映射. 则下列条件等价:

(1) 映射 f 是一个从 A 到 B 的 L-模糊满射;

(2) 任意的 $a \in J(L)$, $f_{(a)}$ 是一个从 $A_{(a)}$ 到 $B_{(a)}$ 的满射;

(3) 任意的 $a \in P(L)$, $f^{(a)}$ 是一个从 $A^{(a)}$ 到 $B^{(a)}$ 的满射.

证明 (1) $\Rightarrow$ (2). 设 $a \in J(L)$ 且 $y \in B_{(a)}$. 因为 $B_{(a)} = \bigcup_{a \in \beta^*(b), b \in J(L)} B_{(b)}$, 所以存在 $b \in J(L)$ 使得 $a \in \beta^*(b)$ 而且 $y \in B_{(b)}$. 由 (1) 可知 $y \in f_{[b]}(A_{[b]})$. 从而存在 $x \in A_{[b]} \subseteq A_{(a)}$ 使得

$$(x,y) \in f_{[b]} \subseteq f_{(a)} \subseteq A_{(a)} \times B_{(a)}.$$

因此 $f_{(a)}: A_{(a)} \to B_{(a)}$ 是满射.

(2) $\Rightarrow$ (3). 设 $a \in P(L)$ 而且 $y \in B^{(a)}$. 则 $B(y) \not\leqslant a$. 取 $b \in \beta^*(B(y))$ 使得 $b \not\leqslant a$. 那么显然有 $y \in B_{(b)}$. 由 (2) 可知存在 $x \in A_{(b)} \subseteq A^{(a)}$ 使得

$$(x, y) \in f_{(b)} \subseteq f^{(a)} \subseteq A^{(a)} \times B^{(a)}.$$

因此 $f^{(a)}: A^{(a)} \to B^{(a)}$ 是满射.

(3) $\Rightarrow$ (1). 设 $a \in J(L)$ 且 $y \in B_{(a)}$. 则 $a \in \beta^*(B(y))$. 从而存在 $b \in J(L)$ 使得 $b \leqslant B(y)$ 且 $a \in \beta^*(b)$. 于是对每个 $c \in P(L)$, 当 $c \not\geqslant b$ 时, 都有 $y \in B^{(c)}$. 由 (3) 我们知道对满足 $c \not\geqslant b$ 的每个 $c \in P(L)$, 都存在 $x_c \in A^{(c)}$ 使得 $(x_c, y) \in f^{(c)}$. 这意味着对满足 $c \not\geqslant b$ 的每个 $c \in P(L)$, 都有 $f(x_c, y) \not\leqslant c$. 于是对满足 $c \not\geqslant b$ 的每个 $c \in P(L)$, 都有 $\bigvee_{c \not\geqslant b, c \in P(L)} f(x_c, y) \not\leqslant c$ 成立. 这表明 $\bigvee_{c \not\geqslant b, c \in P(L)} f(x_c, y) \geqslant b$. 再由 $a \in \beta^*(b)$ 我们知道存在 x_c 使得 $f(x_c, y) \geqslant a$, 也就是

$$(x_c, y) \in f_{[a]} \subseteq A_{[a]} \times B_{[a]}.$$

故有 $y \in f_{[a]}(A_{[a]}) \subseteq B_{[a]}$. 这就证明了 $B_{(a)} \subseteq f_{[a]}(A_{[a]}) \subseteq B_{[a]}$. □

定理 2.7.6 设 $A \in L^X, B \in L^Y$ 且 $f: A \to B$ 是 L-模糊映射. 则下列条件等价:

(1) 映射 f 是从 A 到 B 的 L-模糊双射;
(2) 任意的 $a \in J(L)$, $f_{[a]}$ 是从 $A_{[a]}$ 到 $B_{[a]}$ 的双射;
(3) 任意的 $a \in J(L)$, $f_{(a)}$ 是从 $A_{(a)}$ 到 $B_{(a)}$ 的双射;
(4) 任意的 $a \in P(L)$, $f^{(a)}$ 是从 $A^{(a)}$ 到 $B^{(a)}$ 的双射.

证明 (1) $\Rightarrow$ (2). 设 $a \in J(L)$ 且 $y \in B_{[a]}$. 则 $\forall b \in \beta^*(a)$ 我们知道 $y \in B_{(b)} \subseteq f_{[b]}(A_{[b]}) \subseteq B_{[b]}$. 这表明存在 $x_b \in A_{[b]}$ 使得 $(x_b, y) \in f_{[b]}$ 对每一个 $b \in \beta^*(a)$ 都成立. 假如 $b, c \in \beta^*(a)$, 那么因为 $\beta^*(a)$ 是一个定向集可知, 存在 $d \in \beta^*(a)$ 使得 $d \geqslant b \vee c$. 这意味着 $(x_d, y) \in f_{[d]} \subseteq f_{[b]} \cap f_{[c]}$. 从而由 (1) 我们知道必有 $x_d = x_b = x_c$. 令 $x = x_b$. 则

$$(x, y) \in \bigcap_{b \in \beta^*(a)} f_{[b]} = f_{[a]} \subseteq A_{[a]} \times B_{[a]}.$$

因此 $f_{[a]}: A_{[a]} \to B_{[a]}$ 是满射. 这样就证明了 (2) 是成立的. (2) $\Rightarrow$ (3) $\Rightarrow$ (4) $\Rightarrow$ (1) 由定理 2.7.4 和定理 2.7.5 可得. □

定理 2.7.7 设 $A \in L^X, B \in L^Y, C \in L^Z$, 且 $f: A \to B$ 与 $g: B \to C$ 是两个 L-模糊映射. 则 $g \circ f$ 是从 A 到 C 的 L-模糊映射.

证明 由定理 2.7.2, 我们只需要证明 $\forall a \in P(L), (g \circ f)^{(a)}$ 是从 $A^{(a)}$ 到 $C^{(a)}$ 的映射, 这归结为证明 $(g \circ f)^{(a)} = g^{(a)} \circ f^{(a)}$ 对每个 $a \in P(L)$ 都成立. 事实上此由下面蕴含式可证.

$$(x,z) \in (g \circ f)^{(a)} \Leftrightarrow (g \circ f)(x,z) \not\leqslant a$$
$$\Leftrightarrow \exists y \in Y \text{ 使得 } f(x,y) \wedge g(y,z) \not\leqslant a$$
$$\Leftrightarrow \exists y \in Y \text{ 使得 } f(x,y) \not\leqslant a \text{ 且 } g(y,z) \not\leqslant a$$
$$\Leftrightarrow \exists y \in Y \text{ 使得 } (x,y) \in f^{(a)} \text{ 且 } (y,z) \in g^{(a)}$$
$$\Leftrightarrow (x,z) \in g^{(a)} \circ f^{(a)}. \qquad \square$$

下面我们来考虑 L-模糊映射和 Zadeh 型函数的关系.

引理 2.7.8 设 $0 \in P(L), A \in L^X, B \in L^Y$ 且 $f \leqslant A \times B$. 如果 f 是从 A 到 B 的 L-模糊映射, 那么 $f^{(0)}$ 是从 $A^{(0)}$ 到 $B^{(0)}$ 的映射.

证明 假设 $x \in A^{(0)}$. 那么 $A(x) \neq 0$. 从而存在 $a \in J(L)$ 使得 $a \leqslant A(x)$, 也就是 $x \in A_{[a]}$. 因为 f 是从 A 到 B 的 L-模糊映射, 所以存在 $y \in B_{[a]} \subseteq B^{(0)}$ 使得 $(x,y) \in f_{[a]} \subseteq f^{(0)}$. 如果另外存在一个 $z \in B^{(0)}$ 使得 $(x,z) \in f^{(0)}$, 那么自然地有 $f(x,y) \neq 0$ 和 $f(x,z) \neq 0$. 由 $0 \in P(L)$ 可知必有 $f(x,y) \wedge f(x,z) \neq 0$. 取 $b \in J(L)$ 使得 $f(x,y) \wedge f(x,z) \geqslant b$. 于是我们可以得到 $(x,y) \in f_{[b]}$ 与 $(x,z) \in f_{[b]}$. 再由 f 是从 A 到 B 的 L-模糊映射可知 $y = z$. 这表明 $f^{(0)}$ 是从 $A^{(0)}$ 到 $B^{(0)}$ 的映射. $\qquad \square$

定理 2.7.9 设 $0 \in P(L), A \in L^X, B \in L^Y$ 且 $f \leqslant A \times B$. 则 f 是从 A 到 B 的 L-模糊映射当且仅当 $f = f^{(0)} \wedge (A \times Y)$ 且 $f^{(0)}$ 是从 $A^{(0)}$ 到 $B^{(0)}$ 的映射.

证明 ($\Rightarrow$) 由引理 2.7.8, 我们只需要证明 $f = f^{(0)} \wedge (A \times Y)$.

$f \leqslant f^{(0)} \wedge (A \times Y)$ 是显然的. 为了证明 $f \geqslant f^{(0)} \wedge (A \times Y)$, 我们假设 $(f^{(0)} \wedge (A \times Y))(x,y) \geqslant a$, 这里 $(x,y) \in X \times Y, a \in J(L)$. 那么 $(x,y) \in f^{(0)}$ 且 $A(x) \geqslant a$. 因为 f 是从 A 到 B 的 L-模糊映射, 所以存在 $z \in B_{[a]}$ 使得 $(x,z) \in f_{[a]} \subseteq f^{(0)}$. 由引理 2.7.8 可以得到 $y = z$. 因此 $f(x,y) \geqslant a$. 这表明 $f(x,y) \geqslant (f^{(0)} \wedge (A \times Y))(x,y)$ 对于任意的 $(x,y) \in X \times Y$ 都成立. 于是我们得到 $f \geqslant f^{(0)} \wedge (A \times Y)$ 的证明.

($\Leftarrow$) 由定理 2.7.2 我们只需要证明对每个 $a \in P(L), f^{(a)}$ 是从 $A^{(a)}$ 到 $B^{(a)}$ 的映射. 设 $a \in P(L)$ 且 $x \in A^{(a)}$. 因为 $A^{(a)} \subseteq A^{(0)}$ 而且 $f^{(0)}$ 是从 $A^{(0)}$ 到 $B^{(0)}$ 的映射, 所以存在 $y \in B^{(0)}$ 使得 $(x,y) \in f^{(0)}$. 从而

$$f(x,y) = (f^{(0)} \wedge (A \times Y))(x,y) = A(x) \not\leqslant a.$$

这表明 $(x,y) \in f^{(a)} \subseteq A^{(a)} \times B^{(a)}$.

另外, 如果再存在一个 $z \in B^{(a)}$ 使得 $(x,z) \in f^{(a)}$, 那么显然有 $y = z$. 因此 $f^{(a)}$ 是从 $A^{(a)}$ 到 $B^{(a)}$ 的映射. □

引理 2.7.10 设 $0 \in P(L)$ 且 $f : A \to B$ 是一个 L-模糊映射. 则 $f^{(0)}(A) \leqslant B$.

证明 由定理 2.5.4 可知为了证明 $f^{(0)}(A) \leqslant B$, 我们只需要证明 $f^{(0)}(A^{(a)}) \subseteq B^{(a)}$ 对每个 $a \in P(L)$ 都成立. 事实上, 如果 $y \in f^{(0)}(A^{(a)})$, 那么存在 $x \in A^{(a)}$ 使得 $(x,y) \in f^{(0)}$. 因为 f 是从 A 到 B 的 L-模糊映射, 所以由定理 2.7.2 可知存在 $z \in B^{(a)}$ 使得 $(x,z) \in f^{(a)} \subseteq f^{(0)}$. 再由引理 2.7.8, 我们能够得到 $y = z$. 这表明 $y \in B^{(a)}$. 因此 $f^{(0)}(A^{(a)}) \subseteq B^{(a)}$. □

定理 2.7.11 设 $0 \in P(L)$ 且 $f : A \to B$ 是一个 L-模糊映射. 则
(1) f 是 L-模糊单射当且仅当 $f^{(0)}$ 是从 $A^{(0)}$ 到 $B^{(0)}$ 的单射;
(2) f 是 L-模糊满射当且仅当 $f^{(0)}(A) = B$.

证明 (1) 是显然的. 从定理 2.5.4 和定理 2.7.5 可知 (2) 成立. □

注 2.7.12 从定理 2.7.9 可知一个模糊映射可以看作一个支撑集上的 Zadeh 型函数, 因此在考虑两个模糊集之间的模糊映射时, 只需要考虑 Zadeh 型函数即可.

习 题 2

1. 对于一族 L-模糊集 $\{A_i \mid i \in \Omega\} \subseteq L^X$ 和 $a \in L$, 下面条件是否成立? 在什么条件下成立?

(1) $\left(\bigwedge_{i \in \Omega} A_i\right)_{(a)} = \bigcap_{i \in \Omega} (A_i)_{(a)}$; (2) $\left(\bigvee_{i \in \Omega} A_i\right)_{[a]} = \bigcup_{i \in \Omega} (A_i)_{[a]}$;

(3) $\left(\bigwedge_{i \in \Omega} A_i\right)^{(a)} = \bigcap_{i \in \Omega} (A_i)^{(a)}$; (4) $\left(\bigvee_{i \in \Omega} A_i\right)^{[a]} = \bigcup_{i \in \Omega} (A_i)^{[a]}$.

2. 试证明定理 2.1.3 中的 (1) 和 (4).
3. 试证明 $(\mathscr{P}(X)^L, \leqslant, \vee, \wedge)$ 是完全分配格.
4. 试证明定理 2.2.6.
5. 试证明例 2.2.9 中的两个结论.
6. 请详细写出定理 2.2.13 的证明.
7. 试证明例 2.3.2 中的结论.
8. 试证明定理 2.3.4.
9. 试证明定理 2.3.5.
10. 试证明定理 2.3.6.
11. 试证明定理 2.3.7.
12. 试证明定理 2.3.8.

13. 在定理 2.4.3 中, 我们证明了 $(A \times B)^{(a)} \subseteq A^{(a)} \times B^{(a)}$, 试证明当 $a \in P(L)$ 时, 恒有 $(A \times B)^{(a)} = A^{(a)} \times B^{(a)}$ 成立.

14. 试证明定理 2.4.5.

15. 试证明定理 2.4.9.

16. 设 $\{X_i \mid i \in \Omega\}$ 是一族集合, 对每个 $i \in \Omega$, $A_i \in L^{X_i}$. 则 $\forall a \in L$, 下面条件成立:

(1) $\left(\prod_{i \in \Omega} A_i\right)_{(a)} \subseteq \prod_{i \in \Omega}(A_i)_{(a)} \subseteq \prod_{i \in \Omega}(A_i)_{[a]} = \left(\prod_{i \in \Omega} A_i\right)_{[a]}$;

(2) $\left(\prod_{i \in \Omega} A_i\right)^{(a)} \subseteq \prod_{i \in \Omega}(A_i)^{(a)} \subseteq \prod_{i \in \Omega}(A_i)^{[a]} = \left(\prod_{i \in \Omega} A_i\right)^{[a]}$;

(3) $\prod_{i \in \Omega} A_i = \bigwedge_{a \in L}\left\{a \vee \left(\prod_{i \in \Omega}(A_i)^{[a]}\right)\right\} = \bigwedge_{a \in P(L)}\left\{a \vee \left(\prod_{i \in \Omega}(A_i)^{[a]}\right)\right\}$;

(4) $\prod_{i \in \Omega} A_i = \bigwedge_{a \in L}\left\{a \vee \left(\prod_{i \in \Omega}(A_i)^{(a)}\right)\right\} = \bigwedge_{a \in P(L)}\left\{a \vee \left(\prod_{i \in \Omega}(A_i)^{(a)}\right)\right\}$;

(5) $\prod_{i \in \Omega} A_i = \bigvee_{a \in L}\left\{a \wedge \left(\prod_{i \in \Omega}(A_i)_{(a)}\right)\right\} = \bigvee_{a \in J(L)}\left\{a \wedge \left(\prod_{i \in \Omega}(A_i)_{(a)}\right)\right\}$;

(6) $\prod_{i \in \Omega} A_i = \bigvee_{a \in L}\left\{a \wedge \left(\prod_{i \in \Omega}(A_i)_{[a]}\right)\right\} = \bigvee_{a \in J(L)}\left\{a \wedge \left(\prod_{i \in \Omega}(A_i)_{[a]}\right)\right\}$.

17. 试证明定理 2.5.3.

18. 详细给出定理 2.5.5 的证明.

19. 在定理 2.5.5 中, 我们指出 $f_L^{\rightarrow}\left(\bigwedge_{i \in \Omega} A_i\right) \subseteq \bigwedge_{i \in \Omega} f_L^{\rightarrow}(A_i)$ 成立, 那么逆包含是否成立? 试证明之.

20. 请给出定理 2.6.2 的详细证明.

21. 在定理 2.6.2 中, 试给出 (1) $\Rightarrow$ (4) 成立的充分条件.

22. 在 2.6 节中, 我们仅仅给出了 L-模糊偏序的定义和刻画, 能否进一步定义模糊格等概念? 请读者思考.

23. 设 $0 \in P(L), A \in L^X, B \in L^Y$ 且 $f \leqslant A \times B$. 如果 f 是从 A 到 B 的 L-模糊映射, 那么 $f_{(0)}$ 是从 $A_{(0)}$ 到 $B_{(0)}$ 的映射, 试证之.

24. 设 $0 \in P(L), A \in L^X, B \in L^Y$ 且 $f \leqslant A \times B$. 则 f 是从 A 到 B 的 L-模糊映射当且仅当 $f = f_{(0)} \wedge (A \times Y)$ 且 $f_{(0)}$ 是从 $A_{(0)}$ 到 $B_{(0)}$ 的映射.

25. 设 $0 \in P(L)$ 且 $f : A \to B$ 是一个 L-模糊映射. 则 $f_{(0)}(A) \leqslant B$.

26. 设 $0 \in P(L)$ 且 $f : A \to B$ 是一个 L-模糊映射. 则

(1) f 是 L-模糊单射当且仅当 $f_{(0)}$ 是从 $A_{(0)}$ 到 $B_{(0)}$ 的单射;

(2) f 是 L-模糊满射当且仅当 $f_{(0)}(A) = B$.

27. 设 $A \in L^X, B \in L^Y$ 且 $f \leqslant A \times B$. 试问下列条件是否等价:

(1) f 是从 A 到 B 的 L-模糊映射;

(2) $\forall a \in P(L)$, $f^{[a]}$ 是从 $A^{[a]}$ 到 $B^{[a]}$ 的映射;

(3) $\forall a \in L$, $f^{(a)}$ 是从 $A^{(a)}$ 到 $B^{(a)}$ 的映射.

第 3 章 (L,M)-模糊拓扑与 (L,M)-模糊凸结构

模糊拓扑的概念是由 Chang C L 于 1968 年在文献 [58] 中率先提出的, 开始是针对模糊集的值取自于单位区间 [0,1]. 随后又被 Goguen 推广到了取值为完备格的情形[77]. 这时的模糊拓扑指的是由一族模糊集构成的分明族. 后来它又被 Kubiak 和 Šostak 分别做了进一步的推广 (见 [85,148]), 在这种更加一般化的模糊拓扑定义中, 拓扑是由模糊集作为元素的模糊族构成的. 因为本书主要是介绍模糊代数和模糊凸结构方面的一些工作, 所以关于模糊拓扑的工作不做太多介绍, 感兴趣的同学可以参看专著 [20,97], 两本专著对 L-拓扑空间的工作介绍是相当全面的. 之所以介绍一些模糊拓扑的知识, 是因为它和模糊凸结构有着密切的联系和非常多的相似性. 另外, 作为集合套的应用, 也让读者熟悉它们的应用过程. 对不熟悉拓扑的读者可以跳过这一部分, 这样做并不会影响进一步的阅读. 下面我们根据本书的需要介绍一些相关模糊拓扑的最基本概念.

3.1 拓扑的模糊化

定义 3.1.1 设 X 是一个非空集合, 2^X 的一个子族 $\mathcal{T}$ 称为一个拓扑结构. 如果它满足下列三个条件:

(1) $\varnothing, X \in \mathcal{T}$;
(2) 若 $A, B \in \mathcal{T}$, 则 $A \cap B \in \mathcal{T}$;
(3) 若 $\{A_i \mid i \in \Omega\} \subseteq \mathcal{T}$, 则 $\bigcup_{i \in \Omega} A_i \in \mathcal{T}$.

若 $\mathcal{T}$ 是一个拓扑, 则称 $(X, \mathcal{T})$ 是一个拓扑空间.

把上述定义拓扑中的集合换成 L-模糊集, 作为集族的拓扑也换为 M-模糊集, 则得到下面一般性的 (L,M)-模糊拓扑的定义.

定义 3.1.2 设 X 是一个非空集合, L, M 是两个完备格. 映射 $\mathcal{T}: L^X \to M$ 称为一个 (L,M)-模糊拓扑, 如果它满足下列三个条件:

(1) $\mathcal{T}(\varnothing) = \mathcal{T}(X) = 1$;
(2) 对任意的 $A, B \in L^X$, 都有 $\mathcal{T}(A \wedge B) \geqslant \mathcal{T}(A) \wedge \mathcal{T}(B)$;
(3) 对任意的 $\{A_i \mid i \in \Omega\} \subseteq L^X$, 都有 $\mathcal{T}\left(\bigvee_{i \in \Omega} A_i\right) \geqslant \bigwedge_{i \in \Omega} \mathcal{T}(A_i)$.

若 $\mathcal{T}$ 是一个 (L,M)-模糊拓扑, 则称 $(X, \mathcal{T})$ 是一个 (L,M)-模糊拓扑空间.

一个 $(L,\mathbf{2})$-模糊拓扑空间也叫做一个 L-拓扑空间, 一个 $(\mathbf{2},M)$-模糊拓扑空间也叫做一个 M-模糊化拓扑空间, 一个 (L,L)-模糊拓扑空间也叫做一个 L-模糊拓扑空间.

当 $M = \mathbf{2} = \{0,1\}$ 时, L-拓扑空间的定义可简化为下面情形.

定义 3.1.3 设 X 是一个非空集合, L 是一个完备格. $\mathscr{T} \subseteq L^X$ 叫做一个 L-拓扑, 如果它满足下列三个条件:

(1) $\varnothing, X \in \mathscr{T}$;

(2) 对任意的 $A, B \in \mathscr{T}$, $A \wedge B \in \mathscr{T}$;

(3) 对任意的 $\{A_i \mid i \in \Omega\} \subseteq \mathscr{T}$, 都有 $\bigvee_{i \in \Omega} A_i \in \mathscr{T}$.

当 $L = \mathbf{2} = \{0,1\}$ 时, M-模糊化拓扑空间的定义可简化为下面情形.

定义 3.1.4 设 X 是一个非空集合, M 是一个完备格. 一个映射 $\mathscr{T} : \mathbf{2}^X \to M$ 称为一个 M-模糊化拓扑, 如果它满足下列三个条件:

(1) $\mathscr{T}(\varnothing) = \mathscr{T}(X) = 1$;

(2) 对任意的 $A, B \in \mathbf{2}^X$, 都有 $\mathscr{T}(A \cap B) \geqslant \mathscr{T}(A) \wedge \mathscr{T}(B)$;

(3) 对任意的 $\{A_i \mid i \in \Omega\} \subseteq \mathbf{2}^X$, 都有 $\mathscr{T}\left(\bigcup_{i \in \Omega} A_i\right) \geqslant \bigwedge_{i \in \Omega} \mathscr{T}(A_i)$.

两个拓扑空间之间的连续映射也可以被推广到两个 (L,M)-模糊拓扑空间之间, 下面是它的一般形式.

定义 3.1.5 设 $(X, \mathscr{T})$ 和 $(Y, \mathscr{U})$ 是两个 (L,M)-模糊拓扑空间. 映射 $f : X \to Y$ 称为

(1) (L,M)-模糊连续映射, 若任给 $B \in L^Y$, 有 $\mathscr{U}(B) \leqslant \mathscr{T}(f_L^{\leftarrow}(B))$;

(2) (L,M)-模糊开映射, 若任给 $A \in L^X$, 有 $\mathscr{T}(A) \leqslant \mathscr{U}(f_L^{\rightarrow}(A))$;

(3) (L,M)-模糊同胚映射, 如果 f 是双射的、(L,M)-模糊连续的和 (L,M)-模糊开的.

为了简便, (L,L)-模糊连续映射也称为 L-模糊连续映射, $(L,\mathbf{2})$-模糊连续映射也称为 L-连续映射, $(\mathbf{2},M)$-模糊连续映射称为 M-模糊化连续映射.

注 3.1.6 M-模糊化拓扑和 M-模糊化连续映射的概念来源于文献 [164].

借助两种截集, 我们可以给出 (L,M)-模糊拓扑和 (L,M)-模糊连续映射的刻画 (见下面两个定理), 它们的证明是简单的, 我们把它们留给读者.

定理 3.1.7 设 $\mathscr{T} : L^X \to M$ 是一个映射. 则下列条件等价:

(1) $\mathscr{T}$ 是一个 (L,M)-模糊拓扑;

(2) $\forall a \in M$, $\mathscr{T}_{[a]}$ 是一个 L-拓扑;

(3) $\forall a \in M$, $\mathscr{T}^{[a]}$ 是一个 L-拓扑.

推论 3.1.8 设 $\mathscr{T} : \mathbf{2}^X \to M$ 是一个映射. 则下列条件等价:

(1) $\mathscr{T}$ 是一个 M-模糊化拓扑;
(2) $\forall a \in M$, $\mathscr{T}_{[a]}$ 是一个拓扑;
(3) $\forall a \in M$, $\mathscr{T}^{[a]}$ 是一个拓扑.

定理 3.1.9 设 $(X, \mathscr{T})$ 和 $(Y, \mathscr{U})$ 是两个 (L, M)-模糊拓扑空间. $f: X \to Y$ 是一个映射. 则下列条件等价:
(1) $f: (X, \mathscr{T}) \to (Y, \mathscr{U})$ 是 (L, M)-模糊连续映射;
(2) $\forall a \in M$, $f: (X, \mathscr{T}_{[a]}) \to (Y, \mathscr{U}_{[a]})$ 是 L-连续映射;
(3) $\forall a \in M$, $f: (X, \mathscr{T}^{[a]}) \to (Y, \mathscr{U}^{[a]})$ 是 L-连续映射.

推论 3.1.10 设 $(X, \mathscr{T})$ 和 $(Y, \mathscr{U})$ 是两个 M-模糊化拓扑空间. $f: X \to Y$ 是一个映射. 则下列条件等价:
(1) $f: (X, \mathscr{T}) \to (Y, \mathscr{U})$ 是 M-模糊化连续映射;
(2) $\forall a \in M$, $f: (X, \mathscr{T}_{[a]}) \to (Y, \mathscr{U}_{[a]})$ 是连续映射;
(3) $\forall a \in M$, $f: (X, \mathscr{T}^{[a]}) \to (Y, \mathscr{U}^{[a]})$ 是连续映射.

3.2 (L, M)-模糊内部算子

在这一节中, 我们将给出 (L, M)-模糊内部算子的概念, 并证明它和 (L, M)-模糊拓扑的一一对应关系. 本节和下一节内容主要源于文献 [132].

对于一个拓扑空间 $(X, \mathcal{T})$ 而言, 它的内部算子 $i: \mathbf{2}^X \to \mathbf{2}^X$ 满足下面条件:
(1) $i(X) = X$;
(2) $i(A) \leqslant A$;
(3) $i(A \wedge B) = i(A) \wedge i(B)$;
(4) $i(A) \leqslant i(i(A))$.

基于这个事实, 我们将推广这样的内部算子到 (L, M)-模糊拓扑空间, 为此先给出模糊点的概念.

定义 3.2.1 一个 L-模糊点就是 L^X 中的模糊集 x_λ, 它被定义为

$$x_\lambda(y) = \begin{cases} \lambda, & y = x, \\ 0, & y \neq x. \end{cases}$$

对另外一个 L-模糊集 A 而言, 如果 $x_\lambda \leqslant A$, 那么我们就称 x_λ 属于 A, 否则, 就称 x_λ 不属于 A.

一个单点集 $\{x\}$ 可以看作高为 1 的模糊点. 另外, 我们记 $J(L^X) = \{x_\lambda \mid \lambda \in J(L)\}$.

定义 3.2.2 X 上的一个 (L, M)-模糊内部算子就是满足下面条件的映射 $\text{Int}: L^X \to M^{J(L^X)}$.

(I1) $\mathrm{Int}(A)(x_\lambda) = \bigwedge_{\mu \prec \lambda} \mathrm{Int}(A)(x_\mu)$, $\forall x_\lambda \in J(L^X)$;

(I2) $\mathrm{Int}(X)(x_\lambda) = 1$, $\forall x_\lambda \in J(L^X)$;

(I3) $\mathrm{Int}(A)(x_\lambda) = 0$, $\forall x_\lambda \not\leqslant A$;

(I4) $\mathrm{Int}(A \wedge B) = \mathrm{Int}(A) \wedge \mathrm{Int}(B)$;

(I5) $\forall a \in L \backslash \{1\}$, $(\mathrm{Int}(A))^{(a)} \subseteq \left(\mathrm{Int}\left(\bigvee (\mathrm{Int}(A))^{(a)}\right)\right)^{(a)}$.

$\mathrm{Int}(A)(x_\lambda)$ 叫做 x_λ 属于 A 的程度.

一个带有 (L,M)-模糊内部算子 Int 的集合 X (表示为 (X, Int)) 叫做一个 (L,M)-模糊内部空间.

两个 (L,M)-模糊内部空间 (X, Int_X) 和 (Y, Int_Y) 之间的一个映射 $f: X \to Y$ 叫做连续的, 如果 $\forall x_\lambda \in J(L^X)$, $\forall U \in L^Y$, 都有下面不等式成立.

$$\mathrm{Int}_Y(U)(f_L^\to(x_\lambda)) \leqslant \mathrm{Int}_X(f_L^\leftarrow(U))(x_\lambda).$$

下面的定理给出了 (L,M)-模糊拓扑与 (L,M)-模糊内部算子之间的关系.

定理 3.2.3 设 $\mathcal{T}$ 是 X 上的一个 (L,M)-模糊拓扑. 定义一个映射 $\mathrm{Int}^\mathcal{T}: L^X \to M^{J(L^X)}$ 使得

$$\mathrm{Int}^\mathcal{T}(A)(x_\lambda) = \bigvee_{x_\lambda \leqslant V \leqslant A} \mathcal{T}(V).$$

那么 $\mathrm{Int}^\mathcal{T}$ 是一个 (L,M)-模糊内部算子.

证明 (I1), (I2) 和 (I3) 是显然的. 另外, 容易看出 $\mathrm{Int}^\mathcal{T}$ 是保序的, 因此为了证明 (I4), 我们只需要证明

$$\mathrm{Int}(A \wedge B)(x_\lambda) \geqslant \mathrm{Int}(A)(x_\lambda) \wedge \mathrm{Int}(B)(x_\lambda).$$

为此我们任取 $a \in M$ 使得 $a \prec \mathrm{Int}(A)(x_\lambda) \wedge \mathrm{Int}(B)(x_\lambda)$, 则 $a \prec \mathrm{Int}(A)(x_\lambda)$ 且 $a \prec \mathrm{Int}(B)(x_\lambda)$. 这意味着存在满足 $x_\lambda \leqslant V \leqslant A$ 和 $x_\lambda \leqslant W \leqslant B$ 的 V, W 使得 $a \prec \mathcal{T}(V)$ 和 $a \prec \mathcal{T}(W)$. 于是我们可以得到 $x_\lambda \leqslant V \wedge W \leqslant A \wedge B$ 且

$$a \leqslant \mathcal{T}(V) \wedge \mathcal{T}(W) \leqslant \mathcal{T}(V \wedge W) \leqslant \bigvee_{x_\lambda \leqslant U \leqslant A \wedge B} \mathcal{T}(U) = \mathrm{Int}(A \wedge B)(x_\lambda).$$

这样就证明了 $\mathrm{Int}(A \wedge B)(x_\lambda) \geqslant \mathrm{Int}(A)(x_\lambda) \wedge \mathrm{Int}(B)(x_\lambda)$.

为了证明 (I5), 设 $x_\lambda \in (\mathrm{Int}^\mathcal{T}(A))^{(a)}$, 则

$$\mathrm{Int}^\mathcal{T}(A)(x_\lambda) = \bigvee_{x_\lambda \leqslant U \leqslant A} \mathcal{T}(U) \not\leqslant a.$$

3.2 (L,M)-模糊内部算子

从而存在 $U \in L^X$ 使得 $x_\lambda \leqslant U \leqslant A$ 且 $\mathcal{T}(U) \not\leqslant a$. 这意味着对任何 $y_\mu \prec U$, 皆有 $y_\mu \in \left(\operatorname{Int}^{\mathcal{T}}(U)\right)^{(a)}$ 成立. 由 (I4) 可知

$$x_\lambda \leqslant U = \bigvee\{y_\mu : y_\mu \prec U\} \leqslant \bigvee \left(\operatorname{Int}^{\mathcal{T}}(U)\right)^{(a)} \leqslant \bigvee \left(\operatorname{Int}^{\mathcal{T}}(A)\right)^{(a)}.$$

因此可得到

$$\operatorname{Int}^{\mathcal{T}}\left(\bigvee(\operatorname{Int}^{\mathcal{T}}(A))^{(a)}\right)(x_\lambda) = \bigvee_{x_\lambda \leqslant U \leqslant \bigvee(\operatorname{Int}^{\mathcal{T}}(A))^{(a)}} \mathcal{T}(U) \not\leqslant a.$$

因此有 $x_\lambda \in \left(\operatorname{Int}^{\mathcal{T}}\left(\bigvee(\operatorname{Int}^{\mathcal{T}}(A))^{(a)}\right)\right)^{(a)}$. 这证明 (I5) 成立. $\square$

定理 3.2.4 如果 $f : (X, \mathcal{T}_1) \to (Y, \mathcal{T}_2)$ 关于 (L,M)-模糊拓扑 $\mathcal{T}_1$ 和 $\mathcal{T}_2$ 是连续的, 那么 $f : (X, \operatorname{Int}^{\mathcal{T}_1}) \to (Y, \operatorname{Int}^{\mathcal{T}_2})$ 关于 (L,M)-模糊内部算子 $\operatorname{Int}^{\mathcal{T}_1}$ 和 $\operatorname{Int}^{\mathcal{T}_2}$ 是连续的.

证明 如果 $f : (X, \mathcal{T}_1) \to (Y, \mathcal{T}_2)$ 是连续的, 那么 $\forall U \in L^Y$, 都有 $\mathcal{T}_2(U) \leqslant \mathcal{T}_1(f_L^\leftarrow(U))$, 这意味着 $\forall x_\lambda \in J(L^X)$,

$$\operatorname{Int}^{\mathcal{T}_2}(U)(f_L^\to(x_\lambda)) = \bigvee_{f_L^\to(x_\lambda) \leqslant V \leqslant U} \mathcal{T}_2(V)$$
$$\leqslant \bigvee_{x_\lambda \leqslant f_L^\leftarrow(V) \leqslant f_L^\leftarrow(U)} \mathcal{T}_1(f_L^\leftarrow(V))$$
$$\leqslant \operatorname{Int}^{\mathcal{T}_1}(f_L^\leftarrow(U))(x_\lambda).$$

因此 $f : (X, \operatorname{Int}^{\mathcal{T}_1}) \to (Y, \operatorname{Int}^{\mathcal{T}_2})$ 是连续的. $\square$

定理 3.2.5 设 $\operatorname{Int} : L^X \to L^{J(L^X)}$ 是一个 (L,M)-模糊内部算子. $\forall x_\lambda \in J(L^X)$, 定义 $\mathcal{T}^{\operatorname{Int}} : L^X \to L$ 使得

$$\mathcal{T}^{\operatorname{Int}}(U) = \bigwedge_{x_\lambda \prec U} \operatorname{Int}(U)(x_\lambda).$$

则 $\mathcal{T}^{\operatorname{Int}}$ 是一个 (L,M)-模糊拓扑且 $\operatorname{Int}^{(\mathcal{T}^{\operatorname{Int}})} = \operatorname{Int}$.

证明 (1) $\mathcal{T}^{\operatorname{Int}}(\varnothing) = \mathcal{T}^{\operatorname{Int}}(X) = 1$ 是显然的.

(2) 为了证明 $\forall A, B \in L^X, \mathcal{T}^{\operatorname{Int}}(A \wedge B) \geqslant \mathcal{T}^{\operatorname{Int}}(A) \wedge \mathcal{T}^{\operatorname{Int}}(B)$, 设 $a \prec \mathcal{T}^{\operatorname{Int}}(A) \wedge \mathcal{T}^{\operatorname{Int}}(B)$, 则 $a \prec \mathcal{T}^{\operatorname{Int}}(A)$ 且 $a \prec \mathcal{T}^{\operatorname{Int}}(B)$, 于是 $\forall x_\lambda \prec A \wedge B$, 有

$$a \prec \operatorname{Int}(A)(x_\lambda) \wedge \operatorname{Int}(B)(x_\lambda) = \operatorname{Int}(A \wedge B)(x_\lambda).$$

这表明 $a \leqslant \bigwedge_{x_\lambda \prec A \wedge B} \operatorname{Int}(A \wedge B)(x_\lambda) = \mathcal{T}^{\operatorname{Int}}(A \wedge B)$. 这就证明了 $\mathcal{T}^{\operatorname{Int}}(A \wedge B) \geqslant \mathcal{T}^{\operatorname{Int}}(A) \wedge \mathcal{T}^{\operatorname{Int}}(B)$.

(3) 为了证明对于一族 L-模糊集 $\{A_i \mid i \in \Omega\}$, $\mathcal{T}^{\mathrm{Int}}\left(\bigvee_{i \in \Omega} A_i\right) \geqslant \bigwedge_{i \in \Omega} \mathcal{T}^{\mathrm{Int}}(A_i)$, 设 $a \leqslant \bigwedge_{i \in \Omega} \mathcal{T}^{\mathrm{Int}}(A_i)$, 则 $\forall i \in \Omega$, $a \leqslant \mathcal{T}^{\mathrm{Int}}(A_i)$, 这意味着 $\forall x_\lambda \prec A_i$, 有 $a \leqslant \mathrm{Int}(A_i)(x_\lambda)$. 因为 $x_\lambda \prec \bigvee_{i \in \Omega} A_i \Leftrightarrow \exists i_0 \in \Omega, x_\lambda \prec A_{i_0}$, 所以有

$$a \leqslant \mathrm{Int}(A_{i_0})(x_\lambda) \leqslant \mathrm{Int}\left(\bigvee_{i \in \Omega} A_i\right)(x_\lambda).$$

进一步, 我们可以得到

$$a \leqslant \bigwedge_{x_\lambda \prec \bigvee_{i \in \Omega} A_i} \mathrm{Int}\left(\bigvee_{i \in \Omega} A_i\right)(x_\lambda) = \mathcal{T}^{\mathrm{Int}}\left(\bigvee_{i \in \Omega} A_i\right).$$

最后, 为了证明 $\mathrm{Int}^{(\mathcal{T}^{\mathrm{Int}})} = \mathrm{Int}$, 我们先证明 $\mathrm{Int}^{(\mathcal{T}^{\mathrm{Int}})}(A) \leqslant \mathrm{Int}(A)$, $\forall A \in L^X$. 为此设 $a \in M$, x_λ 是任意一个模糊点且 $a \prec \mathrm{Int}^{(\mathcal{T}^{\mathrm{Int}})}(A)(x_\lambda)$, 则存在 $V \in L^X$ 使得 $x_\lambda \leqslant V \leqslant A$ 且 $a \prec \mathcal{T}^{\mathrm{Int}}(V)$. 于是 $\forall y_\mu \prec V$, 都有 $a \prec \mathrm{Int}(V)(y_\mu)$. 这意味着 $x_\lambda \leqslant V \leqslant \bigvee \mathrm{Int}(V)_{(a)} \leqslant \bigvee \mathrm{Int}(A)_{(a)}$. 借助于 (I1) 可得 $a \leqslant \mathrm{Int}(A)(x_\lambda)$. 这表明 $\mathrm{Int}^{(\mathcal{T}^{\mathrm{Int}})}(A) \leqslant \mathrm{Int}(A)$.

反之, 设 $\mathrm{Int}(A)(x_\lambda) \not\leqslant a$, 则存在 $b \in \alpha(a)$ 使得 $\mathrm{Int}(A)(x_\lambda) \not\leqslant b$, 于是 $x_\lambda \in \mathrm{Int}(A)^{(b)} \subseteq \mathrm{Int}\left(\bigvee \mathrm{Int}(A)^{(b)}\right)^{(b)}$. 令 $V = \bigvee \mathrm{Int}(A)^{(b)}$, 则 $x_\lambda \leqslant V \leqslant A$ 且 $\forall y_\mu \prec V$, 皆有 $\mathrm{Int}(V)(y_\mu) \not\leqslant b$. 这样我们便可以得到 $\mathcal{T}^{\mathrm{Int}}(V) = \bigwedge_{y_\mu \prec V} \mathrm{Int}(V)(y_\mu) \not\leqslant a$. 从而 $\mathrm{Int}^{(\mathcal{T}^{\mathrm{Int}})}(A)(x_\lambda) \not\leqslant a$. 这样就证明了 $\mathrm{Int}^{(\mathcal{T}^{\mathrm{Int}})}(A) \geqslant \mathrm{Int}(A)$. □

定理 3.2.6 如果 $f : (X, \mathrm{Int}_X) \to (Y, \mathrm{Int}_Y)$ 关于 (L, M)-模糊内部算子 Int_X 和 Int_Y 是连续的, 那么 $f : (X, \mathcal{T}^{\mathrm{Int}_X}) \to (Y, \mathcal{T}^{\mathrm{Int}_Y})$ 关于 (L, M)-模糊拓扑 $\mathcal{T}^{\mathrm{Int}_X}$ 和 $\mathcal{T}^{\mathrm{Int}_Y}$ 是连续的.

证明 如果 $f : (X, \mathrm{Int}_X) \to (Y, \mathrm{Int}_Y)$ 是连续的, 那么 $\forall U \in L^Y, \forall x_\lambda \in J(L^X)$,

$$\mathrm{Int}_Y(U)(f^\to(x_\lambda)) \leqslant \mathrm{Int}_X(f^\leftarrow(U))(x_\lambda),$$

这意味着

$$\mathcal{T}^{\mathrm{Int}_Y}(U) = \bigwedge_{y_\mu \prec U} \mathrm{Int}_Y(U)(y_\mu)$$

$$\leqslant \bigwedge_{f_L^\to(x_\lambda) \prec U} \mathrm{Int}_Y(U)(f_L^\to(x_\lambda))$$

$$\leqslant \bigwedge_{x_\lambda \prec f_L^\leftarrow(U)} \mathrm{Int}_Y(U)(f_L^\rightarrow(x_\lambda))$$

$$\leqslant \bigwedge_{x_\lambda \prec f_L^\leftarrow(U)} \mathrm{Int}_X(f_L^\leftarrow(U))(x_\lambda) = \mathcal{T}^{\mathrm{Int}_X}(f_L^\leftarrow(U)).$$

因此 $f:(X, \mathcal{T}^{\mathrm{Int}_X}) \to (Y, \mathcal{T}^{\mathrm{Int}_Y})$ 是连续的. □

3.3 (L,M)-模糊闭包算子

在这一节中, 我们引入 (L,M)-模糊闭包算子的概念, 并研究它和 (L,M)-模糊拓扑之间的一一对应关系.

对于一个拓扑空间 $(X, \mathcal{T})$ 而言, 它的闭包算子 $c: \mathbf{2}^X \to \mathbf{2}^X$ 满足下面条件:
(1) $c(\varnothing) = \varnothing$;
(2) $c(A) \geqslant A$;
(3) $c(A \vee B) = c(A) \vee c(B)$;
(4) $c(A) \geqslant c(c(A))$.

下面我们推广这个闭包算子到一般模糊情形如下:

定义 3.3.1 一个集合 X 上的 (L,M)-模糊闭包算子就是一个满足下面条件的映射 $\mathrm{Cl}: L^X \to M^{J(L^X)}$:

(Cl1) $\mathrm{Cl}(A)(x_\lambda) = \bigwedge_{\mu \prec \lambda} \mathrm{Cl}(A)(x_\mu), \forall x_\lambda \in J(L^X)$;

(Cl2) $\mathrm{Cl}(\varnothing)(x_\lambda) = 0, \forall x_\lambda \in J(L^X)$;

(Cl3) $\mathrm{Cl}(A)(x_\lambda) = 1, \forall x_\lambda \leqslant A$;

(Cl4) $\mathrm{Cl}(A \vee B) = \mathrm{Cl}(A) \vee \mathrm{Cl}(B)$;

(Cl5) $\forall a \in L \backslash \{0\}, \left(\mathrm{Cl}\left(\bigvee (\mathrm{Cl}(A))_{[a]}\right)\right)_{[a]} \subseteq (\mathrm{Cl}(A))_{[a]}$.

$\mathrm{Cl}(A)(x_\lambda)$ 叫做 x_λ 属于 A 的闭包的程度.

一个带有 (L,M)-模糊闭包算子 Cl 的集合 X (也被记为 (X, Cl)) 叫做一个 (L,M)-模糊闭包空间.

两个 (L,M)-模糊闭包空间 (X, Cl_X) 和 (Y, Cl_Y) 之间的映射 $f: X \to Y$ 叫做连续的, 如果 $\forall x_\lambda \in J(L^X), \forall A \in L^X$, 都有

$$\mathrm{Cl}_X(A)(x_\lambda) \leqslant \mathrm{Cl}_Y(f_L^\rightarrow(A))(f_L^\rightarrow(x_\lambda)).$$

下面的定理显然成立.

定理 3.3.2 两个 (L,M)-模糊闭包空间 (X, Cl_X) 和 (Y, Cl_Y) 之间的映射 $f: X \to Y$ 是连续的当且仅当 $\forall x_\lambda \in J(L^X), \forall B \in L^Y$, 都有

$$\mathrm{Cl}_X(f_L^\leftarrow(B))(x_\lambda) \leqslant \mathrm{Cl}_Y(B)(f_L^\rightarrow(x_\lambda)).$$

定理 3.3.3 设 $\mathcal{T}$ 是 X 上的一个 (L,M)-模糊拓扑. 定义一个映射 $\mathrm{Cl}^{\mathcal{T}}$: $L^X \to M^{J(L^X)}$ 使得 $\forall x_\lambda \in J(L^X), \forall A \in L^X$,

$$\mathrm{Cl}^{\mathcal{T}}(A)(x_\lambda) = \bigwedge_{x_\lambda \not\leq D \geq A} (\mathcal{T}(D'))',$$

那么 $\mathrm{Cl}^{\mathcal{T}}$ 是一个 (L,M)-模糊闭包算子.

证明 (Cl1), (Cl2) 和 (Cl3) 是显然的. 下面证明 (Cl4) 和 (Cl5).

(Cl4) 为了证明 (Cl4) 成立, 我们只需要证明下面不等式成立即可.

$$\mathrm{Cl}^{\mathcal{T}}(A \vee B)(x_\lambda) \leq \mathrm{Cl}^{\mathcal{T}}(A)(x_\lambda) \vee \mathrm{Cl}^{\mathcal{T}}(B)(x_\lambda).$$

为此我们设 $a \in J(M), a \leq \mathrm{Cl}^{\mathcal{T}}(A \vee B)(x_\lambda)$, 则对满足 $x_\lambda \not\leq F \geq A \vee B$ 的任何 $F \in L^X$, 皆有 $a \leq (\mathcal{T}(F'))'$. 假如 $a \not\leq \mathrm{Cl}^{\mathcal{T}}(A)(x_\lambda) \vee \mathrm{Cl}^{\mathcal{T}}(B)(x_\lambda)$, 则存在满足条件 $x_\lambda \not\leq D \geq A$ 和 $x_\lambda \not\leq E \geq B$ 的 $D, E \in L^X$ 使得 $a \not\leq (\mathcal{T}(D'))'$ 且 $a \not\leq (\mathcal{T}(E'))'$. 于是

$$a \not\leq (\mathcal{T}(D'))' \vee (\mathcal{T}(E'))' = (\mathcal{T}(D') \wedge \mathcal{T}(E'))' \geq (\mathcal{T}((D \vee E)'))'.$$

令 $F = D \vee E$, 则 $a \not\leq (\mathcal{T}(F'))'$, 这与 $a \leq (\mathcal{T}(F'))'$ 矛盾. 这样我们就证明了

$$\mathrm{Cl}^{\mathcal{T}}(A \vee B)(x_\lambda) \leq \mathrm{Cl}^{\mathcal{T}}(A)(x_\lambda) \vee \mathrm{Cl}^{\mathcal{T}}(B)(x_\lambda).$$

(Cl5) 为了证明 (Cl5) 成立, 我们设 $a \in L \setminus \{0\}$ 且 $x_\lambda \notin (\mathrm{Cl}^{\mathcal{T}}(A))_{[a]}$. 则存在满足 $x_\lambda \not\leq D \geq A$ 的 D 使得 $a \not\leq \mathcal{T}(D')'$. 这意味着 $\forall y_\mu \not\leq D$, 皆有

$$a \not\leq \bigwedge_{y_\mu \not\leq E \geq D} \mathcal{T}(E')' = \mathrm{Cl}^{\mathcal{T}}(D)(y_\mu).$$

于是对任何的 $y_\mu \not\leq D$, 我们都能得到 $y_\mu \notin \left(\mathrm{Cl}^{\mathcal{T}}(D)\right)_{[a]}$. 这说明

$$x_\lambda \not\leq D \geq \bigvee \left(\mathrm{Cl}^{\mathcal{T}}(D)\right)_{[a]} \geq \bigvee \left(\mathrm{Cl}^{\mathcal{T}}(A)\right)_{[a]}.$$

因此可得

$$\mathrm{Cl}^{\mathcal{T}}\left(\bigvee (\mathrm{Cl}^{\mathcal{T}}(A))_{[a]}\right)(x_\lambda) = \bigwedge_{x_\lambda \not\leq D \geq \bigvee (\mathrm{Cl}^{\mathcal{T}}(A))_{[a]}} \mathcal{T}(D')' \not\geq a.$$

这样我们就证明了 $x_\lambda \notin \left(\mathrm{Cl}^{\mathcal{T}}\left(\bigvee (\mathrm{Cl}^{\mathcal{T}}(A))_{[a]}\right)\right)_{[a]}$. □

3.3 (L,M)-模糊闭包算子

定理 3.3.4 如果 $f:(X,\mathcal{T}_1)\to(Y,\mathcal{T}_2)$ 是连续的, 那么 $f:(X,\text{Cl}^{\mathcal{T}_1})\to(Y,\text{Cl}^{\mathcal{T}_2})$ 也是连续的.

证明 如果 $f:(X,\mathcal{T}_1)\to(Y,\mathcal{T}_2)$ 是连续的, 那么 $\forall B\in L^Y$, 皆有 $\mathcal{T}_2(B)\leqslant \mathcal{T}_1(f_L^\leftarrow(B))$, 这意味着 $\forall x_\lambda\in J(L^X)$, 下面不等式成立:

$$\text{Cl}^{\mathcal{T}_2}(B)(f_L^\rightarrow(x_\lambda)) = \bigwedge_{f_L^\rightarrow(x_\lambda)\not\leqslant D\geqslant B} (\mathcal{T}_2(D'))'$$
$$\geqslant \bigwedge_{x_\lambda\not\leqslant f_L^\leftarrow(D)\geqslant f_L^\leftarrow(B)} (\mathcal{T}_1(f_L^\leftarrow(D')))'$$
$$\geqslant \text{Cl}^{\mathcal{T}_1}(f_L^\leftarrow(B))(x_\lambda).$$

因此 $f:(X,\text{Cl}^{\mathcal{T}_1})\to(Y,\text{Cl}^{\mathcal{T}_2})$ 是连续的. $\square$

定理 3.3.5 设 $\text{Cl}:L^X\to M^{J(L^X)}$ 是一个 (L,M)-模糊闭包算子. 定义 $\mathcal{T}^{\text{Cl}}:L^X\to M$ 使得

$$\mathcal{T}^{\text{Cl}}(U) = \bigwedge_{x_\lambda\not\leqslant U'}(\text{Cl}(U')(x_\lambda))'.$$

那么 $\mathcal{T}^{\text{Cl}}$ 是一个 (L,M)-模糊拓扑且 $\text{Cl}^{(\mathcal{T}^{\text{Cl}})}=\text{Cl}$.

证明 (1) 不难验证 $\mathcal{T}^{\text{Cl}}(\varnothing)=\mathcal{T}^{\text{Cl}}(X)=1$.

(2) 我们来证明 $\mathcal{T}^{\text{Cl}}(A\wedge B)\geqslant \mathcal{T}^{\text{Cl}}(A)\wedge \mathcal{T}^{\text{Cl}}(B)$. 假设 $b\in P(M)$ 且 $\mathcal{T}^{\text{Cl}}(A)\wedge \mathcal{T}^{\text{Cl}}(B)\not\leqslant b$, 则存在 $a\in\alpha^*(b)$ 使得 $\mathcal{T}^{\text{Cl}}(A)\wedge \mathcal{T}^{\text{Cl}}(B)\not\leqslant a$, 于是 $\mathcal{T}^{\text{Cl}}(A)\not\leqslant a$ 且 $\mathcal{T}^{\text{Cl}}(B)\not\leqslant a$. 这意味着 $\forall x_\lambda\not\leqslant A'$, $(\text{Cl}(A')(x_\lambda))'\not\leqslant a$ 且 $\forall y_\mu\not\leqslant B'$, $(\text{Cl}(B')(y_\mu))'\not\leqslant a$. 从而对于任意的 $x_\lambda\not\leqslant A'\vee B'$, 皆有

$$(\text{Cl}(A'\vee B')(x_\lambda))' = (\text{Cl}(A')(x_\lambda))'\wedge(\text{Cl}(B')(x_\lambda))'\not\leqslant a.$$

此时显然有

$$\mathcal{T}^{\text{Cl}}(A\wedge B) = \bigwedge_{x_\lambda\not\leqslant(A\wedge B)'}(\text{Cl}(A'\vee B')(x_\lambda))'$$
$$= (\text{Cl}(A')(x_\lambda))'\wedge(\text{Cl}(B')(x_\lambda))'\not\leqslant b.$$

这样我们就证明了 $\mathcal{T}^{\text{Cl}}(A\wedge B)\geqslant \mathcal{T}^{\text{Cl}}(A)\wedge \mathcal{T}^{\text{Cl}}(B)$.

(3) 我们再来证明 $\mathcal{T}^{\text{Cl}}\left(\bigvee_{i\in\Omega}A_i\right)\geqslant \bigwedge_{i\in\Omega}\mathcal{T}^{\text{Cl}}(A_i)$.

设 $b\in M$ 且 $\bigwedge_{i\in\Omega}\mathcal{T}^{\text{Cl}}(A_i)\not\leqslant b$, 则存在 $a\in\alpha^*(b)$ 使得 $\bigwedge_{i\in\Omega}\mathcal{T}^{\text{Cl}}(A_i)\not\leqslant a$. 于是 $\forall i\in\Omega$, 都有 $\mathcal{T}^{\text{Cl}}(A_i)\not\leqslant a$. 从而 $\forall x_\lambda\not\leqslant A_i'$, 皆有 $(\text{Cl}(A_i')(x_\lambda))'\not\leqslant a$. 由于

$$\mathrm{Cl}\left(\bigwedge_{i\in\Omega}A'_i\right)(x_\lambda) \leqslant \mathrm{Cl}(A'_i)(x_\lambda), \text{ 所以 } \left(\mathrm{Cl}\left(\bigwedge_{i\in\Omega}A'_i\right)(x_\lambda)\right)' \not\leqslant a. \text{ 这样我们就可以}$$
得到

$$\mathcal{T}^{\mathrm{Cl}}\left(\bigvee_{i\in\Omega}A_i\right) = \bigwedge_{x_\lambda \not\leqslant \bigwedge_{i\in\Omega}A'_i}\left(\mathrm{Cl}\left(\bigwedge_{i\in\Omega}A'_i\right)(x_\lambda)\right)' \not\leqslant b.$$

故有 $\mathcal{T}^{\mathrm{Cl}}\left(\bigvee_{i\in\Omega}A_i\right) \geqslant \bigwedge_{i\in\Omega}\mathcal{T}^{\mathrm{Cl}}(A_i)$.

综上可知 $\mathcal{T}^{\mathrm{Cl}}$ 是一个 (L,M)-模糊拓扑.

$\mathrm{Cl}^{(\mathcal{T}^{\mathrm{Cl}})} = \mathrm{Cl}$ 的证明就留给读者了. □

定理 3.3.6 如果 $f:(X,\mathrm{Cl}_X) \to (Y,\mathrm{Cl}_Y)$ 关于 (L,M)-模糊闭包算子 Cl_X 和 Cl_Y 是连续的,那么 $f:(X,\mathcal{T}^{\mathrm{Cl}_X}) \to (Y,\mathcal{T}^{\mathrm{Cl}_Y})$ 关于 (L,M)-模糊拓扑 $\mathcal{T}^{\mathrm{Cl}_X}$ 和 $\mathcal{T}^{\mathrm{Cl}_Y}$ 也是连续的.

证明 如果 $f:(X,\mathrm{Cl}_X) \to (Y,\mathrm{Cl}_Y)$ 是连续的,那么 $\forall B \in L^Y, \forall x_\lambda \in J(L^X)$,皆有

$$\mathrm{Cl}_Y(B)(f^\to(x_\lambda)) \geqslant \mathrm{Cl}_X(f^\leftarrow(B))(x_\lambda),$$

这意味着

$$\mathcal{T}^{\mathrm{Cl}_Y}(B) = \bigwedge_{y_\mu \not\leqslant B'}(\mathrm{Cl}_Y(B')(y_\mu))'$$
$$\leqslant \bigwedge_{f_L^\to(x_\lambda)\not\leqslant B'}(\mathrm{Cl}_Y(B')(f_L^\to(x_\lambda)))'$$
$$= \bigwedge_{x_\lambda \not\leqslant f_L^\leftarrow(B')}(\mathrm{Cl}_Y(B')(f_L^\to(x_\lambda)))'$$
$$\leqslant \bigwedge_{x_\lambda \not\leqslant f_L^\leftarrow(B')}(\mathrm{Cl}_X(f_L^\leftarrow(B'))(x_\lambda))'$$
$$= \mathcal{T}^{\mathrm{Cl}_X}(f_L^\leftarrow(B)).$$

因此 $f:(X,\mathcal{T}^{\mathrm{Cl}_X}) \to (Y,\mathcal{T}^{\mathrm{Cl}_Y})$ 是连续的. □

3.4 诱导拓扑空间

在 L-拓扑学中,诱导空间起着非常重要的作用,诱导空间中的开集恰是由下半连续映射的全体构成的. 关于诱导空间中的闭包、内部和导集等运算在文献 [25]

3.4 诱导拓扑空间

中已提供了关系表达式, 但是其证明依赖于文献 [23]. 作为集合套的一个重要应用, 本节将介绍诱导空间的概念及其相关结果, 我们的结果相比文献 [23] 要简单明了.

以下 L 总表示一个具有逆序对合对应的完全分配格.

我们首先考虑一个分明拓扑空间诱导一个 L-拓扑空间的方法.

定理 3.4.1 设 $(X, \mathscr{T})$ 是一个拓扑空间, c 表示 $(X, \mathscr{T})$ 中的闭包算子. 映射 $\widetilde{c}: L^X \to L^X$ 被定义如下:

$$\forall A \in L^X, \quad \widetilde{c}(A) = \bigvee_{a \in L} (a \wedge c(A_{(a)})),$$

则

(1) $\forall A \in L^X$, $\widetilde{c}(A) = \bigvee_{a \in L} (a \wedge c(A_{[a]}))$;

(2) $\forall A \in L^X$, $\widetilde{c}(A) = \bigwedge_{a \in L} (a \vee c(A^{[a]}))$;

(3) $\forall A \in L^X$, $\widetilde{c}(A) = \bigwedge_{a \in L} (a \vee c(A^{(a)}))$;

(4) $\widetilde{c}$ 是一个 L-闭包算子.

证明 (1) $\forall A \in L^X$, 令 $\overline{c}(A) = \bigvee_{a \in L} (a \wedge c(A_{[a]}))$. 则显然有 $\widetilde{c}(A) \leqslant \overline{c}(A)$. 下证 $\widetilde{c}(A) \geqslant \overline{c}(A)$. 对于任意的 $x \in X$, 设 $b \prec \overline{c}(A)(x)$. 则存在 $a \in L$ 使得 $b \prec a$ 且 $x \in c(A_{[a]})$. 取 $e \in L$ 使得 $b \prec e \prec a$, 则 $x \in c(A_{[a]}) \subseteq c(A_{(e)})$, 这意味着 $b \leqslant \bigvee_{e \in L} (e \wedge c(A_{(e)})) = \widetilde{c}(A)(x)$. 因此得到了 $\widetilde{c}(A) \geqslant \overline{c}(A)$ 的证明.

类似 (1), 借助定理 2.2.6 可证

$$\forall A \in L^X, \quad \bigwedge_{a \in L} (a \vee c(A^{[a]})) = \bigwedge_{a \in L} (a \vee c(A^{(a)})).$$

为了证明 (2) 和 (3), 我们现在设 $\underline{c}(A) = \bigwedge_{a \in L} (a \vee c(A^{(a)}))$. 则由定理 2.2.5 和定理 2.2.6 可知下面两个包含式成立

$$\widetilde{c}(A)_{(a)} \subseteq c(A_{(a)}) \subseteq c(A_{[a]}) \subseteq \widetilde{c}(A)_{[a]},$$

$$\underline{c}(A)^{(a)} \subseteq c(A^{(a)}) \subseteq c(A^{[a]}) \subseteq \underline{c}(A)^{[a]}.$$

而由定理 2.1.4 知

$$\widetilde{c}(A)^{(a)} = \bigcup_{b \not\leqslant a} \widetilde{c}(A)_{(b)} \subseteq \bigcup_{b \not\leqslant a} c(A_{(b)}) \subseteq c\left(\bigcup_{b \not\leqslant a} A_{(b)}\right) = c(A^{(a)}) \subseteq \underline{c}(A)^{[a]}.$$

故 $\widetilde{c}(A) \leqslant \underline{c}(A)$. 另一方面, 由

$$\underline{c}(A)^{(a)} = \bigcup_{b \not\leqslant a} \underline{c}(A)_{(b)} \subseteq \bigcup_{b \not\leqslant a} c(A_{(b)}) \subseteq \bigcup_{b \not\leqslant a} \widetilde{c}(A)_{[b]} = \widetilde{c}(A)^{(a)}$$

可知 $\widetilde{c}(A) \geqslant \underline{c}(A)$, 这表明 (2) 和 (3) 成立.

(4) 显然 $\widetilde{c}(\chi_\varnothing) = \chi_\varnothing$ 且 $\widetilde{c}(A) \geqslant A$. 而 $\widetilde{c}(A \vee B) = \widetilde{c}(A) \vee \widetilde{c}(B)$ 由下式可证.

$$\begin{aligned}
\widetilde{c}(A \vee B) &= \bigvee_{a \in L} \left(a \wedge c\left((A \vee B)_{(a)}\right)\right) \\
&= \bigvee_{a \in L} \left(a \wedge c\left(A_{(a)} \cup B_{(a)}\right)\right) \\
&= \bigvee_{a \in L} \left(a \wedge c\left(A_{(a)}\right)\right) \vee \bigvee_{a \in L} \left(a \wedge c\left(B_{(a)}\right)\right) \\
&= \widetilde{c}(A) \vee \widetilde{c}(B).
\end{aligned}$$

最后来证明 $\widetilde{c}(\widetilde{c}(A)) \leqslant \widetilde{c}(A)$. 而这由下面不等式可证.

$$\begin{aligned}
\widetilde{c}(\widetilde{c}(A)) &= \bigvee_{a \in L} \left(a \wedge c\left(\widetilde{c}(A)_{(a)}\right)\right) \\
&\leqslant \bigvee_{a \in L} \left(a \wedge c\left(c(A_{(a)})\right)\right) \\
&\leqslant \bigvee_{a \in L} \left(a \wedge c\left(A_{(a)}\right)\right) = \widetilde{c}(A). \qquad \square
\end{aligned}$$

定义 3.4.2 设 $(X, \mathscr{T})$ 是一个拓扑空间, c 表示 $(X, \mathscr{T})$ 中的闭包算子, 则由 L-闭包算子 $\widetilde{c}: L^X \to L^X$ 可以诱导出 X 上的一个 L-拓扑, 记为 $\omega(\mathscr{T})$, 称 $(X, \omega(\mathscr{T}))$ 为由 $(X, \mathscr{T})$ 诱导的 L-拓扑空间, 而称 $\omega(\mathscr{T})$ 为一个诱导拓扑. 可以验证每个常值模糊集属于 $\omega(\mathscr{T})$.

根据定理 3.4.1 的证明不难证明下面定理.

定理 3.4.3 设 $(X, \mathscr{T})$ 是一个拓扑空间, i 表示 $(X, \mathscr{T})$ 中的内部算子. 映射 $\widetilde{i}: L^X \to L^X$ 定义如下:

$$\forall A \in L^X, \quad \widetilde{i}(A) = \bigvee_{a \in L} (a \wedge i(A_{(a)})),$$

则

(1) $\forall A \in L^X$, $\widetilde{i}(A) = \bigvee_{a \in L} (a \wedge i(A_{[a]}));$

3.4 诱导拓扑空间

(2) $\forall A \in L^X$, $\widetilde{i}(A) = \bigwedge_{a \in L} (a \vee i(A^{[a]}))$;

(3) $\forall A \in L^X$, $\widetilde{i}(A) = \bigwedge_{a \in L} (a \vee i(A^{(a)}))$;

(4) $\widetilde{i}$ 是一个 L-内部算子.

定理 3.4.4 设 $(X, \mathscr{T})$ 是一个拓扑空间, c, i 分别表示 $(X, \mathscr{T})$ 中的闭包算子和内部算子. 则它们诱导的 L-拓扑相同.

证明 由定理 3.4.1(3) 和定理 3.4.3(1) 可知 $\forall A \in L^X$,

$$\widetilde{c}(A)' = \left(\bigwedge_{a \in L}(a \vee c(A^{(a)}))\right)' = \bigvee_{a \in L}(a' \wedge c(A^{(a)})')$$
$$= \bigvee_{a \in L}(a' \wedge i((A^{(a)})')) = \bigvee_{a \in L}(a' \wedge i((A')_{[a']})) = \widetilde{i}(A').$$

因此 c, i 诱导的 L-拓扑相同. □

定理 3.4.5 设 $(X, \omega(\mathscr{T}))$ 为由拓扑空间 $(X, \mathscr{T})$ 诱导的 L-拓扑空间且 $A \in L^X$, 那么下面条件等价:

(1) $A \in \omega(\mathscr{T})$;
(2) $\forall a \in L$, $A^{(a)} \in \mathscr{T}$;
(3) $\forall a \in L$, $A_{(a)} \in \mathscr{T}$.

证明 (1) $\Rightarrow$ (2). 设 $A \in \omega(\mathscr{T})$, 则 $A \leqslant \widetilde{i}(A)$, 这意味着 $\forall a \in L$,

$$A^{(a)} \subseteq \widetilde{i}(A)^{(a)} = \bigcup_{b \in \alpha(a)} i(A^{(b)}) \subseteq i(A^{(a)}).$$

(2) $\Rightarrow$ (1). 设 $\forall a \in L$, $A^{(a)} \in \mathscr{T}$, 则 $i(A^{(a)}) = A^{(a)}$, 于是

$$\widetilde{i}(A) = \bigvee_{a \in L}(a \wedge i(A_{(a)})) = \bigvee_{a \in L}(a \wedge A_{(a)}) = A,$$

即 $A \in \omega(\mathscr{T})$.

(1) $\Rightarrow$ (3) 的证明类似于 (1) $\Rightarrow$ (2). (3) $\Rightarrow$ (1) 的证明类似于 (2) $\Rightarrow$ (1). □

弱诱导空间的概念最早是由张德学和刘应明于文献 [7] 中提出的, 定义如下:

定义 3.4.6 设 $(X, \mathscr{T})$ 是一个 L-拓扑空间. $[\mathscr{T}]$ 表示 $\mathscr{T}$ 中所有分明集的全体, 则可验证 $[\mathscr{T}]$ 是一个分明拓扑, 如果对任意的 $A \in \mathscr{T}$ 和任意的 $a \in L$, 均有 $A_{(a)} \in [\mathscr{T}]$, 则称 $(X, \mathscr{T})$ 为一个弱诱导空间. 如果每个常值模糊集属于 $\mathscr{T}$, 那么称 $(X, \mathscr{T})$ 为一个诱导空间.

从定理 3.4.5 可知, 定义 3.4.2 中的诱导空间等价于定义 3.4.6 中的相应概念.

3.5 Hutton 一致空间和 Erceg 伪度量空间

定义 3.5.1 如果一个映射 $f: L^X \to L^X$ 满足下面两条件:
(1) 对任意的 $A \in L^X$, $f(A) \geqslant A$;
(2) 对 L^X 的任意子集族 $\{A_i \mid i \in \Omega\}$, $f\left(\bigvee_{i\in\Omega} A_i\right) = \bigvee_{i\in\Omega} f(A_i)$,

那么就称 f 是 L^X 上的一个保并增值自映射.

定义 3.5.2 设 $\mathscr{D}$ 是 L^X 上的一族保并增值自映射, 其称为 L^X 上的一个一致结构, 如果满足下面条件:

(U1) $f \in \mathscr{D}, f \leqslant g \Rightarrow g \in \mathscr{D}$, 这里 g 也是 L^X 上的保并增值自映射;
(U2) $f, g \in \mathscr{D} \Rightarrow f \wedge g \in \mathscr{D}$, 这里 $(f \wedge g)(A) = \bigvee_{x \in A}(f(x) \wedge g(x))$;
(U3) $f \in \mathscr{D} \Rightarrow \exists g \in \mathscr{D}, g \circ g \leqslant f$, 这里 $g \circ g$ 表示两个映射的复合;
(U4) $f \in \mathscr{D} \Rightarrow f^{-1} \in \mathscr{D}$, 这里 $f^{-1}(A) = \bigwedge\{B \mid f(B') \leqslant A'\}$.

一个一致结构 $\mathscr{D}$ 的子集 $\mathscr{E}$ 称为 $\mathscr{D}$ 的一个基, 如果 $\forall f \in \mathscr{D}$, 存在一个 $g \in \mathscr{E}$ 使得 $g \leqslant f$. 当 $\mathscr{D}$ 有可数基时, 就称 $(X, \mathscr{D})$ 是一个伪度量空间.

注 3.5.3 在一般拓扑中, (伪) 度量是通过距离函数给出的, 上面的伪度量实际上是一个等价定义, 至于它们如何等价等问题, 我们就不详细论述了. 正是因为上述伪度量不涉及距离函数, 所以 Erceg 给出了一种基于模糊集之间的距离函数, 但是其条件过于复杂, 这里也就不介绍了. 不过后来梁基华[18]给出了它的一个等价刻画如下.

定理 3.5.4 L^X 上的一个 Erceg 伪度量等价于满足下列条件的一族映射 $\{f_r \mid f_r: L^X \to L^X, r > 0\}$.

(E1) $\forall A \in L^X$, $f_r(A) \geqslant A$;
(E2) $f_r\left(\bigvee_{i\in\Omega} A_i\right) = \bigvee_{i\in\Omega} f_r(A_i)$;
(E3) $f_r \circ f_s \leqslant f_{r+s}$;
(E4) $f_r = \bigvee_{s<r} f_s$;
(E5) $f^{-1} = f$.

定理 3.5.5 设 f 是 2^X 上的保并增值自映射. $\forall A \in L^X$, 规定

$$\widetilde{f}(A) = \bigvee_{a \in L} a \wedge f(A_{(a)}).$$

则

(1) $\widetilde{f}$ 是 L^X 上的保并增值自映射;

3.5 Hutton 一致空间和 Erceg 伪度量空间

(2) $\widetilde{f}(A) = \bigvee_{a \in L} (a \wedge f(A_{[a]}))$;

(3) $\widetilde{f}(A) = \bigwedge_{a \in L} (a \vee f(A^{(a)})) = \bigwedge_{a \in L} (a \vee f(A^{[a]}))$.

证明 (1) 是显然的.

(2) $\forall A \in L^X$, 令 $\overline{f}(A) = \bigvee_{a \in L}(a \wedge f(A_{[a]}))$, 则由定理 2.2.5 知, 对于任意的 $a \in L$, 有

$$\overline{f}(A)_{(a)} = \bigcup_{a \in \beta(b)} f(A_{[b]}) = f\left(\bigcup_{a \in \beta(b)} A_{[b]}\right)$$
$$= f(A_{(a)}) = \bigcup_{a \in \beta(b)} f(A_{(b)}) = \widetilde{f}(A)_{(a)},$$

于是 $\overline{f}(A) = \widetilde{f}(A)$.

(3) 类似 (2) 并借助定理 2.2.6 可证第二个等号成立. 现在设

$$\overline{f}(A) = \bigwedge_{a \in L}(a \vee f(A^{(a)})),$$

则由定理 2.2.5 知

$$\overline{f}(A)^{(a)} = \bigcup_{b \in \alpha(a)} f(A^{(b)}) = f\left(\bigcup_{b \in \alpha(a)} A^{(b)}\right) = f(A^{(a)}).$$

而由定理 2.1.4 与定理 2.2.3 知

$$\widetilde{f}(A)^{(a)} = \bigcup_{b \not\leq a} \left(\widetilde{f}(A)\right)_{(b)} \subseteq \bigcup_{b \not\leq a} f(A_{(b)}) = f\left(\bigcup_{b \not\leq a} A_{(b)}\right) = f(A^{(a)}).$$

故 $\widetilde{f}(A) \leqslant \overline{f}(A)$. 另一方面, 由

$$f(A^{(a)}) = f\left(\bigcup_{b \not\leq a} A_{(b)}\right) \subseteq \bigcup_{b \not\leq a} f(A_{[b]}) = \bigcup_{b \not\leq a} \left(\widetilde{f}(A)\right)_{[b]} = \widetilde{f}(A)^{(a)}$$

可知 $\widetilde{f}(A) \geqslant \overline{f}(A)$. □

推论 3.5.6 设 f 是 2^X 上的保并增值自映射, 则 $\forall A \in L^X$, $\widetilde{f}(A)_{(a)} = f(A_{(a)})$.

定理 3.5.7 设 f, g 是 2^X 上的保并增值自映射, 则
(1) $\widetilde{f \wedge g} \leqslant \tilde{f} \wedge \tilde{g}$;
(2) $\widetilde{g \circ g} \leqslant \tilde{g} \circ \tilde{g}$;
(3) $\left(\tilde{f}\right)^{-1} = \widetilde{f^{-1}}$.

证明 (1) 是显然的. (2) 由 $\tilde{g}(A)_{(a)} = g(A_{(a)})$ 易证.
(3) 因为

$$\left(\tilde{f}\right)^{-1}(A) \leqslant B \Leftrightarrow \tilde{f}(B') \leqslant A'$$
$$\Leftrightarrow \forall a \in L, \tilde{f}(B')_{(a)} \subseteq (A')_{(a)}$$
$$\Leftrightarrow \forall a \in L, f(B'_{(a)}) \subseteq (A')_{(a)}$$
$$\Leftrightarrow \forall a \in L, f((B^{[a']})') \subseteq (A^{[a']})'$$
$$\Leftrightarrow \forall a \in L, f^{-1}(A^{[a']}) \subseteq B^{[a']}$$
$$\Leftrightarrow \widetilde{f^{-1}}(A) \leqslant B,$$

所以 $\left(\tilde{f}\right)^{-1} = \widetilde{f^{-1}}$. □

定理 3.5.8 设 $(X, \mathscr{D})$ 是分明 (拟) 一致空间, 这里 $\mathscr{D}$ 是 2^X 上的保并增值自映射族, 则 $\widetilde{\mathscr{D}} = \{\tilde{f} \mid f \in \mathscr{D}\}$ 是 L^X 上的某 (拟) 一致结构 $\mathscr{P}$ 的基且 $\mathscr{P}$ 导出的 L-拓扑恰是由 $\mathscr{D}$ 导出的分明拓扑诱导的.

证明 设 $\mathscr{F}$ 表示 L^X 上所有保并增值自映射族, 令 $\mathscr{P} = \{g \in \mathscr{F} \mid \exists f \in \mathscr{D}$ 使 $g \geqslant \tilde{f}\}$, 则由定理 3.5.7 知 $\mathscr{P}$ 是 L^X 上的 (拟) 一致结构. 若设 c_1, c_2 分别表示 $\mathscr{T}(\mathscr{P}), \mathscr{T}(\mathscr{D})$ 中的闭包算子, 则 $\forall A \in L^X$,

$$c_1(A)_{[a]} = \left(\bigwedge_{f \in \mathscr{D}} \tilde{f}(A)\right)_{[a]} = \bigcap_{f \in \mathscr{D}} \tilde{f}(A)_{[a]}$$
$$= \bigcap_{f \in \mathscr{D}} \bigcap_{b \in \beta(a)} f(A_{[b]}) = \bigcap_{b \in \beta(a)} \bigcap_{f \in \mathscr{D}} f(A_{[b]})$$
$$= \bigcap_{b \in \beta(a)} c_2(A_{[b]}).$$

由定理 2.1.4 与定理 3.5.5 知 $(X, \mathscr{T}(\mathscr{P}))$ 是诱导空间. □

定理 3.5.9 设 $(X, \mathscr{D})$ 是分明伪度量空间, 这里 $\mathscr{D}$ 是 2^X 上满足 (E1)—(E5) 的自映射族. 则 $(X, \widetilde{\mathscr{D}})$ 是模糊伪度量空间且 $(X, \mathscr{T}(\widetilde{\mathscr{D}}))$ 是由 $(X, \mathscr{T}(\mathscr{D}))$ 所诱导的, 这里 $\mathscr{T}(\widetilde{\mathscr{D}}), \mathscr{T}(\mathscr{D})$ 分别表示以 $\widetilde{\mathscr{D}}, \mathscr{D}$ 为基的 (拟) 一致结构生成的拓扑.

证明 只需证 $\forall f_r \in \mathscr{D}, \widetilde{f}_r = \bigvee_{s<r} \widetilde{f}_s$ 即可. $\forall A \in L^X$, 由

$$\widetilde{f}_r(A)_{(a)} = f_r\left(A_{(a)}\right) = \bigcup_{s<r} f_s\left(A_{(a)}\right)$$

$$= \bigcup_{s<r} \widetilde{f}_s(A)_{(a)} = \left(\left(\bigvee_{s<r} \widetilde{f}_s\right)(A)\right)_{(a)}$$

知 $\widetilde{f}_r = \bigvee_{s<r} \widetilde{f}_s$. 从而由定理 3.5.4 与定理 3.5.8 易证本定理. □

3.6 凸 结 构

在许多应用数学的领域中, 极值问题的研究使得凸性理论受到了人们的广泛关注. 20 世纪 50 年代起, 从公理化出发研究凸性的方法, 即凸空间理论逐步发展并成熟. 凸空间理论是通过抽象欧氏空间中凸集的性质而得到的一门用公理化方法处理集合系统的数学分支. 凸空间或者凸结构存在于许多数学分支中, 比如, 布尔代数和格、度量空间、赋范向量空间、图论、中间代数以及拓扑学等.

欧氏空间中的一个集合是凸的, 如果其中任意两个点的连线还在这个集合中. 这种凸集的全体恰好满足下面三条公理.

命题 3.6.1 欧氏空间中的普通凸集满足下面的性质:

(1) 空集和全集是凸的;

(2) 非空凸集族的任意交是凸的;

(3) 一族凸集的全序并是凸的.

将上面性质抽象推广到一般集合就可以得到下面定义.

定义 3.6.2 设 X 是一个非空集合, 2^X 的一个子族 $\mathscr{C}$ 称为一个凸结构, 如果它满足下列三个条件:

(1) $\varnothing, X \in \mathscr{C}$;

(2) 若 $\{A_i \mid i \in \Omega\} \subseteq \mathscr{C}$ 非空, 则 $\bigcap_{i \in \Omega} A_i \in \mathscr{C}$;

(3) 若 $\{A_i \mid i \in \Omega\} \subseteq \mathscr{C}$ 非空且是全序的, 则 $\bigcup_{i \in \Omega} A_i \in \mathscr{C}$.

若 $\mathscr{C}$ 是一个凸结构, 则称 $(X, \mathscr{C})$ 是一个凸空间.

下面我们分别给出凸空间的一些例子.

例 3.6.3 设 $\mathscr{T}$ 是 X 上的一个 Alexandroff 拓扑, 也就是 $\mathscr{T}$ 满足下面三条件:

(1) $\varnothing, X \in \mathscr{T}$;

(2) $\forall \{A_j\}_{j \in J} \subseteq \mathscr{T}, \bigcap_{j \in J} A_j \in \mathscr{T}$;

(3) $\forall \{A_j\}_{j \in J} \subseteq \mathscr{T}, \bigcup_{j \in J} A_j \in \mathscr{T}$,

那么它是一个凸结构.

例 3.6.4 $\mathbb{R}^n$ 上所有凸集组成的集族是一个凸结构.

例 3.6.5 设 $(X, \leqslant)$ 为偏序集. 称 X 上一个集 A 是序凸的, 若它满足 $\forall x, y \in A$, 当 $x \leqslant z \leqslant y$ 时, $z \in A$. 若 $\mathscr{C}$ 表示 X 上所有序凸集组成的集族, 则 $(X, \mathscr{C})$ 是一个凸空间.

例 3.6.6 设 $(X, \leqslant)$ 为偏序集, $\mathscr{C}_1$ 和 $\mathscr{C}_2$ 分别表示 X 上所有上集和所有下集组成的集族. 容易验证 $\mathscr{C}_1$ 和 $\mathscr{C}_2$ 都是凸结构, 分别称它们为上凸结构和下凸结构.

例 3.6.7 设 L 是一个格, $A \in \mathbf{2}^X$. 如果 $\forall x, y \in L$, 有

$$x, y \in A \Rightarrow x \wedge y, x \vee y \in A,$$

则称 A 是 L 的子格. 如果对任意 $a, b \in A$, 都有 $[a, b] = \{c \mid a \leqslant c \leqslant b\} \subseteq A$, 则称 A 是凸的. 让 $\mathscr{C}$ 表示 L 上所有 (凸) 子格组成的集族, 则 $\mathscr{C}$ 是一个凸结构, $(L, \mathscr{C})$ 是一个凸空间.

事实上, 还有很多来自不同数学对象的集族也满足定义 3.6.2 中的三个条件 (1)—(3), 比如, 代数学中的各种子代数, 如子群、子环、子域、理想、子模等都满足凸结构的三个条件 (当然这里要约定空集是相应的子代数). 布尔代数和格中的凸集族、度量空间 (尤其是赋范向量空间和图) 和中间代数的凸集族等也都如此. 另外, 凸结构也很自然地出现在拓扑中, 尤其是在超紧空间理论中.

定义 3.6.8 设 $(X, \mathscr{C}), (Y, \mathscr{D})$ 是两个凸空间, $f: X \to Y$ 是一个映射. 则 f 称为

(1) 是凸保持的, 如果对任意的 $D \in \mathscr{D}$, 都有 $f^{\leftarrow}(D) \in \mathscr{C}$;

(2) 是凸到凸的, 如果对任意的 $C \in \mathscr{C}$, 都有 $f^{\rightarrow}(C) \in \mathscr{D}$;

(3) 同构映射, 如果它是凸保持的且是凸到凸的双射.

3.7 凸结构的模糊化

1994 年, Rosa[126] 首次将模糊性引入到了凸空间理论, 定义了模糊凸空间 (我们称之为 I-凸空间). 2009 年, Maruyama[104] 借助完全分配格 L 这一更为宽泛的取值格, 提出了 L-凸空间的概念.

定义 3.7.1 设 X 是一个非空集合, L 是一个完全分配格, $\mathscr{C} \subseteq L^X$. 若 $\mathscr{C}$ 满足下列条件:

(1) $\chi_\varnothing, \chi_X \in \mathscr{C}$;

(2) 若 $\{A_i \mid i \in \Omega\} \subseteq \mathscr{C}$ 非空, 则 $\bigwedge_{i\in\Omega} A_i \in \mathscr{C}$;

(3) 若 $\{A_i \mid i \in \Omega\} \subseteq \mathscr{C}$ 非空且是全序的, 则 $\bigvee_{i\in\Omega} A_i \in \mathscr{C}$,

则称 $\mathscr{C}$ 是一个 L-凸结构, 称 $(X,\mathscr{C})$ 是一个 L-凸空间, 称 $\mathscr{C}$ 中的元素是 L-凸集.

注 3.7.2 在定义 3.7.1 中, 作为 L-凸结构的集族 $\mathscr{C}$ 是分明的, 里面的元素是模糊的, 也就是说 $\mathscr{C}$ 是一个由模糊集构成的分明族. 当 $L = I$ 时, L-凸空间就是 [126] 中的模糊凸空间.

2014 年, 从一个完全不同的角度, 我和我的学生修振宇给出了凸空间模糊化的一种新方法, 引入了模糊化凸空间的概念.

定义 3.7.3 设 X 是一个非空集且 M 是一个完备格. 称映射 $\mathscr{C}: \mathbf{2}^X \to M$ 是 X 上的一个 M-模糊化凸结构, 若 $\mathscr{C}$ 满足下面条件:

(1) $\mathscr{C}(\varnothing) = \mathscr{C}(X) = 1$;

(2) 若 $\{A_i : i \in \Omega\} \subseteq \mathbf{2}^X$ 非空, 则 $\mathscr{C}\left(\bigcap_{i\in\Omega} A_i\right) \geqslant \bigwedge_{i\in\Omega} \mathscr{C}(A_i)$;

(3) 若 $\{A_i : i \in \Omega\} \subseteq \mathbf{2}^X$ 非空且是全序的, 则 $\mathscr{C}\left(\bigcup_{i\in\Omega} A_i\right) \geqslant \bigwedge_{i\in\Omega} \mathscr{C}(A_i)$.

若 $\mathscr{C}$ 是 X 上一个 M-模糊化凸结构, 则称 $(X,\mathscr{C})$ 为一个 M-模糊化凸空间. 当 $M = I$ 时, 一个 M-模糊化凸空间可以简记为模糊化凸空间.

如果 $\mathscr{C}$ 仅满足 (1) 和 (2), 则称 $\mathscr{C}$ 是一个 M-模糊化闭包结构, $(X,\mathscr{C})$ 是一个 M-模糊化闭包空间.

定义 3.7.4 设 $(X,\mathscr{C}), (Y,\mathscr{D})$ 是 M-模糊化凸空间, $f: X \to Y$ 是一个映射.

(1) 称 $f: X \to Y$ 是一个 M-模糊化凸保持映射, 如果对任意 $B \in \mathbf{2}^Y$, $\mathscr{C}(f^{\leftarrow}(B)) \geqslant \mathscr{D}(B)$;

(2) 称 $f: X \to Y$ 是一个 M-模糊化凸到凸映射, 如果对任意 $A \in \mathbf{2}^X$, $\mathscr{D}(f^{\rightarrow}(A)) \geqslant \mathscr{C}(A)$;

(3) 称 $f: X \to Y$ 是一个 M-模糊化同构, 如果 f 是一个双射, 并且 f 是一个 M-模糊化凸保持映射和一个 M-模糊化凸到凸映射.

注 3.7.5 一个 M-模糊化凸结构实际上是由一些分明集构成的模糊族. 结合上述两种情况, 我们又提出了更一般的凸结构模糊化方法.

定义 3.7.6 映射 $\mathscr{C}: L^X \to M$ 称为 X 上的 (L,M)-模糊凸结构, 如果它满足以下条件:

(1) $\mathscr{C}(\chi_\varnothing) = \mathscr{C}(\chi_X) = 1_M$;

(2) 如果 $\{A_i \mid i \in \Omega\} \subseteq L^X$ 是非空的, 则 $\bigwedge_{i\in\Omega} \mathscr{C}(A_i) \leqslant \mathscr{C}\left(\bigwedge_{i\in\Omega} A_i\right)$;

(3) 如果 $\{A_i \mid i \in \Omega\} \subseteq L^X$ 是非空和全序的, 则 $\bigwedge_{i\in\Omega} \mathscr{C}(A_i) \leqslant \mathscr{C}\left(\bigvee_{i\in\Omega} A_i\right)$.

为了简便, $(X, \mathscr{C})$ 称为 (L, M)-模糊凸空间, (L, L)-模糊凸空间称为 L-模糊凸空间.

定义 3.7.7 设 $(X, \mathscr{C})$ 和 $(Y, \mathscr{D})$ 是 (L, M)-模糊凸空间. 映射 $f : X \to Y$ 称为

(1) (L, M)-模糊凸保持映射, 如果对于所有的 $B \in L^Y$, 有 $\mathscr{D}(B) \leqslant \mathscr{C}(f_L^{\leftarrow}(B))$;

(2) (L, M)-模糊凸到凸映射, 如果对于所有的 $A \in L^X$, 有 $\mathscr{C}(A) \leqslant \mathscr{D}(f_L^{\rightarrow}(A))$;

(3) (L, M)-模糊同构, 如果 f 是双射的、(L, M)-模糊凸保持和 (L, M)-模糊凸到凸的.

为了简便, (L, L)-模糊凸保持映射称为 L-模糊凸保持映射, (L, L)-模糊凸到凸映射称为 L-模糊凸到凸映射, 同时 (L, L)-模糊同构称为 L-模糊同构.

下面我们分别给出 (L, M)-模糊凸空间、L-凸空间和 M-模糊化凸空间的例子.

例 3.7.8 一个映射 $\mathscr{T} : L^X \to M$ 称为一个 (L, M)-Alexandroff 模糊拓扑, 如果它满足下面三个条件:

(1) $\varnothing, X \in \mathscr{T}$;

(2) $\forall \{A_j\}_{j\in J} \subseteq L^X, \mathscr{T}\left(\bigwedge_{j\in J} A_j\right) \geqslant \bigwedge_{j\in J} \mathscr{T}(A_j)$;

(3) $\forall \{A_j\}_{j\in J} \subseteq L^X, \mathscr{T}\left(\bigvee_{j\in J} A_j\right) \geqslant \bigwedge_{j\in J} \mathscr{T}(A_j)$.

这样一个 (L, M)-模糊拓扑显然是一个 (L, M)-模糊凸结构.

例 3.7.9 一个 $\mathbb{R}^n$ 上 L-模糊集 A 称作一个 L-凸集当且仅当对任意 $x, y \in \mathbb{R}^n$ 和对任意 $r \in [0, 1]$, $A(rx + (1-r)y) \geqslant A(x) \wedge A(y)$. 若 $\mathscr{C}_L$ 表示 $\mathbb{R}^n$ 上所有 L-凸集组成的集族, 则 $(\mathbb{R}^n, \mathscr{C}_L)$ 是一个 L-凸空间.

例 3.7.10 设 L 是一个格, $A \in [0, 1]^L$. 称 A 是 L 的模糊子格, 如果 $\forall x, y \in L$, 均有下面两条件成立:

(1) $A(x \wedge y) \geqslant A(x) \wedge A(y)$;

(2) $A(x \vee y) \geqslant A(x) \wedge A(y)$.

若 $\mathscr{C}$ 表示 L 上所有模糊子格组成的集族 (这里我们约定空集是子格), 则 $\mathscr{C}$ 是一个 $[0,1]$-凸结构, $(L, \mathscr{C})$ 是一个 $[0,1]$-凸空间.

例 3.7.11 设 G 是一个群且 $A \in L^G$. 称 A 为 G 的一个模糊子群, 如果它满足下面两个条件:

(1) $A(xy) \geqslant A(x) \wedge A(y)$;

(2) $A(x^{-1}) \geqslant A(x)$.

3.7 凸结构的模糊化

让 $\mathscr{C}$ 表示 G 上所有模糊子群组成的集族 (这里我们约定空集是子群), 则 $\mathscr{C}$ 是一个 L-凸结构, 于是 $(G, \mathscr{C})$ 就是一个 L-凸空间.

例 3.7.12 定义映射 $\mathscr{C} : L^{\mathbb{R}^n} \to L$ 为: $\forall A \in L^{\mathbb{R}^n}$,

$$\mathscr{C}(A) = \bigwedge_{\lambda \in [0,1]} \bigwedge_{(x,y) \in \mathbb{R}^n \times \mathbb{R}^n} (A(x) \wedge A(y)) \to A(\lambda x + (1-\lambda)y),$$

其中二元运算 $\to$ 定义如下: $\forall a, b \in L, a \to b = \bigvee \{c \in L : a \wedge c \leqslant b\}$. 则 $(X, \mathscr{C})$ 是一个 (L, L)-模糊凸空间.

例 3.7.13 让 X 是一个非空集, 定义一个映射 $\mathscr{C} : I^X \to I$ 为

$$\mathscr{C}(A) = \begin{cases} 1, & A \in \{\chi_\varnothing, \chi_X\}, \\ 0.5, & A \notin \{\chi_\varnothing, \chi_X\}. \end{cases}$$

那么容易验证 $(X, \mathscr{C})$ 是一个 I-模糊凸结构. 如果 $A \in I^X$ 且 $A \notin \{\chi_\varnothing, \chi_X\}$, 那么 A 是一个凸模糊集的程度就是 0.5.

借助于两种截集, 我们可以给出 (L, M)-模糊凸结构和 (L, M)-模糊凸保持映射的刻画 (见下面两个定理), 它们的证明是简单的, 请读者自证.

定理 3.7.14 设 $\mathscr{C} : L^X \to M$ 是一个映射. 则下列条件等价:
(1) $\mathscr{C}$ 是一个 (L, M)-模糊凸结构;
(2) $\forall a \in M, \mathscr{C}_{[a]}$ 是一个 L-凸结构;
(3) $\forall a \in M, \mathscr{C}^{[a]}$ 是一个 L-凸结构.

推论 3.7.15 设 $\mathscr{C} : 2^X \to M$ 是一个映射. 则下列条件等价:
(1) $\mathscr{C}$ 是一个 M-模糊化凸结构;
(2) $\forall a \in M, \mathscr{C}_{[a]}$ 是一个凸结构;
(3) $\forall a \in M, \mathscr{C}^{[a]}$ 是一个凸结构.

定理 3.7.16 设 $(X, \mathscr{C})$ 和 $(Y, \mathscr{D})$ 是两个 (L, M)-模糊凸空间. $f : X \to Y$ 是一个映射. 则下列条件等价:
(1) $f : (X, \mathscr{C}) \to (Y, \mathscr{D})$ 是 (L, M)-模糊凸保持映射;
(2) $\forall a \in M, f : (X, \mathscr{C}_{[a]}) \to (Y, \mathscr{D}_{[a]})$ 是 L-凸保持映射;
(3) $\forall a \in M, f : (X, \mathscr{C}^{[a]}) \to (Y, \mathscr{D}^{[a]})$ 是 L-凸保持映射.

推论 3.7.17 设 $(X, \mathscr{C})$ 和 $(Y, \mathscr{D})$ 是两个 M-模糊化凸空间. $f : X \to Y$ 是一个映射. 则下列条件等价:
(1) $f : (X, \mathscr{C}) \to (Y, \mathscr{D})$ 是 M-模糊化凸保持映射;
(2) $\forall a \in M, f : (X, \mathscr{C}_{[a]}) \to (Y, \mathscr{D}_{[a]})$ 是凸保持映射;
(3) $\forall a \in M, f : (X, \mathscr{C}^{[a]}) \to (Y, \mathscr{D}^{[a]})$ 是凸保持映射.

3.8 (L,M)-模糊凸包算子

在这一节中, 我们将引入 (L,M)-模糊凸包算子的概念, 并借之给出 (L,M)-模糊凸结构的刻画.

在一个凸空间 $(X,\mathscr{C})$ 中, 一个凸包算子 co 和它的凸结构 $\mathscr{C}$ 是一一对应的, 此时的凸包算子 co 满足如下条件:

(C1) $\mathrm{co}(\varnothing) = \varnothing$;

(C2) $\mathrm{co}(A) \supseteq A$;

(C3) 如果 $A \subseteq B$, 那么 $\mathrm{co}(A) \subseteq \mathrm{co}(B)$;

(C4) $\mathrm{co}(\mathrm{co}(A)) \subseteq \mathrm{co}(A)$;

(C5) $\mathrm{co}(A) = \bigcup\{\mathrm{co}(F) \mid F \in \mathbf{2}_{\mathrm{fin}}^{A}\}$, 这里 $\forall A \in \mathbf{2}^X$, $\mathbf{2}_{\mathrm{fin}}^{A}$ 表示 A 的所有有限子集的全体.

这个定义首先被推广到如下模糊情形.

定义 3.8.1 集合 X 上的一个模糊化凸包算子就是一个满足下列条件的映射 $\mathrm{co}: \mathbf{2}^X \to M^X$:

(FC1) $\mathrm{co}(\varnothing) = \varnothing$;

(FC2) $\mathrm{co}(A)(x) = 1, \forall x \in A$;

(FC3) 如果 $A \subseteq B$, 那么 $\mathrm{co}(A) \leqslant \mathrm{co}(B)$ (即 $\mathrm{co}(A)(x) \leqslant \mathrm{co}(B)(x), \forall x \in X$);

(FC4) $\mathrm{co}(A)(x) = \bigwedge\limits_{x \notin B \supseteq A} \bigvee\limits_{y \notin B} \mathrm{co}(B)(y), \forall A \in \mathbf{2}^X, \forall x \in X$;

(FC5) $\mathrm{co}(A) = \bigvee\{\mathrm{co}(F) \mid F \in \mathbf{2}_{\mathrm{fin}}^{A}\}, \forall A \in \mathbf{2}^X$.

接下来, 我们给出上述定义的更进一步推广形式.

定义 3.8.2 集合 X 上的一个 (L,M)-模糊凸包算子就是一个满足下列条件的映射 $\mathrm{co}: L^X \to M^{J(L^X)}$:

(FC0) $\mathrm{co}(A)(x_\lambda) = \bigwedge\limits_{\mu \prec \lambda} \mathrm{co}(A)(x_\mu), \forall A \in L^X, \forall x_\lambda \in J(L^X)$;

(FC1) $\mathrm{co}(\varnothing)(x_\lambda) = 0$;

(FC2) $\mathrm{co}(A)(x) = 1, \forall x_\lambda \leqslant A$;

(FC3) 如果 $A \subseteq B$, 那么 $\mathrm{co}(A) \leqslant \mathrm{co}(B)$ (即 $\mathrm{co}(A)(x_\lambda) \leqslant \mathrm{co}(B)(x_\lambda), \forall x_\lambda \in J(L^X)$);

(FC4) $\mathrm{co}\left(\bigvee (\mathrm{co}(A))_{[a]}\right)_{[a]} \subseteq \mathrm{co}(A)_{[a]}$;

(FC5) $\mathrm{co}(A)(x_\lambda) = \bigvee\{\mathrm{co}(F)(x_\lambda) \mid F \ll A\}$.

一个带有 (L,M)-模糊凸包算子 co 的集合 X(也记为 (X,co)) 叫做一个(L,M)-模糊凸包空间.

3.8 (L,M)-模糊凸包算子

两个 (L,M)-模糊凸包空间 (X,co_X) 和 (Y,co_Y) 之间的映射 $f:X\to Y$ 叫做凸包保持的, 如果 $\forall x_\lambda \in J(L^X), \forall A \in L^X$, 都有

$$\mathrm{co}_X(A)(x_\lambda) \leqslant \mathrm{co}_Y(f_L^\to(A))(f_L^\to(x_\lambda)).$$

当 $M=\mathbf{2}$ 时, 上述定义退化为下面形式.

定义 3.8.3 集合 X 上的一个 L-凸包算子就是一个满足下列条件的映射 $\mathrm{co}:L^X \to L^X$:

(FC1) $\mathrm{co}(\varnothing) = \varnothing$;
(FC2) $A \leqslant \mathrm{co}(A)$;
(FC3) $A \subseteq B$ 意味着 $\mathrm{co}(A) \leqslant \mathrm{co}(B)$;
(FC4) $\mathrm{co}(\mathrm{co}(A)) \subseteq \mathrm{co}(A)$;
(FC5) $\mathrm{co}(A) = \bigvee\{\mathrm{co}(F) \mid F \ll A\}$.

下面定理的证明是简单的, 故略去.

定理 3.8.4 两个 (L,M)-模糊凸包空间 (X,co_X) 和 (Y,co_Y) 之间的映射 $f:X\to Y$ 是凸包保持的当且仅当 $\forall x_\lambda \in J(L^X), \forall B \in L^Y$, 都有

$$\mathrm{co}_X(f_L^\leftarrow(B))(x_\lambda) \leqslant \mathrm{co}_Y(B)(f_L^\to(x_\lambda)).$$

定理 3.8.5 设 $\mathscr{C}$ 是 X 上的一个 (L,M)-模糊凸结构, 定义一个映射 $\mathrm{co}^{\mathscr{C}}:L^X \to M^{J(L^X)}$ 使得 $\forall x_\lambda \in J(L^X), \forall A \in L^X$,

$$\mathrm{co}^{\mathscr{C}}(A)(x_\lambda) = \bigwedge_{x_\lambda \not\leqslant D \geqslant A} (\mathscr{C}(D))',$$

那么 $\mathrm{co}^{\mathscr{C}}$ 是一个 (L,M)-模糊凸包算子.

证明 (FC0) 当 $\mu \prec \lambda$ 时, 由 $x_\mu \not\leqslant D \geqslant A \Rightarrow x_\lambda \not\leqslant D \geqslant A$ 可知 $\mathrm{co}^{\mathscr{C}}(A)(x_\lambda) \leqslant \bigwedge_{\mu \prec \lambda} \mathrm{co}^{\mathscr{C}}(A)(x_\mu)$. 为了证明 $\mathrm{co}^{\mathscr{C}}(A)(x_\lambda) \geqslant \bigwedge_{\mu \prec \lambda} \mathrm{co}^{\mathscr{C}}(A)(x_\mu)$, 设 $a \not\leqslant \mathrm{co}^{\mathscr{C}}(A)(x_\lambda)$. 则 $a \not\leqslant \bigwedge_{x_\lambda \not\leqslant D \geqslant A} (\mathscr{C}(D))'$. 于是存在满足 $x_\lambda \not\leqslant D \geqslant A$ 的 D 使得 $a \not\leqslant (\mathscr{C}(D))'$. 由 $x_\lambda \not\leqslant D \geqslant A$ 可知存在 $\mu \prec \lambda$ 使得 $x_\mu \not\leqslant D \geqslant A$. 因此可得

$$a \not\leqslant \bigwedge_{x_\mu \not\leqslant D \geqslant A} (\mathscr{C}(D))' = \mathrm{co}^{\mathscr{C}}(A)(x_\mu).$$

进一步可得 $a \not\leqslant \bigwedge_{\mu \prec \lambda} \mathrm{co}(A)^{\mathscr{C}}(x_\mu)$. 故 $\mathrm{co}^{\mathscr{C}}(A)(x_\lambda) \geqslant \bigwedge_{\mu \prec \lambda} \mathrm{co}^{\mathscr{C}}(A)(x_\mu)$.

(FC1), (FC2) 和 (FC3) 是显然的. 下面证明 (FC4) 和 (FC5).

为了证明 (FC4) 成立, 我们设 $a \in L\setminus\{0\}$ 且 $x_\lambda \not\in (\text{co}^{\mathscr{C}}(A))_{[a]}$. 则存在满足 $x_\lambda \not\leqslant D \geqslant A$ 的 D 使得 $a \not\leqslant \mathscr{C}(D)'$. 这意味着 $\forall y_\mu \not\leqslant D$, 皆有

$$a \not\leqslant \bigwedge_{y_\mu \not\leqslant E \geqslant D} \mathscr{C}(E)' = \text{co}^{\mathscr{C}}(D)(y_\mu).$$

于是对任何的 $y_\mu \not\leqslant D$, 我们都能得到 $y_\mu \not\in (\text{co}^{\mathscr{C}}(D))_{[a]}$. 这说明

$$x_\lambda \not\leqslant D \geqslant \bigvee (\text{co}^{\mathscr{C}}(D))_{[a]} \geqslant \bigvee (\text{co}^{\mathscr{C}}(A))_{[a]}.$$

因此可得

$$\text{co}^{\mathscr{C}}\left(\bigvee(\text{co}^{\mathscr{C}}(A))_{[a]}\right)(x_\lambda) = \bigwedge_{x_\lambda \not\leqslant D \geqslant \bigvee(\text{co}^{\mathscr{C}}(A))_{[a]}} \mathscr{C}(D)' \not\geqslant a.$$

这样我们就证明了 $x_\lambda \not\in \left(\text{co}^{\mathscr{C}}\left(\bigvee(\text{co}^{\mathscr{C}}(A))_{[a]}\right)\right)_{[a]}$. (FC4) 的证明完成了.

为了证明 (FC5) 成立, 我们设 $\text{co}^{\mathscr{C}}(A)(x_\lambda) \geqslant a$. 则由 (C10) 可知 $\forall \mu \prec \lambda$, 都有 $\text{co}^{\mathscr{C}}(A)(x_\mu) \geqslant a$. 于是对于满足 $x_\mu \not\leqslant D \geqslant A$ 的任何 D, 都有 $\mathscr{C}(D)' \geqslant a$. 下面我们来证明 $\bigvee\{\text{co}(F)(x_\lambda) \mid F \ll A\} \geqslant a$.

假设 $\bigvee\{\text{co}(F)(x_\lambda) \mid F \ll A\} \geqslant a$ 不成立, 那么存在 $b \in \beta(a)$ 使得 $\bigvee\{\text{co}(F)(x_\lambda) \mid F \ll A\} \not\geqslant b$. 这样的话, 对于满足 $F \ll A$ 的每个模糊子集 F, 都存在满足 $x_\mu \not\leqslant D_F \geqslant F$ 的 D_F, 使得 $\mathscr{C}(F_D)' \not\geqslant b$. 在这种情况下, 我们有

$$A = \bigvee\{F \mid F \ll A\} \leqslant \bigvee\{D_F \mid F \ll A\}.$$

令 $D = \bigvee\{D_F \mid F \ll A\}$, 则易验证 $x_\lambda \not\leqslant D \geqslant A$ 而且 $\bigvee_{F \ll A} \mathscr{C}(D_F)' \not\geqslant a$. 由于 $\mathscr{C}(D)' \leqslant \bigvee_{F \ll A} \mathscr{C}(D_F)'$, 所以 $\mathscr{C}(D)' \not\geqslant a$, 这与 $\mathscr{C}(D)' \geqslant a$ 相矛盾. 故有 $\bigvee\{\text{co}(F)(x_\lambda) \mid F \ll A\} \geqslant a$ 成立. 这样就证明了 (FC5). □

定理 3.8.6 如果 $f : (X, \mathscr{C}_1) \to (Y, \mathscr{C}_2)$ 是凸保持的, 那么 $f : (X, \text{co}^{\mathscr{C}_1}) \to (Y, \text{co}^{\mathscr{C}_2})$ 也是凸包保持的.

证明 如果 $f : (X, \mathscr{C}_1) \to (Y, \mathscr{C}_2)$ 是凸保持的, 那么 $\forall B \in L^Y$, 皆有 $\mathscr{C}_2(B) \leqslant \mathscr{C}_1(f_L^{\leftarrow}(B))$, 这意味着 $\forall x_\lambda \in J(L^X)$, 下面不等式成立:

$$\text{co}^{\mathscr{C}_2}(B)(f_L^{\rightarrow}(x_\lambda)) = \bigwedge_{f_L^{\rightarrow}(x_\lambda) \not\leqslant D \geqslant B} (\mathscr{C}_2(D))'$$
$$\geqslant \bigwedge_{x_\lambda \not\leqslant f_L^{\leftarrow}(D) \geqslant f_L^{\leftarrow}(B)} (\mathscr{C}_1(f_L^{\leftarrow}(D)))'$$
$$\geqslant \text{co}^{\mathscr{C}_1}(f_L^{\leftarrow}(B))(x_\lambda).$$

因此 $f:(X,\text{co}^{\mathscr{C}_1}) \to (Y,\text{co}^{\mathscr{C}_2})$ 是凸包保持的. $\square$

定理 3.8.7 设 $\text{co}: L^X \to M^{J(L^X)}$ 是一个 (L,M)-模糊凸包算子. 定义 $\mathscr{C}^{\text{co}}: L^X \to M$ 使得

$$\mathscr{C}^{\text{co}}(U) = \bigwedge_{x_\lambda \not\leqslant U} (\text{co}(U)(x_\lambda))'.$$

那么 $\mathscr{C}^{\text{co}}$ 是一个 (L,M)-模糊凸结构且 $\text{co}^{(\mathscr{C}^{\text{co}})} = \text{co}$.

证明 (1) 不难验证 $\mathscr{C}^{\text{co}}(\varnothing) = \mathscr{C}^{\text{co}}(X) = 1$.

(2) 我们再来证明 $\mathscr{C}^{\text{co}}\left(\bigwedge_{i\in\Omega} A_i\right) \geqslant \bigwedge_{i\in\Omega} \mathscr{C}^{\text{co}}(A_i)$.

设 $b \in M$ 且 $\bigwedge_{i\in\Omega} \mathscr{C}^{\text{co}}(A_i) \not\leqslant b$. 则存在 $a \in \alpha^*(b)$ 使得 $\bigwedge_{i\in\Omega} \mathscr{C}^{\text{co}}(A_i) \not\leqslant a$. 于是 $\forall i \in \Omega$, 都有 $\mathscr{C}^{\text{co}}(A_i) \not\leqslant a$. 从而 $\forall x_\lambda \not\leqslant A_i$, 皆有 $(\text{co}(A_i)(x_\lambda))' \not\leqslant a$. 由于 $\text{co}\left(\bigwedge_{i\in\Omega} A_i\right)(x_\lambda) \leqslant \text{co}(A_i)(x_\lambda)$, 所以 $\forall x_\lambda \not\leqslant \bigwedge_{i\in\Omega} A_i$, 都有 $\left(\text{co}\left(\bigwedge_{i\in\Omega} A_i\right)(x_\lambda)\right)' \not\leqslant a$ 成立. 这样我们就可以得到

$$\mathscr{C}^{\text{co}}\left(\bigwedge_{i\in\Omega} A_i\right) = \bigwedge_{x_\lambda \not\leqslant \bigwedge_{i\in\Omega} A_i} \left(\text{co}\left(\bigwedge_{i\in\Omega} A_i\right)(x_\lambda)\right)' \not\leqslant b.$$

故有 $\mathscr{C}^{\text{co}}\left(\bigwedge_{i\in\Omega} A_i\right) \geqslant \bigwedge_{i\in\Omega} \mathscr{C}^{\text{co}}(A_i)$.

(3) 我们再来证明 $\mathscr{C}^{\text{co}}\left(\bigvee_{i\in\Omega} A_i\right) \geqslant \bigwedge_{i\in\Omega} \mathscr{C}^{\text{co}}(A_i)$, 这里 $\{A_i \mid i \in \Omega\}$ 是全序的.

设 $b \in M$ 且 $\bigwedge_{i\in\Omega} \mathscr{C}^{\text{co}}(A_i) \not\leqslant b$, 则存在 $a \in \alpha^*(b)$ 使得 $\bigwedge_{i\in\Omega} \mathscr{C}^{\text{co}}(A_i) \not\leqslant a$. 于是 $\forall i \in \Omega$, 都有 $\mathscr{C}^{\text{co}}(A_i) \not\leqslant a$. 从而 $\forall x_\lambda \not\leqslant A_i$, 皆有 $(\text{co}(A_i)(x_\lambda))' \not\leqslant a$. 由于 $\text{co}\left(\bigwedge_{i\in\Omega} A_i\right)(x_\lambda) \leqslant \text{co}(A_i)(x_\lambda)$, 所以 $\left(\text{co}\left(\bigvee_{i\in\Omega} A_i\right)(x_\lambda)\right)' \not\leqslant a$. 这样我们就可以得到

$$\mathscr{C}^{\text{co}}\left(\bigvee_{i\in\Omega} A_i\right) = \bigwedge_{x_\lambda \not\leqslant \bigvee_{i\in\Omega} A_i} \left(\text{co}\left(\bigvee_{i\in\Omega} A_i\right)(x_\lambda)\right)'$$

$$= \bigwedge_{x_\lambda \not\leqslant \bigvee_{i\in\Omega} A_i} \left(\bigvee\left\{\text{co}(F)(x_\lambda) \,\middle|\, F \ll \bigvee_{i\in\Omega} A_i\right\}\right)'$$

$$= \bigwedge_{x_\lambda \not\leqslant \bigvee_{i\in\Omega} A_i} \left(\bigvee_{i\in\Omega} \bigvee \{ \mathrm{co}(F)(x_\lambda) \mid F \ll A_i \} \right)'$$

$$= \bigwedge_{i\in\Omega} \bigwedge_{x_\lambda \not\leqslant \bigvee_{i\in\Omega} A_i} \left(\bigvee_{F \ll A_i} \mathrm{co}(F)(x_\lambda) \right)'$$

$$\geqslant \bigwedge_{i\in\Omega} \bigwedge_{x_\lambda \not\leqslant A_i} (\mathrm{co}(A_i)(x_\lambda))' = \bigwedge_{i\in\Omega} \mathscr{C}(A_i).$$

故有 $\mathscr{C}^{\mathrm{co}}\left(\bigvee_{i\in\Omega} A_i \right) \geqslant \bigwedge_{i\in\Omega} \mathscr{C}^{\mathrm{co}}(A_i)$.

综上可知 $\mathscr{C}^{\mathrm{co}}$ 是一个 (L,M)-模糊凸结构.

$\mathrm{co}^{(\mathscr{C}^{\mathrm{co}})} = \mathrm{co}$ 的证明就留给读者吧! □

定理 3.8.8 如果 $f: (X, \mathrm{co}_X) \to (Y, \mathrm{co}_Y)$ 关于 (L,M)-模糊凸包算子 co_X 和 co_Y 是凸包保持的, 那么 $f: (X, \mathscr{C}^{\mathrm{co}_X}) \to (Y, \mathscr{C}^{\mathrm{co}_Y})$ 关于 (L,M)-模糊凸结构 $\mathscr{C}^{\mathrm{co}_X}$ 和 $\mathscr{C}^{\mathrm{co}_Y}$ 是凸保持的.

证明 如果 $f: (X, \mathrm{co}_X) \to (Y, \mathrm{co}_Y)$ 是凸包保持的, 那么 $\forall B \in L^Y$, $\forall x_\lambda \in J(L^X)$, 皆有

$$\mathrm{co}_Y(B)(f^\to(x_\lambda)) \geqslant \mathrm{co}_X(f^\leftarrow(B))(x_\lambda),$$

这意味着

$$\mathscr{C}^{\mathrm{co}_Y}(B) = \bigwedge_{y_\mu \not\leqslant B} (\mathrm{co}_Y(B)(y_\mu))'$$

$$\leqslant \bigwedge_{f_L^\to(x_\lambda) \not\leqslant B} (\mathrm{co}_Y(B)(f_L^\to(x_\lambda)))'$$

$$= \bigwedge_{x_\lambda \not\leqslant f_L^\leftarrow(B)} (\mathrm{co}_Y(B)(f_L^\to(x_\lambda)))'$$

$$\leqslant \bigwedge_{x_\lambda \not\leqslant f_L^\leftarrow(B)} (\mathrm{co}_X(f_L^\leftarrow(B))(x_\lambda))'$$

$$= \mathscr{C}^{\mathrm{co}_X}(f_L^\leftarrow(B)).$$

因此 $f: (X, \mathscr{C}^{\mathrm{co}_X}) \to (Y, \mathscr{C}^{\mathrm{co}_Y})$ 是凸保持的. □

3.9 由 M-模糊化凸结构诱导的 (L,M)-模糊凸结构

在 3.4 节中, 我们介绍了拓扑空间的诱导空间, 其在模糊拓扑的研究中是非常重要的. 在这一节中, 我们引入凸空间的诱导空间, 它的作用也是非常重要的. 首先从一些基本引理出发开始我们的介绍.

引理 3.9.1 设 L 是完全分配格, $c \in L$, 则下列条件等价:
(1) c 的每个极大集是下定向的;
(2) c 有一个极大集 A 是下定向的;
(3) $c \in P(L)$.

证明 (1) $\Rightarrow$ (2) 是显然的.

(2) $\Rightarrow$ (3). 假设 $c \notin P(L)$, 则存在 $a, b \in L$ 使得 $c \geqslant a \wedge b$, 但 $c \not\geqslant a$ 且 $c \not\geqslant b$. 于是存在 $s, t \in A$ 使得 $s \not\geqslant a$ 且 $s \not\geqslant b$. 因为 A 是下定向的, 所以存在 $r \in A$ 使得 $r \not\geqslant a$ 且 $r \not\geqslant b$. 另一方面, 由 $c \geqslant a \wedge b$ 以及 A 是极大集可知, 应该有 $a \geqslant r$ 或者 $b \geqslant r$, 产生了矛盾, 故 $c \in P(L)$.

(3) $\Rightarrow$ (1). 设 $c \in P(L)$ 且 A 是 c 的任意一个极大集. 为了证明 A 是下定向的, 取 $s, t \in A$. 令 $C = \{r \in A \mid r \leqslant s\}$, $D = \{r \mid r \not\leqslant s\}$, 则 $A = C \cup D$ 且 $c = \inf A = \inf C \wedge \inf D$. 因为 c 是素元, 所以 $c = \inf C$ 或者 $c = \inf D$. 如果 $c = \inf D$, 那么由 A 为 c 的极大集以及 $s \in A$ 可知有 $r \in D$ 使得 $r \leqslant s$, 矛盾. 这说明 $c = \inf D$ 是不可能的, 故 $c = \inf C$. 又因为 $t \in A$ 且 A 是 c 的极大集, 所以存在 $r \in C$ 使得 $r \leqslant t$, 从而 $r \in A$ 且 $r \leqslant s, r \leqslant t$, 这样我们就证明了 A 是下定向集. □

引理 3.9.2 对任意的 $A \in L^X$ 和 $c \in P(L)$, 集族 $\{A^{[b]} \mid b \in \alpha^*(c)\}$ 是上定向的.

证明 设 $A^{[a]}, A^{[b]} \in \{A^{[b]} \mid b \in \alpha^*(c)\}$, 则 $a, b \in \alpha^*(c)$. 由于 $\alpha^*(c)$ 是下定向集, 所以存在 $e \in \alpha^*(c)$ 使得 $e \leqslant a \wedge b$. 于是由 $A^{[e]}$ 包含 $A^{[a]}$ 和 $A^{[b]}$ 可知 $\{A^{[b]} : b \in \alpha^*(c)\}$ 是上定向的. □

引理 3.9.3 设 $A \in L^X$ 且 $\mathscr{C}$ 是 X 上的凸结构. 若对任何 $b \in P(L)$, $A^{[b]} \in \mathscr{C}$, 则对任何 $c \in P(L)$, $A^{(c)} \in \mathscr{C}$.

证明 由引理 3.9.2 我们知道 $\{A^{[b]} : b \in \alpha^*(c)\}$ 是上定向的, 从而再由 $A^{(c)} = \bigcup_{b \in \alpha^*(c)} A^{[b]}$ 可知 $A^{(c)} \in \mathscr{C}$. □

定理 3.9.4 设 $(X, \mathscr{C})$ 是一个 M-模糊化凸空间. 定义映射 $\omega(\mathscr{C}) : L^X \to M$ 如下:

$$\forall A \in L^X, \ \omega(\mathscr{C})(A) = \bigwedge_{a \in P(L)} \mathscr{C}(A^{[a]}).$$

则 $\omega(\mathscr{C})$ 是一个 (L,M)-模糊凸结构, 称它为由 $\mathscr{C}$ 诱导的 (L,M)-模糊凸结构, 同时称 $(X,\omega(\mathscr{C}))$ 是由 $(X,\mathscr{C})$ 诱导的 (L,M)-模糊凸空间.

证明 (1) 对任何满足条件 $0 \neq a \in P(L)$ 的任意 a, 容易证明 $a \in \alpha(0)$, 因此 $(\chi_\varnothing)^{[a]} = \varnothing$. 如果 $0 \in P(L)$ 且 $0 \in \alpha(0)$, 那么仍然有 $(\chi_\varnothing)^{[0]} = \varnothing$. 另外, 因为 $\alpha(1) = \varnothing$, 所以 $a \notin \alpha(1)$, 这样 $(\chi_X)^{[a]} = X$. 于是我们能够看出 $\omega(\mathscr{C})(\chi_\varnothing) = \omega(\mathscr{C})(\chi_X) = 1_M$.

(2) 设 $\{A_j\}_{j \in J} \subseteq L^X$. 则

$$\omega(\mathscr{C})\left(\bigwedge_{j \in J} A_j\right) = \bigwedge_{a \in P(L)} \mathscr{C}\left(\left(\bigwedge_{j \in J} A_j\right)^{[a]}\right) = \bigwedge_{a \in P(L)} \mathscr{C}\left(\bigcap_{j \in J}(A_j)^{[a]}\right)$$
$$\geqslant \bigwedge_{a \in P(L)} \bigwedge_{j \in J} \mathscr{C}\left((A_j)^{[a]}\right) \geqslant \bigwedge_{j \in J} \bigwedge_{a \in P(L)} \mathscr{C}\left((A_j)^{[a]}\right)$$
$$= \bigwedge_{j \in J} \omega(\mathscr{C})(A_j).$$

(3) 设 $\{A_j\}_{j \in J} \subseteq L^X$ 是一个定向的 L-模糊集族. 为了证明

$$\omega(\mathscr{C})\left(\bigvee_{j \in J} A_j\right) \geqslant \bigwedge_{j \in J} \omega(\mathscr{C})(A_j),$$

我们假设 $m \in M$ 且

$$m \leqslant \bigwedge_{j \in J} \omega(\mathscr{C})(A_j) = \bigwedge_{j \in J} \bigwedge_{a \in P(L)} \mathscr{C}\left((A_j)^{[a]}\right),$$

则 $\forall j \in J$ 和任意的 $a \in P(L)$, 都有 $(A_j)^{[a]} \in \mathscr{C}_{[m]}$. 由推论 3.7.15 可知 $\mathscr{C}_{[m]}$ 是一个凸结构, 进一步再由引理 3.9.3 可知对任何 $c \in P(L)$, 都有 $(A_j)^{(c)} \in \mathscr{C}_{[m]}$. 从而

$$\omega(\mathscr{C})\left(\bigvee_{j \in J} A_j\right) = \bigwedge_{a \in P(L)} \mathscr{C}\left(\left(\bigvee_{j \in J} A_j\right)^{[a]}\right)$$
$$= \bigwedge_{a \in P(L)} \mathscr{C}\left(\bigcap_{\substack{a \in \alpha^*(c) \\ c \in P(L)}} \left(\bigvee_{j \in J} A_j\right)^{(c)}\right)$$
$$\geqslant \bigwedge_{a \in P(L)} \bigwedge_{\substack{a \in \alpha^*(c) \\ c \in P(L)}} \mathscr{C}\left(\bigcup_{j \in J}(A_j)^{(c)}\right)$$

3.9 由 M-模糊化凸结构诱导的 (L,M)-模糊凸结构

$$\geqslant \bigwedge_{a\in P(L)} \bigwedge_{\substack{a\in \alpha^*(c) \\ c\in P(L)}} \bigwedge_{j\in J} \mathscr{C}\left((A_j)^{(c)}\right) \geqslant m.$$

由 m 的任意性, 可证

$$\omega(\mathscr{C})\left(\bigvee_{j\in J} A_j\right) \geqslant \bigwedge_{j\in J} \omega(\mathscr{C})(A_j). \qquad \square$$

推论 3.9.5 设 $(X,\mathscr{C})$ 是一个凸空间. 则 $\omega(\mathscr{C})$ 是一个 L-凸结构, 称它为由 $\mathscr{C}$ 诱导的 L-凸结构, 这里 $\omega(\mathscr{C}) = \{A \in L^X \mid A^{[a]} \in \mathscr{C}, \forall a \in P(L)\}$.

定理 3.9.6 设 $(X,\mathscr{C})$ 是一个 M-模糊化凸空间. 则

$$\forall A \in L^X, \quad \omega(\mathscr{C})(A) = \bigwedge_{a\in P(L)} \mathscr{C}(A^{(a)}).$$

证明 由引理 3.9.3 可知

$$\omega(\mathscr{C})(A) = \bigwedge_{a\in P(L)} \mathscr{C}(A^{[a]}) \leqslant \bigwedge_{a\in P(L)} \mathscr{C}(A^{(a)}).$$

再由定理 2.1.4(7) 可知

$$\omega(\mathscr{C})(A) = \bigwedge_{a\in P(L)} \mathscr{C}(A^{[a]}) \geqslant \bigwedge_{a\in P(L)} \mathscr{C}(A^{(a)}). \qquad \square$$

推论 3.9.7 设 $(X,\mathscr{C})$ 是一个凸空间, $A\in L^X$. 则 $A\in \omega(\mathscr{C})$ 当且仅当 $\forall a\in P(L), A^{(a)} \in \mathscr{C}$.

定理 3.9.8 设 $(X,\mathscr{C})$ 是一个 M-模糊化凸空间. 则

$$\forall A \in L^X, \quad \omega(\mathscr{C})(A) = \bigwedge_{a\in L} \mathscr{C}(A_{[a]}).$$

证明 从定理 2.1.4(9) 我们可得 $\omega(\mathscr{C})(A) \leqslant \bigwedge_{a\in L} \mathscr{C}(A_{[a]})$. 为了证明 $\omega(\mathscr{C})(A) \geqslant \bigwedge_{a\in L} \mathscr{C}(A_{[a]})$, 设 $m \leqslant \bigwedge_{a\in L} \mathscr{C}(A_{[a]})$. 则对任意的 $a\in L, A_{[a]} \in \mathscr{C}_{[m]}$. 下证对任意的 $b\in P(L), A^{(b)} \in \mathscr{C}_{[m]}$. 由定理 2.1.4(10), 我们知道 $A^{(b)} = \bigcup_{a\not\leqslant b} A_{[a]}$. 因此只需要证明 $\{A_{[a]} \mid a \not\leqslant b\}$ 是定向集即可. 为此, 设 $A_{[c]}, A_{[d]} \in \{A_{[a]} \mid a\not\leqslant b\}$, 则 $c\not\leqslant b$ 且 $d\not\leqslant b$. 因为 b 是素元, 所以 $c\wedge d \not\leqslant b$. 这样我们就有 $A_{[c\wedge d]} \in \{A_{[a]} \mid a\not\leqslant b\}$ 且 $A_{[c\wedge d]}$ 包含 $A_{[c]}$ 和 $A_{[d]}$. 于是得证 $\{A_{[a]} \mid a\not\leqslant b\}$ 是定向集. $\qquad \square$

推论 3.9.9 设 $(X,\mathscr{C})$ 是一个凸空间, $A\in L^X$. 则 $A\in\omega(\mathscr{C})$ 当且仅当 $\forall a\in L, A_{[a]}\in\mathscr{C}$.

定理 3.9.10 设 $(X,\mathscr{C})$ 和 $(Y,\mathscr{D})$ 是 M-模糊化凸空间. 则 $f:(X,\mathscr{C})\to(Y,\mathscr{D})$ 是 M-模糊化凸保持的当且仅当 $f:(X,\omega(\mathscr{C}))\to(Y,\omega(\mathscr{D}))$ 是 (L,M)-模糊凸保持的.

证明 ($\Rightarrow$) 设 $f:(X,\mathscr{C})\to(Y,\mathscr{D})$ 是 M-模糊化凸保持的. 则 $\forall B\in \mathbf{2}^Y$, 都有 $\mathscr{D}(B)\leqslant\mathscr{C}(f^{\leftarrow}(B))$. 于是 $\forall F\in L^Y$,

$$\begin{aligned}\omega(\mathscr{C})(f^{\leftarrow}(F)) &= \bigwedge_{a\in L}\mathscr{C}\left((f^{\leftarrow}(F))_{[a]}\right)=\bigwedge_{a\in L}\mathscr{C}\left(f^{\leftarrow}(F_{[a]})\right)\\ &\geqslant \bigwedge_{a\in L}\mathscr{D}\left(F_{[a]}\right)\geqslant\omega(\mathscr{D})(F).\end{aligned}$$

故 $f:(X,\omega(\mathscr{C}))\to(Y,\omega(\mathscr{D}))$ 是 (L,M)-模糊凸保持的.

($\Leftarrow$) 设 $f:(X,\omega(\mathscr{C}))\to(Y,\omega(\mathscr{D}))$ 是 (L,M)-模糊凸保持的. 因为对任意一个分明集 $B\in\mathbf{2}^Y$, 它可以看作一个 L-模糊集, 所以 $\forall a\in P(L), B^{(a)}=B$, $(f^{\leftarrow}(B))^{(a)}=f^{\leftarrow}(B)$. 于是

$$\begin{aligned}\mathscr{C}(f^{\leftarrow}(B)) &= \bigwedge_{a\in P(L)}\mathscr{C}\left((f^{\leftarrow}(B))^{(a)}\right)=\omega(\mathscr{C})(f^{\leftarrow}(B))\\ &\geqslant \omega(\mathscr{D})(B)=\bigwedge_{a\in P(L)}\mathscr{D}\left(B^{(a)}\right)=\mathscr{D}(B).\end{aligned}$$

这表明 $f:(X,\mathscr{C})\to(Y,\mathscr{D})$ 是 M-模糊化凸保持的. □

在这一节最后, 作为集合套的一个重要应用, 我们将从凸包算子出发, 给出诱导 L-凸空间中 L-凸包算子的表现形式.

定理 3.9.11 设 $(X,\mathscr{C})$ 是一个凸空间, co 表示 $(X,\mathscr{C})$ 中的凸包算子. 映射 $\widetilde{\mathrm{co}}:L^X\to L^X$ 定义如下:

$$\forall A\in L^X,\quad \widetilde{\mathrm{co}}(A)=\bigvee_{a\in L}(a\wedge \mathrm{co}(A_{(a)})),$$

则

(1) $\forall A\in L^X, \widetilde{\mathrm{co}}(A)=\bigvee_{a\in L}(a\wedge \mathrm{co}(A_{[a]}))$;

(2) $\forall A\in L^X, \widetilde{\mathrm{co}}(A)=\bigwedge_{a\in L}(a\vee \mathrm{co}(A^{[a]}))$;

(3) $\forall A\in L^X, \widetilde{\mathrm{co}}(A)=\bigwedge_{a\in L}(a\vee \mathrm{co}(A^{(a)}))$;

(4) $\widetilde{\mathrm{co}}$ 是一个 L-凸包算子.

3.9 由 M-模糊化凸结构诱导的 (L,M)-模糊凸结构

证明 (1) $\forall A \in L^X$, 令 $\overline{\mathrm{co}}(A) = \bigvee_{a \in L}(a \wedge \mathrm{co}(A_{[a]}))$, 则显然有 $\widetilde{\mathrm{co}}(A) \leqslant \overline{\mathrm{co}}(A)$. 下证 $\widetilde{\mathrm{co}}(A) \geqslant \overline{\mathrm{co}}(A)$. 对于任意的 $x \in X$, 设 $b \prec \overline{\mathrm{co}}(A)(x)$. 则存在 $a \in L$ 使得 $b \prec a$ 且 $x \in \mathrm{co}(A_{[a]})$. 取 $e \in L$ 使得 $b \prec e \prec a$, 则 $x \in \mathrm{co}(A_{[a]}) \subseteq \mathrm{co}(A_{(e)})$, 这意味着 $b \leqslant \bigvee_{e \in L}(e \wedge \mathrm{co}(A_{(e)})) = \widetilde{\mathrm{co}}(A)(x)$. 因此得到了 $\widetilde{\mathrm{co}}(A) \geqslant \overline{\mathrm{co}}(A)$ 的证明.

类似 (1), 借助定理 2.2.6 可证

$$\forall A \in L^X, \quad \bigwedge_{a \in L}(a \vee \mathrm{co}(A^{[a]})) = \bigwedge_{a \in L}(a \vee \mathrm{co}(A^{(a)})).$$

为了证明 (2) 和 (3), 我们现在设 $\underline{\mathrm{co}}(A) = \bigwedge_{a \in L}(a \vee \mathrm{co}(A^{(a)}))$, 则由定理 2.2.5 和定理 2.2.6 可知下面两个包含式成立:

$$\widetilde{\mathrm{co}}(A)_{(a)} \subseteq \mathrm{co}(A_{(a)}) \subseteq \mathrm{co}(A_{[a]}) \subseteq \widetilde{\mathrm{co}}(A)_{[a]},$$

$$\underline{\mathrm{co}}(A)^{(a)} \subseteq \mathrm{co}(A^{(a)}) \subseteq \mathrm{co}(A^{[a]}) \subseteq \underline{\mathrm{co}}(A)^{[a]}.$$

而由定理 2.1.4 知

$$\widetilde{\mathrm{co}}(A)^{(a)} = \bigcup_{b \not\leqslant a}\widetilde{\mathrm{co}}(A)_{(b)} \subseteq \bigcup_{b \not\leqslant a}\mathrm{co}(A_{(b)}) \subseteq \mathrm{co}\left(\bigcup_{b \not\leqslant a} A_{(b)}\right) = \mathrm{co}(A^{(a)}) \subseteq \underline{\mathrm{co}}(A)^{[a]}.$$

故 $\widetilde{\mathrm{co}}(A) \leqslant \underline{\mathrm{co}}(A)$. 另一方面, 由

$$\underline{\mathrm{co}}(A)^{(a)} = \bigcup_{b \not\leqslant a}\underline{\mathrm{co}}(A)_{(b)} \subseteq \bigcup_{b \not\leqslant a}\mathrm{co}(A_{(b)}) \subseteq \bigcup_{b \not\leqslant a}\widetilde{\mathrm{co}}(A)_{[b]} = \widetilde{\mathrm{co}}(A)^{(a)}$$

可知 $\widetilde{\mathrm{co}}(A) \geqslant \underline{\mathrm{co}}(A)$, 这表明 (2) 和 (3) 成立.

(4) 显然 $\widetilde{\mathrm{co}}(\chi_\varnothing) = \chi_\varnothing$ 且 $\widetilde{\mathrm{co}}(A) \geqslant A$. 另外, 当 $A \leqslant B$ 时, 容易看到 $\widetilde{\mathrm{co}}(A) \leqslant \widetilde{\mathrm{co}}(B)$. 接下来证明 $\widetilde{\mathrm{co}}(\widetilde{\mathrm{co}}(A)) \leqslant \widetilde{\mathrm{co}}(A)$. 而这由下面不等式可证.

$$\widetilde{\mathrm{co}}(\widetilde{\mathrm{co}}(A)) = \bigvee_{a \in L}(a \wedge \mathrm{co}(\widetilde{\mathrm{co}}(A)_{(a)})) \leqslant \bigvee_{a \in L}(a \wedge \mathrm{co}(\mathrm{co}(A_{(a)})))$$

$$\leqslant \bigvee_{a \in L}(a \wedge \mathrm{co}(A_{(a)})) = \widetilde{\mathrm{co}}(A).$$

最后我们来证明

$$\widetilde{\mathrm{co}}(A) = \bigvee\{\mathrm{co}(F) \mid F \ll A\}.$$

只需证明下面不等式即可：

$$\widetilde{\text{co}}(A) \leqslant \bigvee \{\text{co}(F) \mid F \ll A\}.$$

为此设 $a \in L$ 使得 $x \in \text{co}(A_{(a)})$，则存在 $A_{(a)}$ 的有限子集 E 使得 $x \in \text{co}(E)$. 此时显然有 $a \wedge E \ll A$ 且 $(a \wedge E)_{[a]} = E \subseteq A_{[a]}$. 这意味着 $a \leqslant \widetilde{\text{co}}(a \wedge E)(x)$. 故 $a \leqslant \bigvee \{\widetilde{\text{co}}(F) \mid F \ll A\}$. □

定理 3.9.12 设 $(X, \omega(\mathscr{C}))$ 为由凸空间 $(X, \mathscr{C})$ 诱导的 L-凸空间且 $A \in L^X$，那么 $A \in \omega(\mathscr{C})$ 当且仅当 $\widetilde{\text{co}}(A) = A$.

证明 设 $A \in \omega(\mathscr{C})$. 则由推论 3.9.9 可知 $\forall a \in L$, $A_{[a]} \in \mathscr{C}$，这意味着 $\text{co}(A_{[a]}) = A_{[a]}$. 因此 $\widetilde{\text{co}}(A) = A$. 反之，若 $\widetilde{\text{co}}(A) = A$，则 $\forall a \in L$, $(\widetilde{\text{co}}(A))_{[a]} = A_{[a]}$. 于是 $\text{co}(A_{[a]}) \subseteq (\widetilde{\text{co}}(A))_{[a]} = A_{[a]}$. 这表明 $A_{[a]} \in \mathscr{C}$，故由推论 3.9.9 可知 $A \in \omega(\mathscr{C})$. □

习 题 3

1. 试证明定理 3.1.7.
2. 试证明定理 3.1.9.
3. 试问：在定理 3.1.7 中，两种截集可否用另外两种截集取代？
4. 试问：在定理 3.1.9 中，连续映射可否用开映射取代？
5. 试问：在定理 3.1.9 中，连续映射可否换成闭映射？
6. 如果一个映射 $\text{Int}: L^X \to M^{J(L^X)}$ 满足下面条件 (FI1)—(FI4):

(FI1) $\text{Int}(A)(x_\lambda) = \bigwedge_{\mu \prec \lambda} \text{Int}(A)(x_\mu), \forall x_\lambda \in J(L^X), \forall A \in L^X$;

(FI2) $\text{Int}(X)(x_\lambda) = \top_M, \forall x_\lambda \in J(L^X)$;

(FI3) $\text{Int}(A)(x_\lambda) = \bot_M, \forall x_\lambda \not\leqslant A$;

(FI4) $\text{Int}(A \wedge B) = \text{Int}(A) \wedge \text{Int}(B)$,

那么下面 (FI5), (FI6) 和 (FI7) 是等价的，试证之 (证明源于 [135]).

(FI5) $\text{Int}(A)(x_\lambda) = \bigvee_{x_\lambda \leqslant V \leqslant A} \bigwedge_{y_\mu \prec V} \text{Int}(V)(y_\mu)$;

(FI6) $\forall a \in M \setminus \{\top_M\}, (\text{Int}(A))^{(a)} \subseteq \left(\text{Int}\left(\bigvee(\text{Int}(A))^{(a)}\right)\right)^{(a)}$;

(FI7) $\forall a \in M \setminus \{\bot_M\}, (\text{Int}(A))_{(a)} \subseteq \left(\text{Int}\left(\bigvee(\text{Int}(A))_{(a)}\right)\right)_{(a)}$.

7. 试证明定理 3.3.5 中的公式 $\text{Cl}^{(\mathcal{T}^{\text{Cl}})} = \text{Cl}$.
8. 如果一个映射 $\text{Cl}: L^X \to M^{J(L^X)}$ 满足下面 (FC1)—(FC4):

(FC1) $\text{Cl}(A)(x_\lambda) = \bigwedge_{\mu \prec \lambda} \text{Cl}(A)(x_\mu), \forall x_\lambda \in J(L^X)$;

(FC2) $\text{Cl}(\varnothing)(x_\lambda) = \bot_M, \forall x_\lambda \in J(L^X)$;

(FC3) $\text{Cl}(A)(x_\lambda) = \top_M, \forall x_\lambda \leqslant A$;

(FC4) $\text{Cl}(A \vee B) = \text{Cl}(A) \vee \text{Cl}(B)$,

那么下面条件 (FC5), (FC6) 和 (FC7) 是彼此等价的, 试证之 (证明源于文献 [135]).

(FC5) $\mathrm{Cl}(A)(x_\lambda) = \bigwedge_{x_\lambda \not\leq B \geq A} \bigvee_{y_\mu \not\leq B} (\mathrm{Cl}(B))(y_\mu)$;

(FC6) $\forall a \in M \setminus \{\bot_M\}$, $\left(\mathrm{Cl}\left(\bigvee(\mathrm{Cl}(A))_{[a]}\right)\right)_{[a]} \subseteq (\mathrm{Cl}(A))_{[a]}$;

(FC7) $\forall a \in M \setminus \{\top_M\}$, $\left(\mathrm{Cl}\left(\bigvee(\mathrm{Cl}(A))^{[a]}\right)\right)^{[a]} \subseteq (\mathrm{Cl}(A))^{[a]}$.

9. 设 $(X, \mathscr{T})$ 是一个拓扑空间, d 表示 $(X, \mathscr{T})$ 中的导算子. 也就是每个集合的像是它的导集, 则 $d: \mathbf{2}^X \to \mathbf{2}^X$ 满足下面条件:

(D1) $d(\varnothing) = \varnothing$;
(D2) $\forall x \in X$, $x \notin d(\{x\})$;
(D3) $\forall A, B \in \mathbf{2}^X$, $d(A \cup B) = d(A) \cup d(B)$;
(D4) $d(d(A)) \subseteq A \cup d(A)$.

反之, 如果一个映射 $d: \mathbf{2}^X \to \mathbf{2}^X$ 满足上面条件 (D1)—(D4), 那么就称 d 是一个导算子, 它可以诱导出 X 上的一个拓扑如下:

$$\mathscr{T}_d = \{A \in \mathbf{2}^X \mid d(A) \subseteq A\}.$$

试证明上面的所有结论, 并证明 d 和 $\mathscr{T}_d$ 是一一对应的.

10. 设 $(X, \mathscr{T})$ 是一个拓扑空间, d 表示 $(X, \mathscr{T})$ 中的导算子. 定义映射 $\widetilde{d}: L^X \to L^X$ 如下:

$$\forall A \in L^X, \quad \widetilde{d}(A) = \bigvee_{a \in L} (a \wedge d(A_{[a]})),$$

则

(1) $\forall A \in L^X$, $\widetilde{d}(A) = \bigvee_{a \in L} (a \wedge d(A_{(a)}))$;

(2) $\forall A \in L^X$, $\widetilde{d}(A) = \bigwedge_{a \in L} (a \vee d(A^{[a]}))$;

(3) $\forall A \in L^X$, $\widetilde{d}(A) = \bigwedge_{a \in L} (a \vee d(A^{(a)}))$;

(4) $\omega(\mathscr{T}) = \{A \in L^X \mid \widetilde{d}(A) \leq A\}$.

11. X 上的一个拟一致结构就是 $\mathbf{2}^{X \times X}$ 的满足下面条件 (1)—(4) 的一个非空子集 $\mathcal{U}$.

(1) $U \in \mathcal{U} \Rightarrow \Delta \subseteq U$, 这里 $\Delta = \{(x, x) \mid x \in X\}$;
(2) $U \in \mathcal{U}, U \subseteq V \Rightarrow V \in \mathcal{U}$;
(3) $U, V \in \mathcal{U} \Rightarrow U \cap V \in \mathcal{U}$;
(4) $U \in \mathcal{U} \Rightarrow \exists V \in \mathcal{U}$ 使得 $V \circ V \subseteq U$.

对满足 $\Delta \subseteq U$ 的每个 $U \in \mathbf{2}^{X \times X}$, 定义映射 $f_U: X \to \mathbf{2}^X$ 使得

$$\forall x \in X, \quad f_U(x) = \{y \in X \mid (x, y) \in U\},$$

那么 $x \in f_U(x)$, 再定义 $f_U: \mathbf{2}^X \to \mathbf{2}^X$ 使得

$$f_U(A) = \bigcup \{f_U(x) \mid x \in A\},$$

那么 f_U 是一个保并增值自映射. 反之, 任给满足条件: $\forall x \in X, x \in f(x)$ 的映射 $f: X \to \mathbf{2}^X$, 可以定义 $U_f \in \mathbf{2}^{X \times X}$ 使得

$$U_f = \{(x,y) \in X \times X \mid y \in f(x)\}.$$

不难验证 $\Delta \subseteq U_f, U = U_{f_U}$ 且 $f = f_{U_f}$. 另外, 如果 $U, V \in \mathcal{U}$ 而且 U, V 包含 Δ, 那么对每个 $x \in X$, 试证明:

(F1) $f_{U \cap V}(x) = f_U(x) \cap f_V(x)$;

(F2) $f_{U \circ V}(x) = \bigcup \{f_V(y) \mid y \in f_U(x)\}$;

(F3) $U \subseteq V \Rightarrow f_U(x) \subseteq f_V(x)$.

详细可见文献 [10].

12. 试证明例 3.6.4.

13. 试证明例 3.6.5.

14. 试证明例 3.6.6.

15. 试证明例 3.6.7.

16. 试证明一个群的所有子群加上空集构成一个凸结构.

17. 试证明一个群的所有正规子群加上空集构成一个凸结构.

18. 试证明一个环的所有子环加上空集构成一个凸结构.

19. 试证明一个环的所有理想加上空集构成一个凸结构.

20. 试证明一个域的所有子域加上空集构成一个凸结构.

21. 试证明一个模的所有子模加上空集构成一个凸结构.

22. 设 $f: G_1 \to G_2$ 是一个群同态映射, 且 $\mathscr{C}_1, \mathscr{C}_2$ 分别是 G_1, G_2 上由所有子群构成的凸结构. 证明 $f: (G_1, \mathscr{C}_1) \to (G_2, \mathscr{C}_2)$ 是凸空间之间的凸保持映射和凸到凸映射.

23. 设 $f: R_1 \to R_2$ 是一个环同态映射, 且 $\mathscr{C}_1, \mathscr{C}_2$ 分别是 R_1, R_2 上由所有子环构成的凸结构. 证明 $f: (G_1, \mathscr{C}_1) \to (G_2, \mathscr{C}_2)$ 是凸空间之间的凸保持映射和凸到凸映射.

24. 设 $f: R_1 \to R_2$ 是一个环同态映射, 且 $\mathscr{C}_1, \mathscr{C}_2$ 分别是 R_1, R_2 上由所有理想构成的凸结构. 证明 $f: (G_1, \mathscr{C}_1) \to (G_2, \mathscr{C}_2)$ 是凸空间之间的凸保持映射和凸到凸映射.

25. 试证明例 3.7.9.

26. 试证明例 3.7.10.

27. 试证明例 3.7.11.

28. 试证明例 3.7.12.

29. 试证明例 3.7.13.

30. 试证明定理 3.7.14.

31. 试证明定理 3.7.16.

32. 试证明定理 3.8.7 中的公式 $\mathrm{co}^{(\mathscr{C}^{\mathrm{co}})} = \mathrm{co}$.

33. 试证明 $\mathscr{C}^{\mathrm{co}\,\mathscr{C}} = \mathscr{C}$.

34. 设 $(X, \mathscr{C})$ 是一个 $[0,1]$-模糊化凸空间 (即模糊化凸空间). 试证明下面等式成立.

$$\forall A \in L^X, \quad \omega(\mathscr{C})(A) = \bigwedge_{a \in L} \mathscr{C}(A_{(a)}).$$

习 题 3

35. 设 $(X, \mathscr{C})$ 是一个 M-模糊化凸空间, 这里 $M \neq [0,1]$. 那么下面等式是否成立? 如果不成立, 可否举出例子说明之?
$$\forall A \in L^X, \quad \omega(\mathscr{C})(A) = \bigwedge_{a \in L} \mathscr{C}(A_{(a)}).$$

36. 设 $(X, \mathscr{C})$ 是一个 M-模糊化凸空间, 这里 $M \neq [0,1]$. 如果下面等式不成立, 请问 M 满足什么条件成立?
$$\forall A \in L^X, \quad \omega(\mathscr{C})(A) = \bigwedge_{a \in L} \mathscr{C}(A_{(a)}).$$

37. 设 $(X, \mathscr{C})$ 和 $(Y, \mathscr{D})$ 是 M-模糊化凸空间. 试证明 $f: (X, \mathscr{C}) \to (Y, \mathscr{D})$ 是 M-模糊化凸到凸的当且仅当 $f: (X, \omega(\mathscr{C})) \to (Y, \omega(\mathscr{D}))$ 是 (L, M)-模糊凸到凸的.

38. (见定理 3.9.11) 设 $(X, \mathscr{C})$ 是一个凸空间, co 表示 $(X, \mathscr{C})$ 中的凸包算子, 试证明
(1) $\forall A \in L^X, \widetilde{\mathrm{co}}(A) = \bigvee_{a \in J(M)} (a \wedge \mathrm{co}(A_{(a)}));$
(2) $\forall A \in L^X, \widetilde{\mathrm{co}}(A) = \bigvee_{a \in J(M)} (a \wedge \mathrm{co}(A_{[a]}));$
(3) $\forall A \in L^X, \widetilde{\mathrm{co}}(A) = \bigwedge_{a \in P(M)} (a \vee \mathrm{co}(A^{[a]}));$
(4) $\forall A \in L^X, \widetilde{\mathrm{co}}(A) = \bigwedge_{a \in P(M)} (a \vee \mathrm{co}(A^{(a)})).$

第 4 章 L-模糊子群

本章旨在引入 L-模糊子半群度、L-模糊子群度、L-模糊正规子群度的概念,并借助四种截集给出它们的等价刻画,进一步讨论它们与 L-模糊凸结构的关系,指出群同态可以看成 L-模糊凸保持映射和 L-模糊凸到凸的映射等.

4.1 群上的 L-模糊同余关系

其实, 模糊子群的概念很早就被提出了, 在模糊群的早期定义中, 有两种稍有不同的定义. 那么, 使用哪种定义比较合理呢? 之所以在这一节中来介绍群中的 L-模糊同余关系, 就是为了说明哪种定义比较合理, 以便于在今后章节中为进一步推广提供依据. 模糊群最早被 Rosenfeld 提出时是使用 $[0,1]$ 作为模糊集的取值格[127] 的, 后来被推广到一般格值形式.

定义 4.1.1 设 G 是一个群且 $A \in L^G$. 称 A 为 G 的一个模糊子群, 如果它满足下面两个条件:

(1) $A(xy) \geqslant A(x) \wedge A(y)$;

(2) $A(x^{-1}) \geqslant A(x)$.

后来, Anthony 和 Sherwood 在文献 [48] 中增加了一个条件, 变成了下述情形.

定义 4.1.2 设 G 是一个群且 $A \in L^G$. 称 A 为 G 的一个模糊子群, 如果它满足下面三个条件:

(1) $A(xy) \geqslant A(x) \wedge A(y)$;

(2) $A(x^{-1}) \geqslant A(x)$;

(3) $A(1) = 1$.

接下来, 我们将从等价关系的角度, 来说明定义 4.1.2 的合理性. 本节的主要结果源于 [131].

定义 4.1.3 一个群 G 上的 L-模糊等价关系 R 叫做一个 L-模糊左 (或右) 同余关系, 如果 R 满足下面条件 (L1)(或 (R1)).

(L1) $\forall a, x, y \in G, R(ax, ay) = R(x, y)$;

(R1) $\forall a, x, y \in G, R(xa, ya) = R(x, y)$.

进一步来说, R 叫做 G 上的一个 L-模糊同余关系, 如果它既是左同余关系又是右同余关系.

下面定理的证明是直接的.

4.1 群上的 L-模糊同余关系

定理 4.1.4 设 R 是 G 上的一个 L-模糊等价关系. 那么下列条件等价:

(1) R 是 G 上的一个 L-模糊 (左, 右) 同余关系;
(2) $\forall a \in L$, $R_{[a]}$ 是 G 上的一个 (左, 右) 同余关系;
(3) $\forall a \in M(L)$, $R_{[a]}$ 是 G 上的一个 (左, 右) 同余关系;
(4) $\forall a \in L$, $R^{[a]}$ 是 G 上的一个 (左, 右) 同余关系;
(5) $\forall a \in P(L)$, $R^{[a]}$ 是 G 上的一个 (左, 右) 同余关系;
(6) $\forall a \in P(L)$, $R^{(a)}$ 是 G 上的一个 (左, 右) 同余关系.

下面我们给出 L-模糊同余关系和 L-模糊子群的相互诱导关系.

定理 4.1.5 设 R 是群 G 上的一个 L-模糊左 (或者右) 同余关系. 定义 $A \in L^G$ 使得

$$\forall x \in G, \quad A(x) = R(1, x), \quad \text{这里 } 1 \text{ 是 } G \text{ 的单位元}.$$

那么 A 满足下面条件:

(G1) $\forall x, y \in G$, $A(xy) \geqslant A(x) \wedge A(y)$;
(G2) $\forall x \in G$, $A(x^{-1}) = A(x)$;
(G3) $A(1) = 1$.

也就是说, A 是 G 的一个 L-模糊子群.

证明 (G3) 是显然的. (G1) 和 (G2) 可以从下面事实得到.

$$A(xy) = R(1, xy) \geqslant R(1, x) \wedge R(x, xy) = A(x) \wedge A(y)$$

和

$$A(x^{-1}) = R(1, x^{-1}) = R(x, 1) = R(1, x) = A(x). \qquad \square$$

定理 4.1.6 设 A 是一个群 G 的 L-模糊子群. 定义 $R \in L^{G \times G}$ 使得 $\forall x, y \in G$,

$$R_1(x, y) = A(x^{-1}y), \quad R_2(x, y) = A(xy^{-1}).$$

那么 R_1 是一个 L-模糊左同余关系且 R_2 是一个 L-模糊右同余关系.

证明 显然 $\forall x, y \in G$, $R_1(x, x) = A(1) = 1$ 且 $R_1(x, y) = R_1(y, x)$. 另外, $R_1 \circ R_1 \subset R_1$ 由下面事实可得.

$$R_1(x, z) = A(x^{-1}z)$$
$$= \bigvee_{y \in G} A(x^{-1}yy^{-1}z)$$
$$\geqslant \bigvee_{y \in G} \{A(x^{-1}y) \wedge A(y^{-1}z)\}$$

$$= (R_1 \circ R_1)(x,z).$$

这表明 R_1 是一个 L-模糊等价关系. 其他证明是显然的. □

与上面定理的证明类似地可证明下面两个定理.

定理 4.1.7 设 R 是一个群 G 上的 L-模糊同余关系. 定义 $A \in L^G$ 使得

$$\forall x \in G, \quad A(x) = R(1,x),$$

那么 A 满足 (G1)—(G3) 和 (G4), 其中

(G4) $\forall x,y \in G, A(xyx^{-1}) = A(y)$.

也就是说, A 是一个 L-模糊正规子群.

定理 4.1.8 设 A 是群 G 的一个 L-模糊正规子群. 定义 $R \in L^{G \times G}$ 使得 $\forall x,y \in G$,

$$R(x,y) = A(x^{-1}y),$$

那么 R 是一个 L-模糊同余关系.

另外, 借助于几种截集, 我们能够给出 L-模糊子群和正规子群的等价刻画如下.

定理 4.1.9 设 G 是一个群且 $A \in L^G$. 那么下面条件等价:

(1) A 是 G 的一个 L-模糊 (正规) 子群;
(2) $\forall a \in L, A_{[a]}$ 是 G 的一个 (正规) 子群;
(3) $\forall a \in M(L), A_{[a]}$ 是 G 的一个 (正规) 子群;
(4) $\forall a \in L, A^{[a]}$ 是 G 的一个 (正规) 子群;
(5) $\forall a \in P(L), A^{[a]}$ 是 G 的一个 (正规) 子群;
(6) $\forall a \in P(L), A^{(a)}$ 是 G 的一个 (正规) 子群.

证明 $\forall x,y \in G$, 令 $R(x,y) = A(x^{-1}y)$. 则 $\forall a \in L$,

$$(x,y) \in R_{[a]} \Leftrightarrow x^{-1}y \in A_{[a]},$$

$$(x,y) \in R^{[a]} \Leftrightarrow x^{-1}y \in A^{[a]},$$

$$(x,y) \in R^{(a)} \Leftrightarrow x^{-1}y \in A^{(a)}.$$

从而由同余关系与子群的对应关系可得到证明. □

4.2 L-模糊子半群度

在这一节中, 我们介绍 L-模糊子半群的一个新定义, 使得一个半群中的每个 L-模糊子集在一定程度上都是一个 L-模糊子半群, 并借助四种截集分别给出它们的等价刻画.

4.2 L-模糊子半群度

定义 4.2.1 设 S 是一个半群且 $A \in L^S$. 称 A 为

(1) S 的一个 L-模糊子半群, 如果 $\forall x, y \in S, A(xy) \geqslant A(x) \wedge A(y)$;

(2) S 的一个 L-模糊左 (右) 理想, 如果 $\forall x, y \in S, A(xy) \geqslant A(y)$ $(A(xy) \geqslant A(x))$;

(3) S 的一个 L-模糊理想, 如果它既是 S 的一个 L-模糊左理想又是 S 的一个 L-模糊右理想.

由半群 S 中 L-模糊子半群的定义可知, 一个 L-模糊子集要么是 S 的 L-模糊子半群, 要么不是, 二者必居其一且仅居其一.

下面我们通过利用 L 的蕴含算子, 给出一个 L-模糊子集是 S 的 L-模糊子半群的程度的概念, 也就是对 L-模糊子半群进一步模糊化. 此外, 我们还可以定义半群中一个 L-模糊子集是一个 L-模糊理想的程度的概念, 相关性质等放在习题里. 本节主要结果来源于 [59].

定义 4.2.2 设 S 是一个半群, $A \in L^S$, 令

$$\mathscr{S}(A) = \bigwedge_{x,y \in S} ((A(x) \wedge A(y)) \mapsto A(xy)).$$

称 $\mathscr{S}(A)$ 为 A 的 L-模糊子半群度.

显然, A 是 S 的 L-模糊子半群当且仅当 $\mathscr{S}(A) = 1$. 一般地, $\mathscr{S}(A)$ 总可以看作 A 为 L-模糊子半群的程度.

以下为了方便起见, 我们总是约定 $\varnothing$ 是 S 的子半群.

例 4.2.3 设 S 是由所有正奇数组成的集合, 易得其对整数的乘法 "·" 是封闭的, 且乘法满足结合律, 因此 $(S, \cdot)$ 是一个半群. 定义三个模糊集 $A, B, C \in [0,1]^S$ 如下:

$$A(x) = \begin{cases} 0, & \text{当 } x = 15 \text{ 时,} \\ 1, & \text{其他,} \end{cases}$$

$$B(x) = \begin{cases} 0.5, & \text{当 } x = 15 \text{ 时,} \\ 1, & \text{其他,} \end{cases}$$

$$C(x) = 0.5, \quad \forall x \in S.$$

容易验证

$$\mathscr{S}(A) = \bigwedge_{x,y \in S} ((A(x) \wedge A(y)) \mapsto A(xy)) = (A(3) \wedge A(5)) \mapsto A(15) = 0.$$

$$\mathscr{S}(B) = \bigwedge_{x,y \in S} ((B(x) \wedge B(y)) \mapsto B(xy)) = (A(3) \wedge A(5)) \mapsto A(15) = 0.5.$$

$$\mathscr{S}(C) = \bigwedge_{x,y \in S} ((C(x) \wedge C(y)) \mapsto C(xy)) = 0.5 \mapsto 0.5 = 1.$$

为了给出 L-模糊子半群度的等价表达式, 我们先给出下面的引理, 它的证明是显而易见的.

引理 4.2.4 设 S 是一个半群且 $A \in L^S$. 则 $\forall a \in L, \mathscr{S}(A) \geqslant a$ 当且仅当 $\forall x, y \in S$,

$$A(x) \wedge A(y) \wedge a \leqslant A(xy).$$

由引理 4.2.4, 可以很容易地得到 L-模糊子半群度的如下刻画.

定理 4.2.5 设 S 是一个半群且 $A \in L^S$. 则

$$\mathscr{S}(A) = \bigvee \{a \in L \mid A(x) \wedge A(y) \wedge a \leqslant A(xy), \forall x, y \in S\}.$$

下面定理用四个截集刻画了 L-模糊子半群度.

定理 4.2.6 设 S 是一个半群且 $A \in L^S$. 则

(1) $\mathscr{S}(A) = \bigvee \{a \in L \mid \forall b \leqslant a, A_{[b]}$ 是 S 的子半群$\}$;

(2) $\mathscr{S}(A) = \bigvee \{a \in L \mid \forall b \notin \alpha(a), A^{[b]}$ 是 S 的子半群$\}$;

(3) $\mathscr{S}(A) = \bigvee \{a \in L \mid \forall b \in P(L), a \nleqslant b, A^{(b)}$ 是 S 的子半群$\}$;

(4) 如果对于任意的 $a, b \in L$, 都有 $\beta(a \wedge b) = \beta(a) \cap \beta(b)$, 那么 $\mathscr{S}(A) = \bigvee \{a \in L \mid \forall b \in \beta(a), A_{(b)}$ 是 S 的子半群$\}$.

证明 (1) 先证明下面不等式:

$$\mathscr{S}(A) \leqslant \bigvee \{a \in L \mid \forall b \leqslant a, A_{[b]} \text{ 是 } S \text{ 的子半群}\}.$$

$\forall x, y \in S$, 令 $A(x) \wedge A(y) \wedge a \leqslant A(xy)$, 则 $\forall b \leqslant a, \forall x, y \in A_{[b]}$, 有

$$A(xy) \geqslant A(x) \wedge A(y) \wedge a \geqslant A(x) \wedge A(y) \wedge b = b,$$

这说明 $xy \in A_{[b]}$, 于是 $A_{[b]}$ 是 S 的子半群. 因此我们得到

$$\mathscr{S}(A) \leqslant \bigvee \{a \in L \mid \forall b \leqslant a, A_{[b]} \text{ 是 } S \text{ 的子半群}\}.$$

下面我们来证明

$$\bigvee \{a \in L \mid \forall b \leqslant a, A_{[b]} \text{ 是 } S \text{ 的子半群}\} \leqslant \mathscr{S}(A).$$

设 $a \in L$ 且 $\forall b \leqslant a, A_{[b]}$ 是半群 S 的一个子半群, 我们要证明 $\forall x, y \in S$,

$$A(x) \wedge A(y) \wedge a \leqslant A(xy).$$

为此设 $b = A(x) \wedge A(y) \wedge a$, 则 $b \leqslant a$ 且 $x, y \in A_{[b]}$, 于是 $xy \in A_{[b]}$, 这意味着 $A(xy) \geqslant b = A(x) \wedge A(y) \wedge a$. 于是,

$$\bigvee \{a \in L \mid \forall b \leqslant a, A_{[b]} \text{ 是 } S \text{ 的子半群}\} \leqslant \mathscr{S}(A).$$

综上可得

$$\mathscr{S}(A) = \bigvee \{a \in L \mid \forall b \leqslant a, A_{[b]} \text{ 是 } S \text{ 的子半群}\}.$$

(2) 先证明不等式

$$\mathscr{S}(A) \leqslant \bigvee \{a \in L \mid \forall b \notin \alpha(a), A^{[b]} \text{ 是 } S \text{ 的子半群}\}.$$

设 a 满足条件: 对任意的 $x, y \in S$, $A(x) \wedge A(y) \wedge a \leqslant A(xy)$, 下面我们来证明 $\forall b \notin \alpha(a)$, $A^{[b]}$ 是 S 的子半群, 为此任取 $x, y \in A^{[b]}$, 则 $b \notin \alpha(A(x))$, $b \notin \alpha(A(y))$, 从而

$$b \notin \alpha(A(x)) \cup \alpha(A(y)) \cup \alpha(a) = \alpha(A(x) \wedge A(y) \wedge a).$$

于是由 $A(x) \wedge A(y) \wedge a \leqslant A(xy)$ 可知

$$\alpha(A(xy)) \subseteq \alpha(A(x) \wedge A(y) \wedge a),$$

这意味着 $b \notin \alpha(A(xy))$, 也就是 $xy \in A^{[b]}$. 这说明 $A^{[b]}$ 是半群 S 的子半群, 且

$$a \in \{a \in L \mid \forall b \notin \alpha(a), A^{[b]} \text{ 是 } S \text{ 的子半群}\}.$$

因此我们得到

$$\mathscr{S}(A) = \bigvee \{a \in L \mid \forall x, y \in S, A(x) \wedge A(y) \wedge a \leqslant A(xy)\}$$
$$\leqslant \bigvee \{a \in L \mid \forall b \notin \alpha(a), A^{[b]} \text{ 是 } S \text{ 的子半群}\}.$$

接下来证明

$$\mathscr{S}(A) \geqslant \bigvee \{a \in L \mid \forall b \notin \alpha(a), A^{[b]} \text{ 是 } S \text{ 的子半群}\}.$$

为此, 对于任意的 $a \in L$ 和任意的 $b \notin \alpha(a)$, 设 $A^{[b]}$ 是半群 S 的子半群. 下面只需证 $\forall x, y \in S$,

$$A(x) \wedge A(y) \wedge a \leqslant A(xy).$$

为此设 $b \notin \alpha(A(x) \wedge A(y) \wedge a)$, 由

$$\alpha(A(x) \wedge A(y) \wedge a) = \alpha(A(x)) \cup \alpha(A(y)) \cup \alpha(a)$$

我们知道 $b \notin \alpha(a)$ 且 $x, y \in A^{[b]}$, 因此由 $A^{[b]}$ 是半群 S 的子半群可知 $xy \in A^{[b]}$, 那就是 $b \notin \alpha(A(xy))$, 这意味着 $A(x) \wedge A(y) \wedge a \leqslant A(xy)$. 这正好说明

$$\mathscr{S}(A) = \bigvee \{a \in L \mid \forall x, y \in S, A(x) \wedge A(y) \wedge a \leqslant A(xy)\}$$

$$\geqslant \bigvee \{a \in L \mid \forall b \notin \alpha(a), A^{[b]} \text{ 是 } S \text{ 的子半群}\}.$$

因此, (2) 成立.

(3) 先证明下面不等式

$$\mathscr{S}(A) \leqslant \bigvee\{a \in L \mid \forall b \in P(L) \text{且} b \not\geqslant a, A^{(b)} \text{ 是 } S \text{ 的子半群}\}.$$

为此设 $a \in L$ 满足: $\forall x, y \in S, A(x) \wedge A(y) \wedge a \leqslant A(xy)$. 令 $b \in P(L)$, $a \not\leqslant b$ 且 $x, y \in A^{(b)}$. 接下来证明 $xy \in A^{(b)}$. 假设 $xy \notin A^{(b)}$, 则 $A(xy) \leqslant b$, 于是 $A(x) \wedge A(y) \wedge a \leqslant A(xy) \leqslant b$. 由 $b \in P(L)$ 且 $x, y \in A^{(b)}$ 可得 $a \leqslant b$, 它和 $a \not\leqslant b$ 矛盾. 因此, $xy \in A^{(b)}$, 即 $A^{(b)}$ 是半群 S 的子半群. 因此,

$$\mathscr{S}(A) = \bigvee\{a \in L \mid \forall x, y \in S, A(x) \wedge A(y) \wedge a \leqslant A(xy)\}$$

$$\leqslant \bigvee\{a \in L \mid \forall x \not\leqslant y, A^{(b)} \text{ 是 } S \text{ 的子半群}\}.$$

接下来证明

$$\mathscr{S}(A) \geqslant \bigvee\{a \in L \mid \forall b \in P(L) \text{且} b \not\geqslant a, A^{(b)} \text{ 是 } S \text{ 的子半群}\}.$$

设 $a \in L$ 满足 $\forall b \in P(L) \text{且} b \not\geqslant a$, 有 $A^{(b)}$ 是 S 的子半群. 下面只需证 $\forall x, y \in S$, 有

$$A(x) \wedge A(y) \wedge a \leqslant A(xy).$$

设 $b \in P(L)$ 且 $A(x) \wedge A(y) \wedge a \not\leqslant b$, 则 $A(x) \not\leqslant b$, $A(y) \not\leqslant b$ 且 $a \not\leqslant b$, 于是 $x, y \in A^{(b)}$. 对一个子半群 $A^{(b)}$ 来说, 这意味着 $xy \in A^{(b)}$, 也就是 $A(xy) \not\leqslant b$ 成立. 这说明 $A(x) \wedge A(y) \wedge a \leqslant A(xy)$. 这就证明了 $\forall x, y \in S$,

$$\mathscr{S}(A) = \bigvee\{a \in L \mid A(x) \wedge A(y) \wedge a \leqslant A(xy)\}$$

$$\geqslant \bigvee\{a \in L \mid \forall b \in P(L) \text{ 且 } b \not\geqslant a, A^{(b)} \text{ 是 } S \text{ 的子半群}\}.$$

因此, (3) 成立.

(4) 先证明下面不等式

$$\mathscr{S}(A) \leqslant \bigvee\{a \in L \mid \forall b \in \beta(a),\ A_{(b)} \text{ 是 } S \text{ 的子半群}\}.$$

4.2 L-模糊子半群度

设 $a \in \{a \in L \mid A(x) \wedge A(y) \wedge a \leqslant A(xy), \forall x, y \in S\}$, 则 $\forall b \in \beta(a)$, $x, y \in A_{(b)}$, 有

$$b \in \beta(A(x)) \cap \beta(A(y)) \cap \beta(a) = \beta(A(x) \wedge A(y) \wedge a) \subseteq \beta(A(xy)),$$

这意味着 $xy \in A_{(b)}$. 因此 $A_{(b)}$ 是半群 S 的一个子半群. 即

$$\mathscr{S}(A) = \bigvee\{a \in L \mid A(x) \wedge A(y) \wedge a \leqslant A(xy), \forall x, y \in S\}$$

$$\leqslant \bigvee\{a \in L \mid \forall b \in \beta(a), A_{(b)} \text{ 是 } S \text{ 的子半群}\}.$$

接下来证明

$$\mathscr{S}(A) \geqslant \bigvee\{a \in L \mid \forall b \in \beta(a), A_{(b)} \text{ 是 } S \text{ 的子半群}\}.$$

设 $a \in L$ 满足 $\forall b \in \beta(a)$, 有 $A_{(b)}$ 是 S 的子半群, 下面证明 $\forall x, y \in S$,

$$A(x) \wedge A(y) \wedge a \leqslant A(xy).$$

为此令 $b \in \beta(A(x) \wedge A(y) \wedge a)$. 则由

$$\beta(A(x) \wedge A(y) \wedge a) = \beta(A(x)) \cap \beta(A(y)) \cap \beta(a)$$

可知 $x, y \in A_{(b)}$, 且 $b \in \beta(a)$. 由 $A_{(b)}$ 是半群 S 的一个子半群知 $xy \in A_{(b)}$, 也就是 $b \in \beta(A(xy))$, 这说明 $A(x) \wedge A(y) \wedge a \leqslant A(xy)$. 所以,

$$\mathscr{S}(A) = \bigvee\{a \in L \mid \forall x, y \in S, A(x) \wedge A(y) \wedge a \leqslant A(xy)\}$$

$$\geqslant \bigvee\{a \in L \mid \forall b \in \beta(a), A_{(b)} \text{ 是 } S \text{ 的子半群}\}.$$

这样就证明了 (4) 成立. $\square$

推论 4.2.7 设 S 是一个半群且 $A \in L^S$. 则下列条件等价:
(1) A 是 S 的 L-模糊子半群;
(2) $\forall b \in L$, $A_{[b]}$ 是 S 的子半群;
(3) $\forall b \in L$, $A^{[b]}$ 是 S 的子半群;
(4) $\forall b \in P(L)$, $A^{(b)}$ 是 S 的子半群;
(5) 如果对于任意的 $a, b \in L$, 都有 $\beta(a \wedge b) = \beta(a) \cap \beta(b)$, 那么 $\forall b \in \beta(1)$, $A_{(b)}$ 是 S 的子半群.

定理 4.2.8 设 $\{A_i\}_{i \in \Omega}$ 是半群 S 的一族 L-模糊子集. 则

$$\bigwedge_{i \in \Omega} \mathscr{S}(A_i) \leqslant \mathscr{S}\left(\bigwedge_{i \in \Omega} A_i\right).$$

证明 设 $\{A_i\}_{i\in\Omega}$ 是半群 S 的一个 L-模糊子集族, 且设 $a \leqslant \bigwedge_{i\in\Omega}\mathscr{S}(A_i)$, 则 $\forall i \in \Omega$, 都有 $a \leqslant \mathscr{S}(A_i)$. 于是 $\forall i \in \Omega, \forall x, y \in S$, 都有 $A_i(x) \wedge A_i(y) \wedge a \leqslant A_i(xy)$. 进一步有

$$\bigwedge_{i\in\Omega} A_i(x) \wedge \bigwedge_{i\in\Omega} A_i(y) \wedge a \leqslant \bigwedge_{i\in\Omega} A_i(xy).$$

由此可知 $a \leqslant \mathscr{S}\left(\bigwedge_{i\in\Omega} A_i\right)$. 得证 $\bigwedge_{i\in\Omega}\mathscr{S}(A_i) \leqslant \mathscr{S}\left(\bigwedge_{i\in\Omega} A_i\right)$. □

接下来, 我们研究 L-模糊子集的同态像和原像的模糊子半群度.

定理 4.2.9 设 $f: S \to S'$ 是半群同态. 则

(1) 若 $A \in L^S$, 则 $\mathscr{S}(f_L^{\to}(A)) \geqslant \mathscr{S}(A)$; 若 f 是单射, 则 $\mathscr{S}(f_S^{\to}(A)) = \mathscr{S}(A)$.

(2) 若 $B \in L^{S'}$, 则 $\mathscr{S}(f_L^{\leftarrow}(B)) \geqslant \mathscr{S}(B)$; 若 f 是满射, 则 $\mathscr{S}(f_L^{\leftarrow}(B)) = \mathscr{S}(B)$.

证明 (1) 事实上,

$$\mathscr{S}(f_L^{\to}(A))$$
$$= \bigvee \{a \in L \mid \forall x', y' \in S', f_L^{\to}(A)(x') \wedge f_L^{\to}(A)(y') \wedge a \leqslant f_L^{\to}(A)(x'y')\}$$
$$= \bigvee \left\{a \in L \,\middle|\, \forall x', y' \in S', \bigvee_{f(x)=x'} A(x) \wedge \bigvee_{f(y)=y'} A(y) \wedge a \leqslant \bigvee_{f(z)=x'y'} A(z)\right\}$$
$$\geqslant \bigvee \{a \in L \mid \forall x, y \in S, A(x) \wedge A(y) \wedge a \leqslant A(xy)\}$$
$$= \mathscr{S}(A).$$

若 f 是单射, 则上面的 $\leqslant$ 可以用 $=$ 代替. 即 $\mathscr{S}(A) = \mathscr{S}(f_L^{\to}(A))$.

(2) 事实上

$$\mathscr{S}(f_L^{\leftarrow}(B)) = \bigwedge_{x,y\in S} ((f_L^{\leftarrow}(B)(x) \wedge f_L^{\leftarrow}(B)(y)) \mapsto f_L^{\leftarrow}(B)(xy))$$
$$= \bigwedge_{x,y\in S} ((B(f(x))) \wedge (B(f(y)))) \mapsto B(f(x)f(y)))$$
$$\geqslant \bigwedge_{x',y'\in S'} ((B(x')) \wedge (B(y'))) \mapsto (B(x'y')))$$
$$= \mathscr{S}(B).$$

若 f 是满射, 则上面的 $\leqslant$ 可用 $=$ 代替. 即 $\mathscr{S}(B) = \mathscr{S}(f_L^{\leftarrow}(B))$. □

4.3 由 L-模糊子半群度确定的 L-模糊凸结构

在定义 2.4.2 中, 我们曾经定义了 L-模糊子集的笛卡儿积, 下面考察 L-模糊子集的笛卡儿积的子半群度问题.

定理 4.2.10 设 $\{S_i\}_{i=1}^n$ 是一族半群且 $\prod_{i=1}^n A_i$ 是 $\{A_i\}_{i=1}^n$ 的乘积, 其中 $A_i \in L^{S_i}$. 则

$$\mathscr{S}\left(\prod_{i=1}^n A_i\right) \geqslant \bigwedge_{i=1}^n \mathscr{S}(A_i).$$

证明 由于 $\prod_{i=1}^n A_i = \bigwedge_{i=1}^n P_i^{-1}(A_i)$, $\forall i \in \{1, 2, \cdots, n\}$, 令 $P_i : \prod_{i=1}^n S_i \to S_i$ 是投影映射, 则 P_i 是半群同态映射. 由此可知

$$\mathscr{S}\left(P_i^{-1}(A_i)\right) \geqslant \mathscr{S}(A_i),$$

于是进一步可得

$$\mathscr{S}\left(\prod_{i=1}^n A_i\right) = \mathscr{S}\left(\bigwedge_{i=1}^n P_i^{-1}(A_i)\right)$$
$$\geqslant \bigwedge_{i=1}^n \mathscr{S}\left(P_i^{-1}(A_i)\right)$$
$$\geqslant \bigwedge_{i=1}^n \mathscr{S}(A_i). \qquad \square$$

4.3 由 L-模糊子半群度确定的 L-模糊凸结构

在这一节中, 我们使用 L-模糊子半群度构造半群上的 L-模糊凸结构. 之后, 研究半群同态和 L-模糊凸保持映射 (L-模糊凸到凸映射) 之间的关系. 给定任意的 $A \in L^S$, 让 $\mathscr{S}(A)$ 与之对应, 则得到一个映射 $\mathscr{S} : L^S \to L$, 下面我们证明 $\mathscr{S}$ 实际上是半群 S 上的一个 L-模糊凸结构.

我们约定 $\varnothing$ 是任何半群的一个子半群.

定理 4.3.1 设 S 为一个半群且 $A \in L^S$. 则映射 $\mathscr{S} : L^S \to L$ 使得 $A \longmapsto \mathscr{S}(A)$ 是半群 S 上的一个 L-模糊凸结构, 称为由 L-模糊子半群度确定的 L-模糊凸结构.

证明 我们将证明 $\mathscr{S}$ 满足 L-模糊凸结构定义的三个条件.
(1) 显然, $\mathscr{S}(\chi_\varnothing) = \mathscr{S}(\chi_S) = 1$;

(2) 假设 $\{A_i\}_{i\in\Omega}$ 是半群 S 中的一族 L-模糊集,则由定理 4.2.8 可知 $\bigwedge_{i\in\Omega}\mathscr{S}(A_i)$ $\leqslant \mathscr{S}\left(\bigwedge_{i\in\Omega} A_i\right)$;

(3) 假设 $\{A_i \mid i \in \Omega\} \neq \varnothing$, $\{A_i : i \in \Omega\} \subseteq L^S$ 是全序的. 为了要证明

$$\bigwedge_{i\in\Omega}\mathscr{S}(A_i) \leqslant \mathscr{S}\left(\bigvee_{i\in\Omega} A_i\right),$$

只需要证明

$$a \leqslant \mathscr{S}\left(\bigvee_{i\in\Omega} A_i\right), \quad 对于任何的 a \leqslant \bigwedge_{i\in\Omega}\mathscr{S}(A_i) 都成立.$$

设 $a \leqslant \bigwedge_{i\in\Omega}\mathscr{S}(A_i)$, 则由引理 4.2.4 可知, $\forall i \in \Omega, \forall x, y \in S$, 有

$$A_i(x) \wedge A_i(y) \wedge a \leqslant A_i(xy).$$

下面我们证明

$$\left(\bigvee_{i\in\Omega} A_i\right)(x) \wedge \left(\bigvee_{i\in\Omega} A_i\right)(y) \wedge a \leqslant \left(\bigvee_{i\in\Omega} A_i\right)(xy).$$

为此假设 $b \in L$ 且其满足

$$b \prec \left(\bigvee_{i\in\Omega} A_i(x)\right) \wedge \left(\bigvee_{i\in\Omega} A_i(y)\right) \wedge a,$$

则进一步可得到

$$b \prec \bigvee_{i\in\Omega} A_i(x), \ b \prec \bigvee_{i\in\Omega} A_i(y) \ 且 \ b \leqslant a.$$

因此, 存在 $i, j \in \Omega$ 使得

$$b \leqslant A_i(x), \ b \leqslant A_j(y) \ 且 \ b \leqslant a.$$

因为 $\{A_i \mid i \in \Omega\}$ 是全序的, 所以不妨假设 $A_j \leqslant A_i$, 于是 $b \leqslant A_i(x) \wedge A_i(y) \wedge a$. 由 $A_i(x) \wedge A_i(y) \wedge a \leqslant A_i(xy)$ 可得 $b \leqslant A_i(xy)$. 于是 $b \leqslant \left(\bigvee_{i\in\Omega} A_i\right)(xy)$. 因此, 可推知

$$\left(\bigvee_{i\in\Omega} A_i(x)\right) \wedge \left(\bigvee_{i\in\Omega} A_i(y)\right) \wedge a \leqslant \left(\bigvee_{i\in\Omega} A_i\right)(xy).$$

结合引理 4.2.4 可得到 $a \leqslant \mathscr{S}\left(\bigvee_{i\in\Omega} A_i\right)$. 从而可推得

$$\bigwedge_{i\in\Omega} \mathscr{S}(A_i) \leqslant \mathscr{S}\left(\bigvee_{i\in\Omega} A_i\right).$$

□

结合定义 3.7.6 和定理 4.2.9, 可得下列定理.

定理 4.3.2 设 $f: S \to S'$ 是半群同态, $\mathscr{S}$ 和 $\mathscr{S}'$ 分别是半群 S 和 S' 上由 L-模糊子半群度确定的 L-模糊凸结构. 则 $f: (S, \mathscr{S}) \to (S', \mathscr{S}')$ 是一个 L-模糊凸保持的和 L-模糊凸到凸的映射.

结合定义 3.7.6、定理 4.2.10 和定理 4.3.2, 用类似的方法可得下列定理.

定理 4.3.3 设 $\{S_i\}_{i=1}^n$ 是一族半群且 $\forall i \in \{1, 2, \cdots, n\}$, $P_i: \prod_{i=1}^n S_i \to S_i$ 是投影映射. $\forall i \in \{1, 2, \cdots, n\}$, 令 $\mathscr{S}_i$ 表示 S_i 上的 L-模糊子半群度, 令 $\mathscr{S}$ 表示 $\prod_{i=1}^n S_i$ 上的 L-模糊子半群度, 则 $P_i: \left(\prod_{i=1}^n S_i, \mathscr{S}\right) \to (S_i, \mathscr{S}_i)$ 是 L-模糊凸保持的和 L-模糊凸到凸的映射.

再次由定义 3.7.6 和定理 4.3.2 可得下列推论.

推论 4.3.4 设 $f: S \to S'$ 为半群同构映射. 则 $f: (S, \mathscr{S}) \to (S', \mathscr{S}')$ 是 L-模糊凸空间之间的同构映射.

4.4　L-模糊子群度

在本节中, 首先给出 L-模糊子群度的定义, 然后给出 L-模糊子群度的等价刻画. 本节和下面两节内容主要来自于文献 [137].

在定义 4.1.2 中, 我们可以看到, 给定一个群的模糊集, 它要么是模糊子群, 要么不是, 两者必居其一, 且仅居其一, 事实上, 这一点是可以进一步模糊化的. 具体定义如下.

定义 4.4.1 设 A 是群 G 上的一个 L-模糊子集. A 的子群度定义为

$$\mathscr{G}(A) = A(1) \wedge \bigwedge_{x,y\in G} \left((A(x) \wedge A(y)) \mapsto A(xy^{-1})\right).$$

很明显, A 是 G 上的一个 L-模糊子群当且仅当 $\mathscr{G}(A) = 1$.

例 4.4.2 设 $\mathbb{Z}$ 是一个整数加群且 $L = \mathbf{2}^{\{1,2,3\}}$. 定义 $A: \mathbb{Z} \to L$ 为

$$A(n) = \begin{cases} \{1,2\}, & \text{如果 } n \text{ 是一个偶数}, \\ \{1,3\}, & \text{如果 } n \text{ 是奇数但不等于 } 7, \\ \{3\}, & \text{如果 } n = 7. \end{cases}$$

下面我们来计算 $\mathscr{G}(A) = A(0) \wedge \bigwedge_{x,y \in G} (A(x) \wedge A(y) \mapsto A(x-y))$.

(1) 当 x, y 都是偶数时,
$$A(x) \wedge A(y) \mapsto A(x-y) = \{1,2\} \wedge \{1,2\} \mapsto \{1,2\} = \{1,2,3\}.$$

(2) 当 x, y 一奇一偶不含 7 时, 若 $x - y = 7$, 则
$$A(x) \wedge A(y) \mapsto A(x-y) = \{1,2\} \wedge \{1,3\} \mapsto \{3\} = \{2,3\}.$$

(3) 当 x, y 一奇一偶不含 7 时, 若 $x - y \neq 7$, 则
$$A(x) \wedge A(y) \mapsto A(x-y) = \{1,2\} \wedge \{1,3\} \mapsto \{1,3\} = \{1,2,3\}.$$

(4) 当 x, y 一奇一偶含 7 时, 则
$$A(x) \wedge A(y) \mapsto A(x-y) = \{1,2\} \wedge \{3\} \mapsto \{1,3\} = \{1,2,3\}.$$

(5) 当 x, y 都是非 7 奇数时,
$$A(x) \wedge A(y) \mapsto A(x-y) = \{1,3\} \mapsto \{1,2\} = \{1,2\}.$$

(6) 当 x, y 都是奇数且含 7 时,
$$A(x) \wedge A(y) \mapsto A(x-y) = \{1,3\} \wedge \{3\} \mapsto \{1,2\} = \{1,2\}.$$

故有 $\mathscr{G}(A) = A(0) \wedge \bigwedge_{x,y \in G} (A(x) \wedge A(y) \mapsto A(x-y)) = \{2\}.$

下面我们来讨论模糊子群度的等价刻画问题.

定理 4.4.3 设 A 是群 G 上的一个 L-模糊子集. 那么
$$\mathscr{G}(A) = A(1) \wedge \bigwedge_{x,y \in G} ((A(x) \wedge A(y)) \mapsto A(xy)) \wedge \bigwedge_{x \in G} (A(x) \mapsto A(x^{-1})).$$

证明 由 $\mathscr{G}(A) = A(1) \wedge \bigwedge_{x,y \in G} ((A(x) \wedge A(y)) \mapsto A(xy^{-1}))$, 可得
$$\mathscr{G}(A) \leqslant A(1), \quad \mathscr{G}(A) \leqslant (A(x) \wedge A(y) \mapsto A(xy^{-1})), \quad \forall x, y \in G.$$

这意味着
$$\mathscr{G}(A) \leqslant A(1), \quad \mathscr{G}(A) \wedge A(x) \wedge A(y) \leqslant A(xy^{-1}), \quad \forall x, y \in G.$$

特别地, 有
$$\mathscr{G}(A) \wedge A(y) \leqslant A(1),$$
$$\mathscr{G}(A) \wedge A(y) \leqslant \mathscr{G}(A) \wedge A(1) \wedge A(y) \leqslant A(y^{-1}),$$
$$\mathscr{G}(A) \wedge A(y) = \mathscr{G}(A) \wedge A(y^{-1}),$$
$$\mathscr{G}(A) \wedge A(x) \wedge A(y) = \mathscr{G}(A) \wedge A(x) \wedge A(y^{-1}).$$

这说明
$$\mathscr{G}(A) \leqslant A(1) \wedge \bigwedge_{x,y \in G} \left(A(x) \wedge A(y) \mapsto A(xy) \wedge A(y^{-1}) \right)$$
$$\leqslant A(1) \wedge \bigwedge_{x,y \in G} \left(A(x) \wedge A(y) \mapsto A(xy) \right) \wedge \bigwedge_{y \in G} \left(A(y) \mapsto A(y^{-1}) \right).$$

类似地, 可以证得
$$\mathscr{G}(A) \geqslant A(1) \wedge \bigwedge_{x,y \in G} \left(A(x) \wedge A(y) \mapsto A(xy) \right) \wedge \bigwedge_{y \in G} \left(A(y) \mapsto A(y^{-1}) \right). \quad \square$$

下面的引理是显然的.

引理 4.4.4 设 A 是群 G 上的一个 L-模糊子集. 那么 $\mathscr{G}(A) \geqslant a$ 当且仅当对于任意的 $x, y \in G$,
$$a \leqslant A(1), \quad A(x) \wedge A(y) \wedge a \leqslant A(xy^{-1}).$$

由引理 4.4.4, 很容易得到 L-模糊子群度的另外一种刻画.

定理 4.4.5 设 A 是群 G 上的一个 L-模糊子集. 那么
$$\mathscr{G}(A) = \bigvee \left\{ a \in L \mid a \leqslant A(1), A(x) \wedge A(y) \wedge a \leqslant A(xy^{-1}), \forall x, y \in G \right\}.$$

接下来利用四种截集给出 L-模糊子群度的等价刻画定理.

定理 4.4.6 设 A 是群 G 上的一个 L-模糊子集. 那么
$$\mathscr{G}(A) = \bigvee \left\{ a \in L \mid \forall b \leqslant a \leqslant A(1),\ A_{[b]} \text{ 是 } G \text{ 上的一个子群} \right\}.$$

证明 假设对于任意的 $x, y \in G$, $A(x) \wedge A(y) \wedge a \leqslant A(xy^{-1})$. 那么对于任意的 $b \leqslant a \leqslant A(1)$ 以及对于任意的 $x, y \in A_{[b]}$, 有
$$A(xy^{-1}) \geqslant A(x) \wedge A(y) \wedge a \geqslant A(x) \wedge A(y) \wedge b \geqslant b,$$

这说明 $xy^{-1} \in A_{[b]}$. 因此, $A_{[b]}$ 是 G 上的一个子群. 所以,

$$\mathscr{G}(A) = \bigvee\{a \in L \mid a \leqslant A(1), \ A(x) \wedge A(y) \wedge a \leqslant A(xy^{-1}), \forall x, y \in G\}$$

$$\leqslant \bigvee\{a \in L \mid \forall b \leqslant a \leqslant A(1), \ A_{[b]} \text{ 是 } G \text{ 上的一个子群}\}.$$

反过来, 假设 $a \in L$ 和 $\forall b \leqslant a \leqslant A(1)$, $A_{[b]}$ 是 G 上的一个子群. 对于任意的 $x, y \in G$, 设 $b = A(x) \wedge A(y) \wedge a$. 那么 $b \leqslant a$ 和 $x, y \in A_{[b]}$, 因此, $xy^{-1} \in A_{[b]}$, 即, $A(xy^{-1}) \geqslant b = A(x) \wedge A(y) \wedge a$. 这说明

$$\mathscr{G}(A) = \bigvee\{a \in L \mid a \leqslant A(1), \ A(x) \wedge A(y) \wedge a \leqslant A(xy^{-1}), \forall x, y \in G\}$$

$$\geqslant \bigvee\{a \in L \mid \forall b \leqslant a \leqslant A(1), \ A_{[a]} \text{ 是 } G \text{ 上的一个子群}\}. \qquad \square$$

定理 4.4.7 设 A 是群 G 上的一个 L-模糊子集. 那么

$$\mathscr{G}(A) = \bigvee\left\{a \in L \mid a \leqslant A(1), \forall b \notin \alpha(a), \ A^{[b]} \text{ 是 } G \text{ 上的一个子群}\right\}.$$

证明 假设对于任意的 $x, y \in G$, $A(x) \wedge A(y) \wedge a \leqslant A(xy^{-1})$, 这里 $a \leqslant A(1)$. 那么对于任意的 $b \notin \alpha(a)$ 和 $x, y \in A^{[b]}$, 有

$$b \notin \alpha(A(x)) \cup \alpha(A(y)) \cup \alpha(a) = \alpha(A(x) \wedge A(y) \wedge a).$$

由

$$A(x) \wedge A(y) \wedge a \leqslant A(xy^{-1}),$$

可得

$$\alpha(A(xy^{-1})) \subseteq \alpha(A(x) \wedge A(y) \wedge a).$$

因此, $b \notin \alpha(A(xy^{-1}))$, 即, $xy^{-1} \in A^{[b]}$. 这说明 $A^{[b]}$ 是 G 的一个子群, 并且

$$a \in \{a \in L \mid a \leqslant A(1), \forall b \notin \alpha(a), \ A^{[b]} \text{ 是 } G \text{ 上的一个子群}\}.$$

于是有

$$\mathscr{G}(A) = \bigvee\{a \in L \mid a \leqslant A(1), A(x) \wedge A(y) \wedge a \leqslant A(xy^{-1}), \forall x, y \in G\}$$

$$\leqslant \bigvee\{a \in L \mid a \leqslant A(1), \forall b \notin \alpha(a), \ A^{[b]} \text{ 是 } G \text{ 上的一个子群}\}.$$

反过来, 假设

$$a \in \{a \in L \mid a \leqslant A(1), \forall b \notin \alpha(a), \ A^{[b]} \text{ 是 } G \text{ 上的一个子群}\}.$$

4.4 L-模糊子群度

现在证明对于任意的 $x,y \in G$ 和 $a \leqslant A(1)$, 有 $A(x) \wedge A(y) \wedge a \leqslant A(xy^{-1})$ 成立. 假设 $b \notin \alpha(A(x) \wedge A(y) \wedge a)$. 由

$$\alpha(A(x) \wedge A(y) \wedge a) = \alpha(A(x)) \cup \alpha(A(y)) \cup \alpha(a),$$

可得 $b \notin \alpha(a)$ 和 $x,y \in A^{[b]}$. 因为 $A^{[b]}$ 是 G 的一个子群, 所以 $xy^{-1} \in A^{[b]}$, 即, $b \notin \alpha(A(xy^{-1}))$. 这说明 $A(x) \wedge A(y) \wedge a \leqslant A(xy^{-1})$. 于是得到

$$\mathscr{G}(A) = \bigvee \{a \in L \mid a \leqslant A(1), A(x) \wedge A(y) \wedge a \leqslant A(xy^{-1}), \forall x,y \in G\}$$

$$\geqslant \bigvee \{a \in L \mid a \leqslant A(1), \forall b \notin \alpha(a),\ A^{[b]} \text{ 是 } G \text{ 上的一个子群}\}. \qquad \square$$

定理 4.4.8 设 A 是群 G 上的一个 L-模糊子集. 那么

$$\mathscr{G}(A) = \bigvee \{a \in L \mid a \leqslant A(1), \forall b \in P(L), b \not\geqslant a,\ A^{(b)} \text{ 是 } G \text{ 上的一个子群}\}.$$

证明 假设对于任意的 $x,y \in G$ 和 $a \leqslant A(1)$, 有 $A(x) \wedge A(y) \wedge a \leqslant A(xy^{-1})$ 成立. 设 $b \in P(L)$, $b \not\geqslant a$ 且 $x,y \in A^{(b)}$. 现在来证明 $xy^{-1} \in A^{(b)}$. 如果 $xy^{-1} \notin A^{(b)}$, 即 $A(xy^{-1}) \leqslant b$, 那么 $A(x) \wedge A(y) \wedge a \leqslant A(xy^{-1}) \leqslant b$. 由 $b \in P(L)$ 和 $x,y \in A^{(b)}$, 有 $a \leqslant b$, 这与 $b \not\geqslant a$ 矛盾. 于是有 $xy^{-1} \in A^{(b)}$. 这表明 $A^{(b)}$ 是 G 上的一个子群. 因此,

$$\mathscr{G}(A) = \bigvee \{a \in L \mid a \leqslant A(1), A(x) \wedge A(y) \wedge a \leqslant A(xy^{-1}), \forall x,y \in G\}$$

$$\leqslant \bigvee \{a \in L \mid a \leqslant A(1), \forall b \in P(L), b \not\geqslant a, A^{(b)} \text{ 是 } G \text{ 上的一个子群}\}.$$

反过来, 假设

$$a \in \{a \in L \mid a \leqslant A(1), \forall b \in P(L), b \not\geqslant a,\ A^{(b)} \text{ 是 } G \text{ 上的一个子群}\}.$$

现在来证明 $\forall x,y \in G$, 有 $A(x) \wedge A(y) \wedge a \leqslant A(xy^{-1})$. 设 $b \in P(L)$ 且 $A(x) \wedge A(y) \wedge a \not\leqslant b$. 那么 $A(x) \not\leqslant b$, $A(y) \not\leqslant b$ 且 $a \not\leqslant b$, 即 $x,y \in A^{(b)}$. 因为 $A^{(b)}$ 是 G 的一个子群, 从而有 $xy^{-1} \in A^{(b)}$, 即 $A(xy^{-1}) \not\leqslant b$. 这说明 $A(x) \wedge A(y) \wedge a \leqslant A(xy^{-1})$. 于是有

$$\mathscr{G}(A) = \bigvee \{a \in L \mid a \leqslant A(1), \forall b \in P(L), b \not\geqslant a,\ A^{(b)} \text{ 是 } G \text{ 上的一个子群}\}$$

$$\geqslant \bigvee \{a \in L \mid a \leqslant A(1), A(x) \wedge A(y) \wedge a \leqslant A(xy^{-1}), \forall x,y \in G\}. \qquad \square$$

定理 4.4.9 设 A 是群 G 上的一个 L-模糊子集. 对于任意的 $a,b \in L$, 如果 $\beta(a \wedge b) = \beta(a) \cap \beta(b)$, 那么

$$\mathscr{G}(A) = \bigvee \{a \in L \mid a \leqslant A(1), \forall b \in \beta(a),\ A_{(b)} \text{ 是 } G \text{ 上的一个子群}\}.$$

证明 假设 $a \in \{a \in L \mid a \leqslant A(1), A(x) \wedge A(y) \wedge a \leqslant A(xy^{-1}), \forall x, y \in G\}$. 那么对于任意的 $b \in \beta(a)$ 以及任意的 $x, y \in A_{(b)}$, 有

$$b \in \beta(A(x)) \cap \beta(A(y)) \cap \beta(a) = \beta(A(x) \wedge A(y) \wedge a) \subseteq \beta(A(xy^{-1})),$$

即 $xy^{-1} \in A_{(b)}$. 这说明 $A_{(b)}$ 是 G 的一个子群. 于是可得

$$\mathscr{G}(A) = \bigvee \{a \in L \mid a \leqslant A(1), A(x) \wedge A(y) \wedge a \leqslant A(xy^{-1}), \forall x, y \in G\}$$
$$\leqslant \bigvee \{a \in L \mid a \leqslant A(1), \forall b \in \beta(a), A_{(b)} \text{ 是 } G \text{ 上的一个子群}\}.$$

反过来, 假设

$$a \in \{a \in L \mid a \leqslant A(1), \forall b \in \beta(a), A_{(b)} \text{ 是 } G \text{ 上的一个子群}\}.$$

现在来证明对于任意的 $x, y \in G$, 我们都有

$$A(x) \wedge A(y) \wedge a \leqslant A(xy^{-1}).$$

设 $b \in \beta(A(x) \wedge A(y) \wedge a)$. 由

$$\beta(A(x) \wedge A(y) \wedge a) = \beta(A(x)) \cap \beta(A(y)) \cap \beta(a),$$

可得 $x, y \in A_{(b)}$ 且 $b \in \beta(a)$. 因为 $A_{(b)}$ 是 G 的一个子群, 所以有 $xy^{-1} \in A_{(b)}$, 即 $b \in \beta(A(xy^{-1}))$. 这说明

$$A(x) \wedge A(y) \wedge a \leqslant A(xy^{-1}).$$

于是得到

$$\mathscr{G}(A) = \bigvee \{a \in L \mid a \leqslant A(1), A(x) \wedge A(y) \wedge a \leqslant A(xy^{-1}), \forall x, y \in G\}$$
$$\geqslant \bigvee \{a \in L \mid a \leqslant A(1), \forall b \in \beta(a), A_{(b)} \text{ 是 } G \text{ 上的一个子群}\}. \qquad \square$$

当 $\mathscr{G}(A) = 1$ 时, 上述几个刻画定理可退化为下面推论形式.

推论 4.4.10 设 G 是一个群且 $A \in L^G$. 则下列条件等价:
(1) A 是 G 的 L-模糊子群;
(2) $\forall b \in L$, $A_{[b]}$ 是 G 的子群;
(3) $\forall b \in L$, $A^{[b]}$ 是 G 的子群;
(4) $\forall b \in P(L)$, $A^{(b)}$ 是 G 的子群;
(5) 如果对于任意的 $a, b \in L$, 都有 $\beta(a \wedge b) = \beta(a) \cap \beta(b)$, 那么 $\forall b \in \beta(1)$, $A_{(b)}$ 是 G 的子群.

4.4 L-模糊子群度

定理 4.4.11 设 $\{A_i\}_{i\in\Omega}$ 是群 G 上的一族 L-模糊子集. 则

$$\mathscr{G}\left(\bigwedge_{i\in\Omega} A_i\right) \geqslant \bigwedge_{i\in\Omega} \mathscr{G}(A_i).$$

证明 这能够从下面蕴含推理可得.

$a \leqslant \bigwedge_{i\in\Omega} \mathscr{G}(A_i) \Rightarrow a \leqslant \mathscr{G}(A_i), \forall i \in \Omega$

$\Rightarrow a \leqslant \bigwedge_{i\in\Omega} A_i(1), A_i(x) \wedge A_i(y) \wedge a \leqslant A_i(xy^{-1}), \forall i \in \Omega, \forall x, y \in G$

$\Rightarrow a \leqslant \bigwedge_{i\in\Omega} A_i(1), \bigwedge_{i\in\Omega} A_i(x) \wedge \bigwedge_{i\in\Omega} A_i(y) \wedge a \leqslant \bigwedge_{i\in\Omega} A_i(xy^{-1})$

$\Rightarrow a \leqslant \mathscr{G}\left(\bigwedge_{i\in\Omega} A_i\right).$ □

下面我们从 L-模糊子集的像和原像的角度来研究模糊子群度的性质.

定理 4.4.12 设 $f: G \to G'$ 是一个群同态.

(1) 如果 A 是群 G 的一个 L-模糊子集, 那么 $\mathscr{G}(f_L^{\rightarrow}(A)) \geqslant \mathscr{G}(A)$; 如果 f 是单射, 那么 $\mathscr{G}(f_L^{\rightarrow}(A)) = \mathscr{G}(A)$.

(2) 如果 B 是群 G' 的一个 L-模糊子集, 那么 $\mathscr{G}(f_L^{\leftarrow}(B)) \geqslant \mathscr{G}(B)$; 如果 f 是满射, 那么 $\mathscr{G}(f_L^{\leftarrow}(B)) = \mathscr{G}(B)$.

证明 (1) 由定理 4.4.5 可得下面不等式.

$\mathscr{G}(f_L^{\rightarrow}(A))$

$= \bigvee\{a \in L \mid a \leqslant f_L^{\rightarrow}(A)(1'), f_L^{\rightarrow}(A)(x') \wedge f_L^{\rightarrow}(A)(y') \wedge a$

$\leqslant f_L^{\rightarrow}(A)(x'y'^{-1}), \forall x', y' \in G'\}$

$= \bigvee\left\{a \in L \,\middle|\, a \leqslant \bigvee_{f(x)=1'} A(x), \bigvee_{f(x)=x'} A(x) \wedge \bigvee_{f(y)=y'} A(y) \wedge a \right.$

$\left. \leqslant \bigvee_{f(z)=x'y'^{-1}} A(z), \forall x', y' \in G'\right\}$

$\geqslant \bigvee\{a \in L \mid a \leqslant A(1), A(x) \wedge A(y) \wedge a \leqslant A(xy^{-1}), \forall x, y \in G\}$

$= \mathscr{G}(A).$

如果 f 是单射,那么上面的 $\leqslant$ 可以被 $=$ 代替. 因此有 $\mathscr{G}(A) = \mathscr{G}(f_L^{\rightarrow}(A))$.

(2) 由定理 4.4.5 可知

$$\mathscr{G}(f_L^{\leftarrow}(B))$$
$$= f_L^{\leftarrow}(B)(1) \wedge \bigwedge_{x,y \in G} \left((f_L^{\leftarrow}(B)(x) \wedge f_L^{\leftarrow}(B)(y)) \mapsto f_L^{\leftarrow}(B)(xy^{-1}) \right)$$
$$= B(f(1)) \wedge \bigwedge_{x,y \in G} \left((B(f(x)) \wedge B(f(y))) \mapsto B\left(f(x)f(y)^{-1}\right) \right)$$
$$\geqslant B(1) \wedge \bigwedge_{x',y' \in G'} \left((B(x') \wedge B(y')) \mapsto B(x'{y'}^{-1}) \right)$$
$$= \mathscr{G}(B).$$

如果 f 是满射,那么上面的 $\leqslant$ 可以被 $=$ 代替. 因此可以得到 $\mathscr{G}(B) = \mathscr{G}(f_L^{\leftarrow}(B))$. □

4.5 L-模糊子群的运算

在这一节中,我们考虑几种 L-模糊子集的乘积的子群度.

在定义 2.4.2 中,曾给出一族 L-模糊集笛卡儿积的概念,下面我们考虑这种乘积在群乘积中的 L-模糊子群度问题.

定理 4.5.1 设 $\{G_i\}_{i=1}^n$ 是一族群且 $\prod_{i=1}^n A_i$ 是 $\{A_i\}_{i=1}^n$ 的乘积,其中对于任意 $i \in \{1, 2, \cdots, n\}$, $A_i \in L^{G_i}$. 则 $\mathscr{G}\left(\prod_{i=1}^n A_i\right) \geqslant \bigwedge_{i=1}^n \mathscr{G}(A_i)$.

证明 **方法 1.** 设 $G = \prod_{i=1}^n G_i$, $x = \{x_i\}_{i=1}^n$, $y = \{y_i\}_{i=1}^n$. 则

$$\mathscr{G}\left(\prod_{i=1}^n A_i\right)$$
$$= \left(\prod_{i=1}^n A_i\right)(1) \wedge \bigwedge_{x,y \in G} \left(\left(\left(\prod_{i=1}^n A_i\right)(x) \wedge \left(\prod_{i=1}^n A_i\right)(y) \right) \mapsto \left(\prod_{i=1}^n A_i\right)(xy^{-1}) \right)$$
$$= \bigwedge_{i=1}^n A_i(1) \wedge \bigwedge_{x,y \in G} \left(\left(\bigwedge_{i=1}^n A_i(x_i) \wedge \bigwedge_{i=1}^n A_i(y_i) \right) \mapsto \bigwedge_{i=1}^n A_i(x_i y_i^{-1}) \right)$$
$$\geqslant \bigwedge_{i=1}^n A_i(1) \wedge \bigwedge_{x,y \in G} \bigwedge_{i=1}^n \left((A_i(x_i) \wedge A_i(y_i)) \mapsto A_i(x_i y_i^{-1}) \right)$$

4.5 L-模糊子群的运算

$$= \bigwedge_{i=1}^{n} \bigwedge_{x_i, y_i \in G_i} A_i(1) \wedge ((A_i(x_i) \wedge A_i(y_i)) \mapsto A_i(x_i y_i^{-1}))$$

$$= \bigwedge_{i=1}^{n} \mathscr{G}(A_i).$$

方法 2. 设 $P_i : \prod_{i=1}^{n} G_i \to G_i$ 是投射, 容易验证 $\prod_{i=1}^{n} A_i = \bigwedge_{i=1}^{n} P_i^{\leftarrow}(A_i)$. 由定理 4.4.11 和定理 4.4.12 可知

$$\mathscr{G}\left(\prod_{i=1}^{n} A_i\right) = \mathscr{G}\left(\bigwedge_{i=1}^{n} P_i^{\leftarrow}(A_i)\right) \geqslant \bigwedge_{i=1}^{n} \mathscr{G}(P_i^{\leftarrow}(A_i)) \geqslant \bigwedge_{i=1}^{n} \mathscr{G}(A_i). \quad \square$$

习题 4 中的练习 7 曾经定义了一个半群中两个 L-模糊集的内乘积, 在一个群中, 我们仍可以使用这个定义, 那就是定义 4.5.2.

定义 4.5.2 设 A 和 B 是群 G 的两个 L-模糊子集. 那么 AB 是 G 上的一个 L-模糊子集, 其定义为

$$(AB)(z) = \bigvee_{z=xy} (A(x) \wedge B(y)),$$

AB 称为 A 和 B 的内乘积.

容易验证: 对于任意的 $a \in P(L)$, $(AB)^{(a)} = A^{(a)} B^{(a)}$.

定理 4.5.3 若 A 为群 G 上的 L-模糊子集, 则

$$\mathscr{G}(A) = A(1) \wedge \bigwedge_{z \in G} ((AA)(z) \mapsto A(z)) \wedge \bigwedge_{z \in G} (A(z) \mapsto A(z^{-1})).$$

证明 首先我们证明

$$\bigwedge_{z \in G} ((AA)(z) \mapsto A(z)) = \bigwedge_{x, y \in G} ((A(x) \wedge A(y)) \mapsto A(xy)).$$

事实上, 若 a 满足对任意 $z \in G$, $(AA)(z) \wedge a \leqslant A(z)$, 则对于任意 $x, y \in G$, 下式成立:

$$A(x) \wedge A(y) \wedge a \leqslant (AA)(xy) \wedge a \leqslant A(xy).$$

反之, 若 a 满足对任意 $x, y \in G$, $A(x) \wedge A(y) \wedge a \leqslant A(xy)$, 则对任意 $z \in G$, 有

$$(AA)(z) \wedge a = \bigvee_{z=xy} (A(x) \wedge A(y)) \wedge a \leqslant \bigvee_{z=xy} A(xy) = A(z).$$

进而

$$\bigwedge_{z\in G}((AA)(z)\mapsto A(z)) = \bigvee\{a\in L \mid (AA)(z)\wedge a \leqslant A(z), \forall z\in G\}$$

$$= \bigvee\{a\in L \mid A(x)\wedge A(y)\wedge a \leqslant A(xy), \forall x,y\in G\}$$

$$= \bigwedge_{x,y\in G}((A(x)\wedge A(y))\mapsto A(xy)).$$

再由定理 4.4.3, 可得下面所证

$$\mathscr{G}(A) = A(1)\wedge \bigwedge_{x,y\in G}((A(x)\wedge A(y))\mapsto A(xy))\wedge \bigwedge_{z\in G}(A(z)\mapsto \mu(z^{-1}))$$

$$= A(1)\wedge \bigwedge_{z\in G}((AA)(z)\mapsto A(z))\wedge \bigwedge_{z\in G}(A(z)\mapsto \mu(z^{-1})). \qquad \square$$

引理 4.5.4 设 A 和 B 是群 G 上的两个 L-模糊子集. 则

(1) $\bigwedge_{z\in G}((AA)(z)\mapsto A(z))\wedge \bigwedge_{z\in G}((BB)(z)\mapsto B(z)) \leqslant \bigwedge_{z\in G}((AA)(BB)(z)\mapsto (AB)(z));$

(2) $\bigwedge_{z\in G}(A(z)\mapsto A(z^{-1}))\wedge \bigwedge_{z\in G}(B(z)\mapsto B(z^{-1})) \leqslant \bigwedge_{z\in G}(AB(z)\mapsto BA(z^{-1})).$

证明 (1) 可由下面的不等式证得.

$$\bigwedge_{z\in G}((AA)(z)\mapsto A(z))\wedge \bigwedge_{z\in G}((BB)(z)\mapsto B(z))$$

$$= \bigvee\{a\in L \mid (AA)(z)\wedge a \leqslant A(z), \forall z\in G\}$$

$$\wedge \bigvee\{a\in L \mid (BB)(z)\wedge a \leqslant B(z), \forall z\in G\}$$

$$= \bigvee\{a\in L \mid (AA)(z)\wedge a \leqslant A(z), (BB)(z)\wedge a \leqslant B(z), \forall z\in G\}$$

$$\leqslant \bigvee\left\{a\in L \;\middle|\; \bigvee_{z=xy}((AA)(x)\wedge(BB)(y))\wedge a \leqslant \bigvee_{z=xy}(A(x)\wedge B(y)), \forall z\in G\right\}$$

$$= \bigvee\{a\in L \mid (AA)(BB)(z)\wedge a \leqslant (AB)(z), \forall z\in G\}$$

$$= \bigwedge_{z\in G}((AA)(BB)(z)\mapsto (AB)(z)).$$

(2) 可由下面的不等式证得.

$$\bigwedge_{z\in G}(A(z)\mapsto A(z^{-1}))\wedge \bigwedge_{z\in G}(B(z)\mapsto B(z^{-1}))$$

$$= \bigvee \{a \in L \mid A(z) \wedge a \leqslant A(z^{-1}), \forall z \in G\}$$
$$\wedge \bigvee \{a \in L \mid B(z) \wedge a \leqslant B(z^{-1}), \forall z \in G\}$$
$$= \bigvee \{a \in L \mid A(z) \wedge a \leqslant A(z^{-1}), B(z) \wedge a \leqslant B(z^{-1}), \forall z \in G\}$$
$$\leqslant \bigvee \left\{a \in L \mid \bigvee_{z=xy}(A(x) \wedge B(y)) \wedge a \leqslant \bigvee_{z=xy}(A(x^{-1}) \wedge B(y^{-1})), \forall z \in G\right\}$$
$$= \bigvee \{a \in L \mid (AB)(z) \wedge a \leqslant (BA)(z^{-1}), \forall z \in G\}$$
$$= \bigwedge_{z \in G}((AB)(z) \mapsto (BA)(z^{-1})). \qquad \square$$

定理 4.5.5 设 A 和 B 是群 G 上的两个 L-模糊子集. 如果 $AB = BA$, 那么 $\mathscr{G}(AB) \geqslant \mathscr{G}(A) \wedge \mathscr{G}(B)$.

证明 由引理 4.5.4 及 $AB = BA$ 可得

$\mathscr{G}(AB)$
$$= (AB)(1) \wedge \bigwedge_{z \in G}((AB)(AB)(z) \mapsto (AB)(z)) \wedge \bigwedge_{z \in G}((AB)(z) \mapsto (AB)(z^{-1}))$$
$$\geqslant A(1) \wedge B(1) \wedge \bigwedge_{z \in G}((AA)(z) \mapsto A(z)) \wedge \bigwedge_{z \in G}((BB)(z) \mapsto B(z))$$
$$\wedge \bigwedge_{z \in G}(A(z) \mapsto A(z^{-1})) \wedge \bigwedge_{z \in G}(B(z) \mapsto B(z^{-1}))$$
$$= \mathscr{G}(A) \wedge \mathscr{G}(B). \qquad \square$$

4.6 L-模糊正规子群度

在这一节中, 我们给出 L-模糊正规子群度的概念及其刻画.

定义 4.6.1 设 A 是群 G 上的一个 L-模糊子集. A 的正规子群度 $\mathscr{G}_N(A)$ 定义为
$$\mathscr{G}_N(A) = \mathscr{G}(A) \wedge \bigwedge_{x,y \in G}(A(y) \mapsto A(xyx^{-1})).$$

很明显, A 是 G 的一个 L-模糊正规子群当且仅当 $\mathscr{G}_N(A) = 1$; 如果 G 是可换群, 那么 $\mathscr{G}_N(A) = \mathscr{G}(A)$. 反之, 如果 $\mathscr{G}_N(A) = \mathscr{G}(A)$, 那么 G 未必是可换群.

比如, 对于一个不是可换群 G, 有 $\mathscr{G}_N(G) = \mathscr{G}(G) = 1$.

例 4.6.2 (见例 4.4.2) 设 $\mathbb{Z}$ 是一个整数加群且 $L = 2^{\{1,2,3\}}$. 定义 $A: \mathbb{Z} \to L$ 为

$$A(n) = \begin{cases} \{1,2\}, & \text{如果 } n \text{ 是一个偶数}, \\ \{1,3\}, & \text{如果 } n \text{ 是奇数但不等于 } 7, \\ \{3\}, & \text{如果 } n = 7, \end{cases}$$

那么 $\mathscr{G}_N(A) = \mathscr{G}(A) = \{2\}$.

类似于引理 4.4.4, 有以下引理成立.

引理 4.6.3 设 A 是群 G 的 L-模糊子集. 那么 $\mathscr{G}_N(A) \geqslant a$ 当且仅当 $\mathscr{G}(A) \geqslant a$, 且对于任意的 $x, y \in G$,

$$A(y) \wedge a \leqslant A(xyx^{-1}).$$

通过引理 4.4.4 和引理 4.6.3, 可以得到以下定理.

定理 4.6.4 设 A 是群 G 的 L-模糊子集. 则

$$\mathscr{G}_N(A) = \bigvee \{a \in L \mid a \leqslant A(1), A(x) \wedge A(y) \wedge a \leqslant A(xy^{-1}),$$
$$A(y) \wedge a \leqslant A(xyx^{-1}), \forall x, y \in G\}.$$

类似于定理 4.4.6—定理 4.4.9, 可以得到下面定理 4.6.5 和定理 4.6.6.

定理 4.6.5 设 A 是群 G 的 L-模糊子集. 则
(1) $\mathscr{G}_N(A) = \bigvee \{a \in L \mid a \leqslant A(1), \forall b \leqslant a, A_{[b]} \text{ 是 } G \text{ 的正规子群}\}$;
(2) $\mathscr{G}_N(A) = \bigvee \{a \in L \mid a \leqslant A(1), \forall b \notin \alpha(a), A^{[b]} \text{ 是 } G \text{ 的正规子群}\}$;
(3) $\mathscr{G}_N(A) = \bigvee \{a \in L \mid a \leqslant A(1), \forall b \in P(L), b \not\geqslant a, A^{(b)} \text{ 是 } G \text{ 的正规子群}\}$.

定理 4.6.6 设 A 是群 G 的 L-模糊子集. 如果对于任意的 $a, b \in L$, $\beta(a \wedge b) = \beta(a) \cap \beta(b)$, 那么

$$\mathscr{G}_N(A) = \bigvee \{a \in L \mid a \leqslant A(1), \forall b \in \beta(a), A_{(b)} \text{ 是 } G \text{ 的正规子群}\}.$$

当 $\mathscr{G}_N(A) = 1$ 时, 上述定理可退化为下面推论形式.

推论 4.6.7 设 G 是一个群且 $A \in L^G$. 则下列条件等价:
(1) A 是 G 的 L-模糊正规子群;
(2) $\forall b \in L$, $A_{[b]}$ 是 G 的正规子群;
(3) $\forall b \in L$, $A^{[b]}$ 是 G 的正规子群;
(4) $\forall b \in P(L)$, $A^{(b)}$ 是 G 的正规子群;
(5) 如果对于任意的 $a, b \in L$, 都有 $\beta(a \wedge b) = \beta(a) \cap \beta(b)$, 那么 $\forall b \in \beta(1)$, $A_{(b)}$ 是 G 的正规子群.

4.6　L-模糊正规子群度

定理 4.6.8　设 $\{A_i\}_{i\in\Omega}$ 是群 G 的一族 L-模糊子集. 则

$$\mathscr{G}_N\left(\bigwedge_{i\in\Omega} A_i\right) \geqslant \bigwedge_{i\in\Omega} \mathscr{G}_N(A_i).$$

证明　这从下面不等式可以得到.

$$\mathscr{G}_N\left(\bigwedge_{i\in\Omega} A_i\right)$$
$$= \mathscr{G}\left(\bigwedge_{i\in\Omega} A_i\right) \wedge \bigwedge_{x,y\in G}\left(\left(\bigwedge_{i\in\Omega} A_i\right)(y) \mapsto \left(\bigwedge_{i\in\Omega} A_i\right)(xyx^{-1})\right)$$
$$= \mathscr{G}\left(\bigwedge_{i\in\Omega} A_i\right) \wedge \bigwedge_{x,y\in G}\bigwedge_{i\in\Omega}\left(\left(\bigwedge_{i\in\Omega} A_i\right)(y) \mapsto A_i(xyx^{-1})\right)$$
$$\geqslant \mathscr{G}\left(\bigwedge_{i\in\Omega} A_i\right) \wedge \bigwedge_{x,y\in G}\bigwedge_{i\in\Omega}\left(A_i(y) \mapsto A_i(xyx^{-1})\right)$$
$$\geqslant \bigwedge_{i\in\Omega}\mathscr{G}(A_i) \wedge \bigwedge_{x,y\in G}\bigwedge_{i\in\Omega}\left(A_i(y) \mapsto A_i(xyx^{-1})\right)$$
$$= \bigwedge_{i\in\Omega}\left(\mathscr{G}(A_i) \wedge \bigwedge_{x,y\in G}\left(A_i(y) \mapsto A_i(xyx^{-1})\right)\right)$$
$$= \bigwedge_{i\in\Omega}\mathscr{G}_N(A_i). \qquad \square$$

定理 4.6.9　设 $f: G \to G'$ 是一个群同态.

(1) 如果 A 是群 G 的一个 L-模糊子集且 f 是满射, 那么 $\mathscr{G}_N(f_L^\to(A)) \geqslant \mathscr{G}_N(A)$; 如果 f 是单射, 那么 $\mathscr{G}_N(f_L^\to(A)) \leqslant \mathscr{G}_N(A)$.

(2) 如果 B 是群 G' 的一个 L-模糊子集, 那么 $\mathscr{G}_N(f_L^\leftarrow(B)) \geqslant \mathscr{G}_N(B)$; 如果 f 是满射, 那么 $\mathscr{G}_N(f_L^\leftarrow(B)) = \mathscr{G}_N(B)$.

证明　(1) 因为 f 是满射, 由定理 4.4.12, 有

$$\mathscr{G}_N(f_L^\to(A))$$
$$= \mathscr{G}(f_L^\to(A)) \wedge \bigvee\{a\in L \mid f_L^\to(A)(y') \wedge a \leqslant f_L^\to(A)(x'y'x'^{-1}), \forall x',y'\in G'\}$$
$$\geqslant \mathscr{G}(A) \wedge \bigvee\left\{a\in L \,\Big|\, \bigvee_{f(y)=y'} A(y) \wedge a \leqslant \bigvee_{f(z)=x'y'x'^{-1}} A(z), \forall x',y'\in G'\right\}$$

$$\geqslant \mathscr{G}(A) \wedge \bigvee\{a \in L \mid A(y) \wedge a \leqslant A(xyx^{-1}), \forall x, y \in G\} = \mathscr{G}_N(A).$$

如果 f 是单射, 由定理 4.4.12, 有

$$\mathscr{G}_N(f_L^{\rightarrow}(A))$$
$$= \mathscr{G}(f_L^{\rightarrow}(A)) \wedge \bigvee \left\{ a \in L \mid f_L^{\rightarrow}(A)(y') \wedge a \leqslant f_L^{\rightarrow}(A)(x'y'x'^{-1}), \forall x', y' \in G' \right\}$$
$$= \mathscr{G}(A) \wedge \bigvee \left\{ a \in L \mid f_L^{\rightarrow}(A)(y') \wedge a \leqslant f_L^{\rightarrow}(A)(x'y'x'^{-1}), \forall x', y' \in G' \right\}$$
$$\leqslant \mathscr{G}(A) \wedge \bigvee\{a \in L \mid f_L^{\rightarrow}(A)(f(y)) \wedge a \leqslant f_L^{\rightarrow}(A)(f(x)f(y)f(x)^{-1}), \forall x, y \in G\}$$
$$= \mathscr{G}(A) \wedge \bigvee\{a \in L \mid \left(f_L^{\leftarrow}(f_L^{\rightarrow}(A))\right)(y) \wedge a \leqslant \left(f_L^{\leftarrow}(f_L^{\rightarrow}(A))\right)(xyx^{-1}), \forall x, y \in G\}$$
$$= \mathscr{G}(A) \wedge \bigvee\{a \in L \mid A(y) \wedge a \leqslant A(xyx^{-1}), \forall x, y \in G\}$$
$$= \mathscr{G}_N(A).$$

(2) 这从下面不等式能够得到.

$$\mathscr{G}_N(f_L^{\leftarrow}(B)) = \mathscr{G}(f_L^{\leftarrow}(B)) \wedge \bigwedge_{x,y \in G} \left(f_L^{\leftarrow}(B)(y) \mapsto f_L^{\leftarrow}(B)(xyx^{-1})\right)$$
$$= \mathscr{G}(f_L^{\leftarrow}(B)) \wedge \bigwedge_{x,y \in G} \left(B(f(y)) \mapsto B(f(x)f(y)f(x)^{-1})\right)$$
$$\geqslant \mathscr{G}(B) \wedge \bigwedge_{x',y' \in G'} \left(B(y') \mapsto B(x'y'x'^{-1})\right)$$
$$= \mathscr{G}_N(B).$$

如果 f 是满射, 由 (1) 可得

$$\mathscr{G}_N(B) = \mathscr{G}_N\left(f_L^{\rightarrow}(f_L^{\leftarrow}(B))\right) \geqslant \mathscr{G}_N\left(f_L^{\leftarrow}(B)\right).$$

因此, 有 $\mathscr{G}_N(B) = \mathscr{G}_N(f_L^{\leftarrow}(B))$. □

定理 4.6.10 设 $\{G_i\}_{i=1}^n$ 是一族群, 且 $\prod_{i=1}^n A_i$ 是 $\{A_i\}_{i=1}^n$ 的乘积, 其中对任意 $i \in \{1, 2, \cdots, n\}$, $A_i \in L^{G_i}$. 则 $\mathscr{G}_N\left(\prod_{i=1}^n A_i\right) \geqslant \bigwedge_{i=1}^n \mathscr{G}_N(A_i)$.

证明 **方法 1**. 设 $G = \prod_{i=1}^n G_i$. 由定理 4.5.1 可得

$$\mathscr{G}_N\left(\prod_{i=1}^n A_i\right)$$

$$= \mathscr{G}\left(\prod_{i=1}^{n} A_i\right) \wedge \bigwedge_{x,y \in G}\left(\left(\prod_{i=1}^{n} A_i\right)(y) \mapsto \left(\prod_{i=1}^{n} A_i\right)(xyx^{-1})\right)$$

$$= \mathscr{G}\left(\prod_{i=1}^{n} A_i\right) \wedge \bigwedge_{x=\{x_i\},y=\{y_i\} \in G}\left(\bigwedge_{i=1}^{n} A_i(y_i) \mapsto \bigwedge_{i=1}^{n} A_i(x_i y_i x_i^{-1})\right)$$

$$\geqslant \bigwedge_{i=1}^{n} \mathscr{G}(A_i) \wedge \bigwedge_{x=\{x_i\},y=\{y_i\} \in G} \bigwedge_{i=1}^{n}\left(A_i(y_i) \mapsto A_i(x_i y_i x_i^{-1})\right)$$

$$= \bigwedge_{i=1}^{n}\left(\mathscr{G}(A_i) \wedge \bigwedge_{x_i,y_i \in G_i}\left(A_i(y_i) \mapsto A_i(x_i y_i x_i^{-1})\right)\right)$$

$$= \bigwedge_{i=1}^{n} \mathscr{G}_N(A_i).$$

方法 2. 设 $P_i : \prod_{i=1}^{n} G_i \to G_i$ 是投射, 容易验证 $\prod_{i=1}^{n} A_i = \bigwedge_{i=1}^{n} P_i^{\leftarrow}(A_i)$. 由定理 4.6.8 和定理 4.6.9 可知

$$\mathscr{G}_N\left(\prod_{i=1}^{n} A_i\right) = \mathscr{G}_N\left(\bigwedge_{i=1}^{n} P_i^{\leftarrow}(A_i)\right) \geqslant \bigwedge_{i=1}^{n} \mathscr{G}_N(P_i^{\leftarrow}(A_i)) \geqslant \bigwedge_{i=1}^{n} \mathscr{G}_N(A_i). \qquad \square$$

定理 4.6.11 设 A 和 B 是 G 上的两个 L-模糊子集. 则

$$\mathscr{G}(AB) \geqslant \mathscr{G}_N(A) \wedge \mathscr{G}(B).$$

证明 由定理 4.4.8 与定理 4.6.5 我们能够得到下面不等式.

$\mathscr{G}(AB)$

$= \bigvee\{a \in L \mid a \leqslant (AB)(1), \forall b \not\leqslant a \text{且} b \in P(L), (AB)^{(b)} \text{ 是 } G \text{ 的子群}\}$

$= \bigvee\{a \in L \mid a \leqslant (AB)(1), \forall b \not\leqslant a \text{且} b \in P(L), A^{(b)} B^{(b)} \text{ 是 } G \text{ 的子群}\}$

$\geqslant \bigvee\{a \in L \mid a \leqslant A(1), a \leqslant B(1), \forall b \not\leqslant a \text{且} b \in P(L),$

$\qquad A^{(b)} \text{ 是正规子群}, B^{(b)} \text{ 是子群}\}$

$= \mathscr{G}_N(A) \wedge \mathscr{G}(B). \qquad \square$

4.7 由 L-模糊子群度确定的 L-模糊凸结构

本节主要研究群上由 L-模糊子群度所确定的 L-模糊凸结构. 此外, 还分析了其对应的 L-模糊凸保持映射以及 L-模糊凸到凸映射.

对于任意 $A \in L^G$,$\mathscr{G}(A)$ 自然地被看作由 $A \longmapsto \mathscr{G}(A)$ 确定的映射 $\mathscr{G}: L^G \to L$ 的像. 接下来的定理说明了 $\mathscr{G}$ 是 G 上的一个 L-模糊凸结构.

定理 4.7.1 设 G 是一个群. 则由 $A \longmapsto \mathscr{G}(A)$ 确定的映射 $\mathscr{G}: L^G \to L$ 是 G 上的一个 L-模糊凸结构, 称其为 G 上的由 L-模糊子群度所确定的 L-模糊凸结构.

证明 (1) 显然有 $\mathscr{G}(\varnothing) = \mathscr{G}(G) = 1$.

(2) 设 $\{A_i \mid i \in \Omega\} \subseteq L^G$ 是非空的, 由定理 4.4.11 可得

$$\bigwedge_{i \in \Omega} \mathscr{G}(A_i) \leqslant \mathscr{G}\left(\bigwedge_{i \in \Omega} A_i\right).$$

(3) 设 $\{A_i \mid i \in \Omega\} \subseteq L^G$ 是非空的全序集. 为了证明

$$\bigwedge_{i \in \Omega} \mathscr{G}(A_i) \leqslant \mathscr{G}\left(\bigvee_{i \in \Omega} A_i\right),$$

需证明对于任意的 $a \leqslant \bigwedge_{i \in \Omega} \mathscr{G}(A_i)$, 有 $a \leqslant \mathscr{G}\left(\bigvee_{i \in \Omega} A_i\right)$. 由引理 4.4.4, 对于任意的 $i \in \Omega, s, t \in G$,

$$a \leqslant A_i(1), \quad A_i(s) \wedge A_i(t) \wedge a \leqslant A_i(st^{-1}),$$

令 $b \in J(L)$ 使得

$$b \prec \left(\bigvee_{i \in \Omega} A_i(s)\right) \wedge \left(\bigvee_{i \in \Omega} A_i(t)\right) \wedge a,$$

则有

$$b \prec \bigvee_{i \in \Omega} A_i(s), \quad b \prec \bigvee_{i \in \Omega} A_i(t) \quad 和 \quad b \leqslant a.$$

由此可知有 $i, j \in \Omega$ 使得 $b \leqslant A_i(s), b \leqslant A_j(t)$ 和 $b \leqslant a$. 因为 $\{A_i \mid i \in \Omega\}$ 是全序的, 假设 $A_j \leqslant A_i$, 则 $b \leqslant A_i(s) \wedge A_i(t) \wedge a$. 由

$$A_i(s) \wedge A_i(t) \wedge a \leqslant A_i(st^{-1})$$

可知 $b \leqslant A_i(st^{-1})$ 且 $b \leqslant \mu_l(z)$. 于是有 $b \leqslant \bigvee_{i \in \Omega} A_i(st^{-1})$ 且 $b \leqslant \bigvee_{l \in \Omega} \mu_l(z)$. 由 b 的任意性可得

$$\left(\bigvee_{i \in \Omega} A_i(s)\right) \wedge \left(\bigvee_{i \in \Omega} A_i(t)\right) \wedge a \leqslant \bigvee_{i \in \Omega} A_i(st^{-1}).$$

结合引理 4.4.4 可得 $a \leqslant \mathscr{G}\left(\bigvee_{i\in\Omega} A_i\right)$. 由 a 的任意性我们知道有 $\bigwedge_{i\in\Omega} \mathscr{G}(A_i) \leqslant \mathscr{G}\left(\bigvee_{i\in\Omega} A_i\right)$ 成立. 故 $\mathscr{G}$ 是 G 上的 L-模糊凸结构. □

用相同的方法可以证明下面结论.

定理 4.7.2 设 G 是一个群. 则由 $A \longmapsto \mathscr{G}(A)$ 的映射确定 $\mathscr{G}_N : L^G \to L$ 是 G 上的一个 L-模糊凸结构, 称其为 G 上的由 L-模糊正规子群度所确定的 L-模糊凸结构.

再由定理 4.4.12, 易得到下面定理.

定理 4.7.3 设 $f : G \to G'$ 是一个群同态映射, $\mathscr{G}$ 和 $\mathscr{G}'$ 分别是 G 和 G' 上由 L-子群度确定的 L-模糊凸结构. 则 $f : (G, \mathscr{G}) \to (G', \mathscr{G}')$ 是一个 L-模糊凸保持映射和一个 L-模糊凸到凸映射.

类似地, 我们可以得到下面定理.

定理 4.7.4 设 $f : G \to G'$ 是一个满的群同态映射, $\mathscr{G}_N$ 和 $\mathscr{G}'_N$ 分别是 G 和 G' 上由 L-正规子群度确定的 L-模糊凸结构. 则 $f : (G, \mathscr{G}_N) \to (G', \mathscr{G}'_N)$ 是一个 L-模糊凸保持映射和一个 L-模糊凸到凸映射.

推论 4.7.5 设 $f : G \to G'$ 为群同构映射. 则 $f : (G, \mathscr{G}) \to (G', \mathscr{G}')$ 是 L-模糊凸空间之间的同构映射.

推论 4.7.6 设 $f : G \to G'$ 为群同构映射. 则 $f : (G, \mathscr{G}_N) \to (G', \mathscr{G}'_N)$ 是 L-模糊凸空间之间的同构映射.

定理 4.7.7 设 $\{G_i\}_{i=1}^n$ 是一族群且 $\forall i \in \{1, 2, \cdots, n\}$, $P_i : \prod_{i=1}^n G_i \to G_i$ 是投影映射. $\forall i \in \{1, 2, \cdots, n\}$, 令 $\mathscr{G}_i$ 表示 G_i 上的 L-模糊子群度, $\mathscr{G}$ 表示 $\prod_{i=1}^n G_i$ 上的 L-模糊子群度, 则 $P_i : \left(\prod_{i=1}^n G_i, \mathscr{G}\right) \to (G_i, \mathscr{G}_i)$ 是 L-模糊凸保持的和 L-模糊凸到凸的映射.

4.8 L-模糊商群

在这一节中, 我们研究一种简单的模糊商群的概念, 这里仅要求正规子群是分明的, 而没有要求是模糊的.

定义 4.8.1 设 G 是一个群, N 是 G 的一个正规子群, $A \in L^G$. 定义商群 R/N 的一个 L-模糊集 A/N 使得

$$(A/N)(xN) = \bigvee_{n \in N} A(xn), \quad \forall x \in G.$$

那么 A/N 称为 A 关于正规子群 N 的 L-模糊商集.

接下来我们给出 L-模糊商集的表示定理.

定理 4.8.2 设 N 是 G 的正规子群, $A \in L^G$. 则

(1) $(A/N)_{(a)} \subseteq A_{(a)}/N \subseteq A_{[a]}/N \subseteq (A/N)_{[a]}, \forall a \in L$;

(2) $A/N = \bigvee_{a \in L} \{a \wedge (A_{[a]}/N)\} = \bigvee_{a \in M(L)} \{a \wedge (A_{[a]}/N)\}$;

(3) $A/N = \bigvee_{a \in L} \{a \wedge (A_{(a)}/N)\} = \bigvee_{a \in M(L)} \{a \wedge (A_{(a)}/N)\}$;

(4) $(A/N)_{[a]} = \bigcap_{b \in \beta(a)} (A_{[b]}/N) = \bigcap_{b \in \beta(a)} (A_{(b)}/N)$;

(5) $(A/N)_{(a)} = \bigcup_{a \in \beta(b)} (A_{[b]}/N) = \bigcup_{a \in \beta(b)} (A_{(b)}/N)$.

证明 (1) 对于任意的 $a \in L$, 首先来证明 $(A/N)_{(a)} \subseteq A_{(a)}/N$, 令 $xN \in (A/N)_{(a)}$, 则由

$$a \in \beta((A/N)(xN)) = \beta\left(\bigvee_{n \in N} A(xn)\right) = \bigcup_{n \in N} \beta(A(xn))$$

可知有 $n \in N$ 使得 $a \in \beta(A(xn))$, 也就是 $xn \in A_{(a)}$. 取 $n^{-1} \in N$, 则有 $x = (xn)n^{-1} \in A_{(a)}N$, 即 $xN \in A_{(a)}/N$. 这表明 $(A/N)_{(a)} \subseteq A_{(a)}/N$.

$A_{(a)}/N \subseteq A_{[a]}/N$ 是显然的.

现在证明 $A_{[a]}/N \subseteq (A/N)_{[a]}$. 假设 $xN \in A_{[a]}/N$, 由

$$A_{[a]}/N = \{xN \mid x \in A_{[a]}\} = \{xnN \mid xn \in A_{[a]}, n \in N\}$$

可知有 $(A/N)(xN) = \bigvee_{n \in N} A(xn) \geqslant a$, 因此可得 $xN \in (A/N)_{[a]}$, 这表明 $A_{[a]}/N \subseteq (A/N)_{[a]}$.

由 (1) 和定理 2.1.4 可证 (2), (3), (4) 和 (5). □

同样利用另外两种截集也可以给出 L-模糊商集的表示定理.

定理 4.8.3 设 N 是群 G 的正规子群, $A \in L^G$. 则

(1) $(A/N)^{(a)} = A^{(a)}/N \subseteq A^{[a]}/N \subseteq (A/N)^{[a]}, \forall a \in L$;

(2) $A/N = \bigwedge_{a \in L} \{a \vee (A^{[a]}/N)\} = \bigwedge_{a \in P(L)} \{a \vee (A^{[a]}/N)\}$;

(3) $A/N = \bigwedge_{a \in L} \{a \vee (A^{(a)}/N)\} = \bigwedge_{a \in P(L)} \{a \vee (A^{(a)}/N)\}$;

(4) $(A/N)^{[a]} = \bigcap_{a \in \alpha(b)} (A^{[b]}/N) = \bigcap_{a \in \alpha(b)} (A^{(b)}/N)$;

(5) $(A/N)^{(a)} = \bigcup_{b \in \alpha(a)} (A^{[b]}/N) = \bigcup_{b \in \alpha(a)} (A^{(b)}/N)$.

4.8 L-模糊商群

证明 (1) 对于任意的 $a \in L$, 首先来证明 $(A/N)^{(a)} = A^{(a)}/N$, 为此假设 $xN \in (A/N)^{(a)}$, 则有

$$xN \in (A/N)^{(a)} \Leftrightarrow (A/N)(xN) \not\leq a$$
$$\Leftrightarrow \bigvee_{n \in N} A(xn) \not\leq a$$
$$\Leftrightarrow \exists n \in N, A(xn) \not\leq a$$
$$\Leftrightarrow \exists n \in N, xn \in A^{(a)}$$
$$\Leftrightarrow xN \in A^{(a)}/N,$$

所以可得 $(A/N)^{(a)} = A^{(a)}/N$.

显然 $A^{(a)}/N \subseteq A^{[a]}/N$.

现在证明 $A^{[a]}/N \subseteq (A/N)^{[a]}$. 假设 $xN \notin (A/N)^{[a]}$, 则 $a \in \alpha((A/N)(xN))$. 由下面蕴含式

$$a \in \alpha\left((A/N)(xN)\right) = \alpha\left(\bigvee_{n \in N} A(xn)\right)$$
$$\subseteq \bigcap_{r \in I} \alpha\left(A(xn)\right)$$
$$\Rightarrow \forall n \in N, a \in \alpha\left(A(xn)\right)$$
$$\Rightarrow \forall n \in N, xn \notin A^{[a]}$$
$$\Rightarrow xN \notin A^{[a]}/N,$$

可得 $A^{[a]}/N \subseteq (A/N)^{[a]}$.

由 (1) 和定理 2.1.4 可证 (2), (3), (4) 和 (5). □

接下来我们研究 L-模糊子群度和 L-模糊商群度的关系.

定理 4.8.4 设 N 是群 G 的正规子群, $R/N = \{xN \mid x \in G\}$ 是 G 关于 N 的商群, $A \in L^G$. 则

$$\mathscr{G}(A) \leqslant \mathscr{G}(A/N).$$

证明 **方法 1.** 假设 $p: R \to R/N$ 是自然投射, 则对任意的 $xN \in G/N$,

$$p(A)(xN) = \bigvee_{p(x)=xN} A(x) = \bigvee_{n \in N} A(xn) = (A/N)(xN),$$

这说明

$$p(A)(xN) = (A/N)(xN).$$

从而由定理 4.4.12 可以得到

$$\mathscr{G}(A) \leqslant \mathscr{G}(A/N).$$

方法 2. 由定理 4.4.8 可得下面不等式.

$\mathscr{G}(A) = \bigvee \{a \in L \mid a \leqslant A(1), \forall b \in P(L), b \not\geqslant a, A^{(b)} \text{是 } G \text{ 的子群}\}$

$\leqslant \bigvee \{a \in L \mid a \leqslant A(1), \forall b \in P(L), b \not\geqslant a, A^{(b)}/N \text{是 } G/N \text{ 的子群}\}$

$= \bigvee \{a \in L \mid a \leqslant A(1), \forall b \in P(L), b \not\geqslant a, (A/N)^{(b)} \text{是 } G/N \text{ 的子群}\}$

$\leqslant \bigvee \{a \in L \mid a \leqslant (A/N)(N), \forall b \in P(L), b \not\geqslant a, (A/N)^{(b)} \text{是 } G/N \text{ 的子群}\}$

$= \mathscr{G}(A/N).$ □

设 N 是群 G 的正规子群,则 $A/N \mapsto \mathscr{G}(A/N)$ 所定义的映射 $\mathscr{G}: L^{G/N} \to L$ 是 G/N 上的 L-模糊凸结构. 另外, 两个群之间的群同态是 L-模糊凸保持映射和 L-模糊凸到凸的映射. 于是我们可以得到下面推论.

定理 4.8.5 设 N 是群 G 的正规子群, $A \in L^G$, 则投射 $p: G \to G/N$ ($A \to A/N$) 是 L-模糊凸保持映射和 L-模糊凸到凸的映射.

习 题 4

1. 证明定理 4.1.4.
2. 证明定理 4.1.7.
3. 证明定理 4.1.8.
4. 设 S 是一个半群, $A \in L^S$. 令

$$\mathcal{L}(A) = \bigwedge_{x,y \in S} ((A(x) \mapsto A(xy)) \wedge (A(y) \mapsto A(xy))).$$

称 $\mathcal{L}(A)$ 为 A 的理想度. A 是 S 的 L-模糊理想当且仅当 $\mathcal{L}(A) = 1$.
试证明下面几个等式成立.

(1) $\mathcal{L}(A) = \bigvee \left\{ a \in L \mid \forall b \leqslant a, A_{[b]} \text{ 是 } S \text{ 的理想} \right\}$;

(2) $\mathcal{L}(A) = \bigvee \left\{ a \in L \mid \forall b \notin \alpha(a), A^{[b]} \text{ 是 } S \text{ 的理想} \right\}$;

(3) $\mathcal{L}(A) = \bigvee \left\{ a \in L \mid \forall b \in P(L), a \not\leqslant b, A^{(b)} \text{ 是 } S \text{ 的理想} \right\}$;

习　题　4

(4) 如果对于任意的 $a, b \in L$, 都有 $\beta(a \wedge b) = \beta(a) \cap \beta(b)$, 那么 $\mathcal{L}(A) = \bigvee \{a \in L \mid \forall b \in \beta(a), A_{(b)}$ 是 S 的理想$\}$.

5. 设 $\{A_i\}_{i \in \Omega}$ 是半群 S 的一族 L-模糊子集. 试证明

$$\bigwedge_{i \in \Omega} \mathcal{L}(A_i) \leqslant \mathcal{L}\left(\bigwedge_{i \in \Omega} A_i\right).$$

6. 设 $f : S \to S'$ 是半群同态. 试证明

(1) 若 $A \in L^S$, 则 $\mathcal{L}(f_L^{\to}(A)) \geqslant \mathcal{L}(A)$; 若 f 是单射, 则 $\mathcal{L}(f_L^{\to}(A)) = \mathcal{L}(A)$.

(2) 若 $B \in L^{S'}$, 则 $\mathcal{L}(f_L^{\leftarrow}(B)) \geqslant \mathcal{L}(B)$; 若 f 是满射, 则 $\mathcal{L}(f_L^{\leftarrow}(B)) = \mathcal{L}(B)$.

7. 设 $(S, \circ)$ 为半群, $a \in P(L)$ 且 $A, B \in L^S$. 定义 $A \circ B \in L^S$ 如下:

$$(A \circ B)(z) = \bigvee_{z = x \circ y} (A(x) \wedge A(y)).$$

证明 $(A \circ B)^{(a)} = A^{(a)} \circ B^{(a)}$.

8. 设 $S = \{l, n, p, q\}$ 是一个半群, 它的运算和例 4.2.3 的相同. 定义 $A, B \in L^S$ 如下:

$$A(l) = 0.4, \quad A(n) = 0.7, \quad A(p) = 0.6, \quad A(q) = 0.5$$

且

$$B(l) = 0.3, \quad B(n) = 0.7, \quad B(p) = 0.8, \quad B(q) = 0.9.$$

请写出模糊集 $A \circ B$.

9. 设 S 是半群且 $A, B \in L^S$. 若 $A \circ B = B \circ A$, 试证明

$$\mathscr{S}(A \circ B) \geqslant \mathscr{S}(A) \wedge \mathscr{S}(B).$$

10. 设 S 为一个半群且 $A \in L^S$. 证明映射 $\mathcal{L} : L^S \to L$ 使得 $A \longmapsto \mathcal{L}(A)$ 是半群 S 上的一个 L-模糊凸结构, 称为由 L-模糊理想度确定的 L-模糊凸结构. (见第 4 题)

11. 设 $f : S \to S'$ 是半群同态, $\mathcal{L}$ 和 $\mathcal{L}'$ 分别是半群 S 和 S' 上由 L-模糊理想度确定的 L-模糊凸结构. 证明 $f : (S, \mathcal{L}) \to (S', \mathcal{L}')$ 是一个 L-模糊凸保持的和 L-模糊凸到凸的映射.

12. 设 $\{S_i\}_{i=1}^n$ 是一族半群且 $\forall i \in \{1, 2, \cdots, n\}$, $P_i : \prod_{i=1}^n S_i \to S_i$ 是投影映射. $\forall i \in \{1, 2, \cdots, n\}$, 令 $\mathcal{L}_i$ 表示 S_i 上的 L-模糊理想度, 令 $\mathcal{L}$ 表示 $\prod_{i=1}^n S_i$ 上的 L-模糊理想度, 则 $P_i : \left(\prod_{i=1}^n S_i, \mathcal{L}\right) \to (S_i, \mathcal{L}_i)$ 是 L-模糊凸保持的和 L-模糊凸到凸的映射.

13. 设 $f : S \to S'$ 为半群同构映射, 则 $f : (S, \mathcal{L}) \to (S', \mathcal{L}')$ 是 L-模糊凸空间之间的同构映射.

14. 试用截集刻画定理证明定理 4.4.11.

15. 试用两个不同的截集刻画定理证明定理 4.4.12 中的 (1) 和 (2).

16. 试从截集刻画定理出发证明定理 4.5.1.

17. 设 A 和 B 是群 G 的两个 L-模糊子集. 则 $\forall a \in L$, 下面条件成立:

(1) $(AB)_{(a)} \subseteq A_{(a)}B_{(a)} \subseteq A_{[a]}B_{[a]} \subseteq (AB)_{[a]}$;

(2) $(AB)^{(a)} \subseteq A^{(a)}B^{(a)} \subseteq A^{[a]}B^{[a]} \subseteq (AB)^{[a]}$;

(3) $AB = \bigwedge_{a \in L} \{a \vee (A^{[a]}B^{[a]})\} = \bigwedge_{a \in P(L)} \{a \vee (A^{[a]}B^{[a]})\}$;

(4) $AB = \bigwedge_{a \in L} \{a \vee (A^{(a)}B^{(a)})\} = \bigwedge_{a \in P(L)} \{a \vee (A^{(a)}B^{(a)})\}$;

(5) $AB = \bigvee_{a \in L} \{a \wedge (A_{[a]}B_{[a]})\} = \bigvee_{a \in J(L)} \{a \wedge (A_{[a]}B_{[a]})\}$;

(6) $AB = \bigvee_{a \in L} \{a \wedge (A_{(a)}B_{(a)})\} = \bigvee_{a \in J(L)} \{a \wedge (A_{(a)}B_{(a)})\}$.

试证之.

18. 试从截集刻画定理出发证明定理 4.5.5.

19. 试用截集刻画定理和分明结论证明定理 4.6.9.

20. 试用截集刻画定理和分明结论证明定理 4.6.10.

21. 设 A 和 B 是群 G 的两个 L-模糊子集. 则

(1)
$$\bigwedge_{z \in G}((AA)(BB)(z) \mapsto (AB)(z)) \wedge \bigwedge_{x,y \in G} A(y) \mapsto A(xyx^{-1})$$
$$\leqslant \bigwedge_{z \in G}((AB)(AB)(z) \mapsto (AB)(z));$$

(2)
$$\bigwedge_{z \in G}((AB)(z) \mapsto (BA)(z^{-1})) \wedge \bigwedge_{x,y \in G}(A(y) \mapsto A(xyx^{-1}))$$
$$\leqslant \bigwedge_{z \in G}((AB)(z) \mapsto (AB)(z^{-1}));$$

(3)
$$\bigwedge_{x,y \in G}(A(y) \mapsto A(xyx^{-1})) \wedge \bigwedge_{x,y \in G}(B(y) \mapsto B(xyx^{-1}))$$
$$\leqslant \bigwedge_{x,y \in G}((AB)(y) \mapsto (AB)(xyx^{-1})).$$

22. 试证明定理 4.7.2.

23. 在定理 4.7.7 中, 如果把子群度换成正规子群度, 那么定理是否正确?

24. 在定义 4.8.1 中, 我们研究的是一种简单的模糊商群, 实际上, 也可以定义另外一种模糊商集, 例如, 设 G 是一个群, N 是 G 的一个 L-模糊正规子群, $\forall x \in G$, 定义一个 G 的 L-模糊集 xN 使得

$$(xN)(y) = N(x^{-1}y), \quad \forall y \in G.$$

试证明:

(1) 对任意的 $a \in L$, $(xN)_{[a]} = xN_{[a]}$;

(2) 对任意的 $a \in L$, $(xN)_{(a)} = xN_{(a)}$;

(3) 对任意的 $a \in L$, $(xN)^{(a)} = xN^{(a)}$;

(4) 对任意的 $a \in L$, $(xN)^{[a]} = xN^{[a]}$.

25. (接上题) 如果 N 是 G 的一个 L-模糊正规子群, 那么商群 $G/(N_{[a]})$ (或者 $G/(N_{(a)})$) 可以构成一个 L_β-集合套, 这样能够得到一个什么样的群的类似物呢？类似地, 也可以构造一个 L_α-集合套, 这样能够得到一个什么样的群的类似物呢？

第 5 章 L-模糊子环与理想

模糊子环的概念是紧随着模糊集的引入而产生的, 在其定义下, 一个环中的模糊子集要么是子环, 要么不是, 二者必居其一, 且仅居其一. 类似子群度的方法, 本章给出 L-模糊子环进一步模糊化的方法, 即给出 L-模糊子环度的概念, 并给出 L-模糊子环度的等价刻画, 同时, 讨论 L-模糊子环度与 L-模糊凸结构的关系, 指出环同态实际上也是 L-模糊凸保持映射和 L-模糊凸到凸的映射. 此外, 研究 L-模糊子环的运算, 给出 L-模糊子环内积、直和的表现定理, 讨论 L-模糊子集内积、外积、直和的 L-模糊子环度. 本章结果主要源于文献 [47, 168].

5.1 L-模糊子环度

类似于 L-模糊子群度的定义, 很自然地可以给出 L-模糊子环度的概念如下.

定义 5.1.1 设 R 是一个环, $A \in L^R$. 则 A 的 L-模糊子环度 $\mathscr{R}(A)$ 定义如下:
$$\mathscr{R}(A) = A(0) \wedge \bigwedge_{x,y \in R} \left[(A(x) \wedge A(y)) \mapsto A(xy) \wedge A(x-y) \right].$$

也可以说 A 是环 R 的 L-模糊子环的程度是 $\mathscr{R}(A)$.

显然, $\mathscr{R}(A) = 1$ 当且仅当 A 满足下面三个条件:

(1) $\forall x, y \in X, A(x-y) \geqslant A(x) \wedge A(y)$;

(2) $\forall x, y \in X, A(xy) \geqslant A(x) \wedge A(y)$;

(3) $A(0) = 1$.

因此定义 5.1.1 是模糊子环定义的进一步模糊化. 对环 R 的任意 L-模糊子集 A 而言, 它在某种程度 $\mathscr{R}(A)$ 上, 一定是一个 L-模糊子环.

下面我们给出模糊子环度几种情况的例子.

例 5.1.2 令 $\mathbb{Z}_3$ 是模 3 的剩余类环, 并定义 $\mathbb{Z}_3$ 上的运算为常规的加法和乘法. 定义三个 L-模糊集 $A, B, C \in L^{\mathbb{Z}_3}$ 如下:

$$A(z) = \begin{cases} 0.8, & z = [0], \\ 0.3, & z = [1], \\ 0.4, & z = [2], \end{cases}$$

5.1 L-模糊子环度

$$B(z) = \begin{cases} 0, & z = [0], \\ 1, & z = [1], \\ 0.4, & z = [2], \end{cases}$$

$$C(z) = \begin{cases} 0.2, & z = [0], \\ 0.2, & z = [1], \\ 0.2, & z = [2]. \end{cases}$$

下面我们来验证

$$\mathscr{R}(A) = 0.3, \quad \mathscr{R}(B) = 0, \quad \mathscr{R}(C) = 0.2.$$

实际上

$$A(x) \wedge A(y) \mapsto A(x-y) \wedge A(xy) = \begin{cases} 0.8 \mapsto 0.8 = 1, & x = [0], y = [0], \\ 0.3 \mapsto 0.4 = 1, & x = [0], y = [1], \\ 0.4 \mapsto 0.3 = 0.3, & x = [0], y = [2], \\ 0.3 \mapsto 0.3 = 1, & x = [1], y = [0], \\ 0.3 \mapsto 0.3 = 1, & x = [1], y = [1], \\ 0.3 \mapsto 0.4 = 1, & x = [1], y = [2], \\ 0.4 \mapsto 0.4 = 1, & x = [2], y = [0], \\ 0.3 \mapsto 0.3 = 1, & x = [2], y = [1], \\ 0.4 \mapsto 0.3 = 0.3, & x = [2], y = [2], \end{cases}$$

所以可以得到 $\mathscr{R}(A) = 0.8 \wedge 1 \wedge 0.3 = 0.3$. 又

$$B(x) \wedge B(y) \mapsto B(x-y) \wedge B(xy) = \begin{cases} 0 \mapsto 0 = 1, & x = [0], y = [0], \\ 0 \mapsto 0 = 1, & x = [0], y = [1], \\ 0 \mapsto 0 = 1, & x = [0], y = [2], \\ 0 \mapsto 0 = 1, & x = [1], y = [0], \\ 1 \mapsto 0 = 0, & x = [1], y = [1], \\ 0.4 \mapsto 0.4 = 1, & x = [1], y = [2], \\ 0 \mapsto 0 = 1, & x = [2], y = [0], \\ 0.4 \mapsto 0.4 = 1, & x = [2], y = [1], \\ 0.4 \mapsto 0 = 0, & x = [2], y = [2], \end{cases}$$

所以可以得到 $\mathscr{R}(B) = 0 \wedge 1 \wedge 0 = 0$.

显然 $\mathscr{R}(C) = 0.2$.

L-模糊子环度也可以表示为如下形式.

定理 5.1.3 设 A 为 R 的 L-模糊子集. 则

$$\mathscr{R}(A) = A(0) \wedge \bigwedge_{x,y \in R} ((A(x) \wedge A(y)) \mapsto A(xy))$$

$$\wedge \bigwedge_{x,y \in R} ((A(x) \wedge A(y)) \mapsto A(x+y)) \wedge \bigwedge_{y \in R} (A(y) \mapsto A(-y)).$$

证明 由定义 5.1.1 可以得到 $\forall x, y \in R$,

$$\mathscr{R}(A) \wedge A(x) \wedge A(y) \leqslant A(xy), \quad \mathscr{R}(A) \wedge A(x) \wedge A(y) \leqslant A(x-y).$$

特别地, 有

$$\mathscr{R}(A) \wedge A(y) \leqslant A(0),$$

$$\mathscr{R}(A) \wedge A(y) \leqslant \mathscr{R}(A) \wedge A(0) \wedge A(y) \leqslant A(-y),$$

$$\mathscr{R}(A) \wedge A(y) = \mathscr{R}(A) \wedge A(-y),$$

$$\mathscr{R}(A) \wedge A(x) \wedge A(y) = \mathscr{R}(A) \wedge A(x) \wedge A(-y).$$

这表明

$$\mathscr{R}(A) \leqslant A(0) \wedge \bigwedge_{x,y \in R} (A(x) \wedge A(y) \mapsto A(xy)),$$

$$\mathscr{R}(A) \leqslant \bigwedge_{x,y \in R} (A(x) \wedge A(y) \mapsto A(x+y)) \wedge \bigwedge_{y \in R} (A(y) \mapsto A(-y)).$$

所以

$$\mathscr{R}(A) \leqslant A(0) \wedge \bigwedge_{x,y \in R} ((A(x) \wedge A(y)) \mapsto A(xy))$$

$$\wedge \bigwedge_{x,y \in R} ((A(x) \wedge A(y)) \mapsto A(x+y)) \wedge \bigwedge_{y \in R} (A(y) \mapsto A(-y)).$$

类似地, 可以证明

$$\mathscr{R}(A) \geqslant A(0) \wedge \bigwedge_{x,y \in R} ((A(x) \wedge A(y)) \mapsto A(xy))$$

$$\wedge \bigwedge_{x,y\in R} ((A(x)\wedge A(y))\mapsto A(x+y)) \wedge \bigwedge_{y\in R} (A(y)\mapsto A(-y)).$$

所以定理得证. □

由以上定义及蕴含运算可知下述引理是显然成立的.

引理 5.1.4 令 A 为环 R 中的 L-模糊子集. 则 $\mathscr{R}(A) \geqslant a$ 当且仅当对于任意的 $x, y \in R$,

$$A(0) \geqslant a, \quad A(x)\wedge A(y)\wedge a \leqslant A(xy), \quad A(x)\wedge A(y)\wedge a \leqslant A(x-y).$$

接下来我们考察 L-模糊子环度的等价刻画问题.

定理 5.1.5 令 A 为环 R 中的 L-模糊子集. 则
(1) $\mathscr{R}(A) = \bigvee\{a\in L \mid a\leqslant A(0), A(x)\wedge A(y)\wedge a\leqslant A(xy)\wedge A(x-y), \forall x,y\in R\}$;
(2) $\mathscr{R}(A) = \bigvee\{a\in L \mid \forall b\leqslant a\leqslant A(0), A_{[b]}$ 是环 R 的子环$\}$;
(3) $\mathscr{R}(A) = \bigvee\{a\in L \mid a\leqslant A(0), \forall b\notin\alpha(a), A^{[b]}$ 是环 R 的子环$\}$;
(4) $\mathscr{R}(A) = \bigvee\{a\in L \mid a\leqslant A(0), \forall b\in P(L), b\not\geqslant a, A^{(b)}$ 是环 R 的子环$\}$;
(5) 当 L 满足条件: $\forall a,b\in L, \beta(a\wedge b) = \beta(a)\cap\beta(b)$ 时, 有

$$\mathscr{R}(A) = \bigvee\{a\in L \mid a\leqslant A(0), \forall b\in\beta(a), A_{(b)} \text{ 是环 } R \text{ 的子环}\}.$$

证明 从定义 5.1.1 和蕴含运算的性质不难得到 (1).

下证 (2). 假设 $\forall x,y\in R$, 有

$$a\leqslant A(0), \quad A(x)\wedge A(y)\wedge a\leqslant A(xy)\wedge A(x-y).$$

则 $\forall b\leqslant a, x,y\in A_{[b]}$, 有

$$A(xy)\geqslant A(x)\wedge A(y)\wedge a\geqslant b, \quad A(x-y)\geqslant A(x)\wedge A(y)\wedge a\geqslant b,$$

这表明 $xy\in A_{[b]}, x-y\in A_{[b]}$. 因此 $A_{[b]}$ 是 R 的子环. 于是由 (1) 可知

$$\mathscr{R}(A) = \bigvee\{a\in L \mid a\leqslant A(0), A(x)\wedge A(y)\wedge a\leqslant A(xy)\wedge A(x-y), \forall x,y\in R\}$$
$$\leqslant \bigvee\{a\in L \mid \forall b\leqslant a\leqslant A(0), A_{[b]} \text{ 是 } R \text{ 的子环}\}.$$

反之, 设 $a\in L$ 使得 $a\leqslant A(0), \forall b\leqslant a$, 都有 $A_{[b]}$ 是 R 的子环. 对于任意的 $x,y\in R$, 令 $b = A(x)\wedge A(y)\wedge a$, 则 $b\leqslant a$, 且 $x,y\in A_{[b]}$, 于是有 $xy, x-y\in A_{[b]}$, 因此

$$A(xy)\geqslant b = A(x)\wedge A(y)\wedge a, \quad A(x-y)\geqslant b = A(x)\wedge A(y)\wedge a.$$

这意味着

$$\mathscr{R}(A) = \bigvee\{a \in L \mid a \leqslant A(0),\ A(x) \wedge A(y) \wedge a \leqslant A(xy) \wedge A(x-y), \forall x,y \in R\}$$
$$\geqslant \bigvee\{a \in L \mid \forall b \leqslant a \leqslant A(0),\ A_{[b]} \text{ 是 } R \text{ 的子环}\}.$$

这就证明了 (2) 成立.

为了证明 (3), 假设 $\forall x, y \in R$, 有

$$a \leqslant A(0), \quad A(x) \wedge A(y) \wedge a \leqslant A(xy) \wedge A(x-y).$$

则 $\forall b \notin \alpha(a), x, y \in A^{[b]}$, 有

$$b \notin \alpha(A(x)) \cup \alpha(A(y)) \cup \alpha(a) = \alpha(A(x) \wedge A(y) \wedge a).$$

由

$$A(x) \wedge A(y) \wedge a \leqslant A(xy), \quad A(x) \wedge A(y) \wedge a \leqslant A(x-y),$$

可以知道

$$\alpha(A(xy)) \subseteq \alpha(A(x) \wedge A(y) \wedge a), \quad \alpha(A(x-y)) \subseteq \alpha(A(x) \wedge A(y) \wedge a),$$

因此 $b \notin \alpha(A(xy))$ 且 $b \notin \alpha(A(x-y))$, 也就是 $xy, x-y \in A^{[b]}$. 这表明 $A^{[b]}$ 是 R 的子环, 并且

$$a \in \{a \in L \mid a \leqslant A(0),\ \forall b \notin \alpha(a), A^{[b]} \text{ 是 } R \text{ 的子环}\}.$$

于是有

$$\mathscr{R}(A) = \bigvee\{a \in L \mid a \leqslant A(0),\ A(x) \wedge A(y) \wedge a \leqslant A(xy) \wedge A(x-y), \forall x,y \in R\}$$
$$\leqslant \bigvee\{a \in L \mid a \leqslant A(0),\ \forall b \notin \alpha(a), A^{[b]} \text{ 是 } R \text{ 的子环}\}.$$

相反地, 设

$$a \in \{a \in L \mid a \leqslant A(0),\ \forall b \notin \alpha(a), A^{[b]} \text{ 是 } R \text{ 的子环}\}.$$

现在证明对于任意的 $x, y \in R$,

$$A(x) \wedge A(y) \wedge a \leqslant A(xy), \quad A(x) \wedge A(y) \wedge a \leqslant A(x-y).$$

假设 $b \notin \alpha(A(x) \wedge A(y) \wedge a)$. 由

$$\alpha(A(x) \wedge A(y) \wedge a) = \alpha(A(x)) \cup \alpha(A(y)) \cup \alpha(a),$$

5.1 L-模糊子环度

可以知道 $b \notin \alpha(a)$, $x,y \in A^{[b]}$. 因为 $A^{[b]}$ 是 R 的子环, 所以 $xy, x-y \in A^{[b]}$, 也就是说 $b \notin \alpha(A(xy))$, $b \notin \alpha(A(x-y))$. 这表明

$$A(x) \wedge A(y) \wedge a \leqslant A(xy), \quad A(x) \wedge A(y) \wedge a \leqslant A(x-y).$$

这就证明了

$$\mathscr{R}(A) = \bigvee\{a \in L \mid a \leqslant A(0), A(x) \wedge A(y) \wedge a \leqslant A(xy) \wedge A(x-y), \forall x, y \in R\}$$
$$\geqslant \bigvee\{a \in L \mid a \leqslant A(0), \forall b \notin \alpha(a), A^{[b]} \text{ 是 } R \text{ 的子环}\}.$$

所以 (3) 是成立的.

为了证明 (4), 假设 $a \in L$, 且对于任意的 $x, y \in R$, 有

$$a \leqslant A(0), \quad A(x) \wedge A(y) \wedge a \leqslant A(xy), \quad A(x) \wedge A(y) \wedge a \leqslant A(x-y).$$

令 $b \in P(L)$, $b \not\geqslant a$ 且 $x, y \in A^{(b)}$. 现在证明 $xy, x-y \in A^{(b)}$. 若 $xy \notin A^{(b)}$ 或者 $x - y \notin A^{(b)}$, 也就是 $A(xy) \leqslant b$ 或者 $A(x-y) \leqslant b$, 则

$$A(x) \wedge A(y) \wedge a \leqslant A(xy) \leqslant b \quad \text{或者} \quad A(x) \wedge A(y) \wedge a \leqslant A(x-y) \leqslant b.$$

由 $b \in P(L)$ 和 $x, y \in A^{(b)}$, 我们得到 $a \leqslant b$, 这与 $b \not\geqslant a$ 矛盾. 因此 $xy, x-y \in A^{(b)}$. 这表明 $A^{(b)}$ 是 R 的子环. 于是

$$\mathscr{R}(A) = \bigvee\{a \in L \mid a \leqslant A(0), A(x) \wedge A(y) \wedge a \leqslant A(xy) \wedge A(x-y), \forall x, y \in R\}$$
$$\leqslant \bigvee\{a \in L \mid a \leqslant A(0), \forall b \in P(L), b \not\geqslant a, A^{(b)} \text{ 是 } R \text{ 的子环}\}.$$

相反地, 假设

$$a \in \{a \in L \mid a \leqslant A(0), \forall b \in P(L), b \not\geqslant a, A^{(b)} \text{ 是 } R \text{ 的子环}\}.$$

现在证明对于任意的 $x, y \in R$,

$$A(x) \wedge A(y) \wedge a \leqslant A(xy), \quad A(x) \wedge A(y) \wedge a \leqslant A(x-y).$$

令 $b \in P(L)$, $A(x) \wedge A(y) \wedge a \not\leqslant b$. 则 $A(x) \not\leqslant b$, $A(y) \not\leqslant b$, $a \not\leqslant b$, 也就是 $x, y \in A^{(b)}$. 因此 $A^{(b)}$ 是 R 的子环, 所以 $xy \in A^{(b)}$, $x - y \in A^{(b)}$, 即 $A(xy) \not\leqslant b$, $A(x-y) \not\leqslant b$. 这表明

$$A(x) \wedge A(y) \wedge a \leqslant A(xy), \quad A(x) \wedge A(y) \wedge a \leqslant A(x-y).$$

因此

$$\mathscr{R}(\mu) = \bigvee\{a \in L \mid a \leqslant A(0), A(x) \wedge A(y) \wedge a \leqslant A(xy) \wedge A(x-y), \forall x, y \in R\}$$

$$\geqslant \bigvee \{a \in L \mid a \leqslant A(0), \forall b \in P(L), b \not\geqslant a, A^{(b)} \text{ 是 } R \text{ 的子环}\}.$$

所以 (4) 是成立的.

为了证明 (5), 假设

$$a \in \{a \in L \mid a \leqslant A(0),\ A(x) \wedge A(y) \wedge a \leqslant A(xy) \wedge A(x-y),\ \forall x, y \in R\}.$$

则 $\forall b \in \beta(a),\ x, y \in A_{(b)}$, 有

$$b \in \beta(A(x)) \cap \beta(A(y)) \cap \beta(a) = \beta(A(x) \wedge A(y) \wedge a) \subseteq \beta(A(xy)),$$

$$b \in \beta(A(x)) \cap \beta(A(y)) \cap \beta(a) = \beta(A(x) \wedge A(y) \wedge a) \subseteq \beta(A(x-y)),$$

即 $xy, x - y \in A_{(b)}$. 所以 $A_{(b)}$ 是 R 的子环. 这表明

$$\mathscr{R}(A) = \bigvee \{a \in L \mid a \leqslant A(0),\ A(x) \wedge A(y) \wedge a \leqslant A(xy) \wedge A(x-y), \forall x, y \in R\}$$

$$\leqslant \bigvee \{a \in L \mid a \leqslant A(0), \forall b \in \beta(a), A_{(b)} \text{ 是 } R \text{ 的子环}\}.$$

相反地, 假设

$$a \in \{a \in L \mid a \leqslant A(0),\ \forall b \in \beta(a), A_{(b)} \text{ 是 } R \text{ 的子环}\}.$$

现在证明对于任意的 $x, y \in R$,

$$A(x) \wedge A(y) \wedge a \leqslant A(xy), \quad A(x) \wedge A(y) \wedge a \leqslant A(x-y).$$

令 $b \in \beta(A(x) \wedge A(y) \wedge a)$. 由

$$\beta(A(x) \wedge A(y) \wedge a) = \beta(A(x)) \cap \beta(A(y)) \cap \beta(a),$$

可以知道 $x, y \in A_{(b)}, b \in \beta(a)$. 因此 $A_{(b)}$ 是 R 的子环, 所以 $xy, x - y \in A_{(b)}$, 即 $b \in \beta(A(xy)), b \in \beta(A(x-y))$. 这表明

$$A(x) \wedge A(y) \wedge a \leqslant A(xy), \quad A(x) \wedge A(y) \wedge a \leqslant A(x-y).$$

因此

$$\mathscr{R}(A) = \bigvee \{a \in L \mid a \leqslant A(0),\ A(x) \wedge A(y) \wedge a \leqslant A(xy) \wedge A(x-y), \forall x, y \in R\}$$

$$\geqslant \{a \in L \mid a \leqslant A(0),\ \forall b \in \beta(a), A_{(b)} \text{ 是 } R \text{ 的子环}\}.$$

所以 (5) 是成立的. □

当 $\mathscr{R}(A) = 1$ 时, 上述定理可以退化为下面推论形式.

推论 5.1.6 令 A 为环 R 中的 L-模糊子集. 则下列条件等价:

(1) A 是环 R 的 L-模糊子环;

(2) $\forall b \in L$, $A_{[b]}$ 是环 R 的子环;

(3) $\forall b \in L$, $A^{[b]}$ 是环 R 的子环;

(4) $\forall b \in P(L)$, $A^{(b)}$ 是环 R 的子环;

(5) 当 L 满足条件: $\forall a, b \in L$, $\beta(a \wedge b) = \beta(a) \cap \beta(b)$ 时, $\forall b \in \beta(1)$, $A_{(b)}$ 是环 R 的子环.

定理 5.1.7 设 $\{A_i\}_{i \in \Omega}$ 是环 R 的 L-模糊子集族, 则

$$\mathscr{R}\left(\bigwedge_{i \in \Omega} A_i\right) \geqslant \bigwedge_{i \in \Omega} \mathscr{R}(A_i).$$

证明 设 $a \leqslant \bigwedge_{i \in \Omega} \mathscr{R}(A_i)$. 则 $\forall i \in \Omega, \forall x, y \in R$, 都有下面不等式成立:

$$a \leqslant A_i(0), \quad A_i(x) \wedge A_i(y) \wedge a \leqslant A_i(xy), \quad A_i(x) \wedge A_i(y) \wedge a \leqslant A_i(x-y).$$

因此

$$a \leqslant \bigwedge_{i \in \Omega} A_i(0), \quad \bigwedge_{i \in \Omega} A_i(x) \wedge \bigwedge_{i \in \Omega} A_i(y) \wedge a \leqslant \bigwedge_{i \in \Omega} A_i(xy),$$

$$\bigwedge_{i \in \Omega} A_i(x) \wedge \bigwedge_{i \in \Omega} A_i(y) \wedge a \leqslant \bigwedge_{i \in \Omega} A_i(x-y).$$

这表明 $a \leqslant \mathscr{R}\left(\bigwedge_{i \in \Omega} A_i\right)$. 所以有 $\mathscr{R}\left(\bigwedge_{i \in \Omega} A_i\right) \geqslant \bigwedge_{i \in \Omega} \mathscr{R}(A_i)$ 成立. □

5.2 L-模糊子环的运算

在这一节中, 我们给出 L-模糊子环内积、直和的表现定理, 然后讨论 L-模糊子环度的一些运算性质.

在定义 4.5.2 中, 我们曾研究了群中 L-模糊子集的运算, 那里曾经引入了 L-模糊集合的笛卡儿积和内积. 下面我们再引入环中 L-模糊子集的内积与直和的概念, 并研究这些运算的性质. 注意此处的内积已经不同于定义 4.5.2, 尽管符号相同.

定义 5.2.1 设 A 和 B 是环 R 的两个 L-模糊子集. $\forall x \in R$, 定义 R 的 L-模糊子集 AB 使得它满足

$$(AB)(x) = \bigvee_{x=\sum\limits_{i\in\Lambda} y_i z_i} \left(\bigwedge_{i\in\Lambda} (A(y_i) \wedge B(z_i)) \right),$$

在上述的表达式中, Λ 表示 Ω 的有限子集, 称 AB 为 A 和 B 在环 R 内的积.

定义 5.2.2 设 $\{A_i \mid i \in \Omega\}$ 是环 R 的一族 L-模糊子集. 定义 R 的 L-模糊子集 $\sum\limits_{i\in\Omega} A_i$ 使得

$$\left(\sum_{i\in\Omega} A_i\right)(x) = \bigvee_{x=\sum\limits_{i\in\Lambda} x_i} \left(\bigwedge_{i\in\Lambda} A_i(x_i) \right),$$

在上述的表达式中, Λ 表示 Ω 的有限子集, 称 $\sum\limits_{i\in\Omega} A_i$ 为 $\{A_i \mid i \in \Omega\}$ 在环 R 内的和.

下面定理给出 L-模糊子环内积的截集表现形式.

定理 5.2.3 设 R 是环, $A, B \in L^R$. 则下列诸条件成立:

(1) $\forall a \in L, (AB)_{(a)} \subseteq A_{(a)} B_{(a)} \subseteq A_{[a]} B_{[a]} \subseteq (AB)_{[a]}$.

(2) $\forall a \in L, (AB)^{(a)} \subseteq A^{(a)} B^{(a)} \subseteq A^{[a]} B^{[a]} \subseteq (AB)^{[a]}$; 特别地, 当 $a \in P(L)$ 时, 有 $(AB)^{(a)} = A^{(a)} B^{(a)}$ 成立.

(3) $AB = \bigvee\limits_{a\in L} \{a \wedge (A_{[a]} B_{[a]})\} = \bigvee\limits_{a\in M(L)} \{a \wedge (A_{[a]} B_{[a]})\}$.

(4) $AB = \bigvee\limits_{a\in L} \{a \wedge (A_{(a)} B_{(a)})\} = \bigvee\limits_{a\in M(L)} \{a \wedge (A_{(a)} B_{(a)})\}$.

(5) $AB = \bigwedge\limits_{a\in L} \{a \vee (A^{[a]} B^{[a]})\} = \bigwedge\limits_{a\in P(L)} \{a \vee (A^{[a]} B^{[a]})\}$.

(6) $AB = \bigwedge\limits_{a\in L} \{a \vee (A^{(a)} B^{(a)})\} = \bigwedge\limits_{a\in P(L)} \{a \vee (A^{(a)} B^{(a)})\}$.

证明 (1) 对于任意的 $a \in L$, 首先证明 $(AB)_{(a)} \subseteq A_{(a)} B_{(a)}$. 由

$$x \in (AB)_{(a)}$$

$$\Rightarrow a \in \beta(AB(x)) = \beta \left(\bigvee_{x=\sum\limits_{i\in\Lambda} y_i z_i} \left(\bigwedge_{i\in\Lambda} (A(y_i) \wedge B(z_i)) \right) \right)$$

5.2 L-模糊子环的运算

$$= \bigcup_{x=\sum\limits_{i\in\Lambda} y_i z_i} \beta\left(\bigwedge_{i\in\Lambda}(A(y_i)\wedge B(z_i))\right)$$

$$\subseteq \bigcup_{x=\sum\limits_{i\in\Lambda} y_i z_i} \bigcap_{i\in\Lambda}(\beta(A(y_i))\cap \beta(B(z_i)))$$

可以知道存在有限集 $\Lambda \subseteq \Omega$ 和 $y_i, z_i \in R$, $\forall i \in \Lambda$ 使得 $x = \sum\limits_{i\in\Lambda} y_i z_i$, $a \in \beta(A(y_i))$ 且 $a \in \beta(B(z_i))$, 即 $y_i \in A_{(a)}$ 且 $z_i \in B_{(a)}$, 所以 $x = \sum\limits_{i\in\Lambda} y_i z_i \in A_{(a)}B_{(a)}$. 这表明 $(AB)_{(a)} \subseteq A_{(a)}B_{(a)}$.

$A_{(a)}B_{(a)} \subseteq A_{[a]}B_{[a]}$ 是显然的.

现在证明 $A_{[a]}B_{[a]} \subseteq (AB)_{[a]}$. 设 $x \in A_{[a]}B_{[a]}$. 则由

$$A_{[a]}B_{[a]} = \left\{x \,\middle|\, 存在\ \Omega\ 的有限子集\ \Lambda\ 和\ y_i \in A_{[a]}, z_i \in B_{[a]}\ 使得\ x = \sum_{i\in\Lambda} y_i z_i\right\}$$

可知

$$(AB)(x) = \bigvee_{x=\sum\limits_{i\in\Lambda} y_i z_i}\left\{\bigwedge_{i\in\Lambda}(A(y_i)\wedge B(z_i))\right\}$$

$$= \bigvee_{x=\sum\limits_{i\in\Lambda} y_i z_i}\left\{\bigwedge_{i\in\Lambda}(a\wedge a)\right\} \geqslant a.$$

这意味着 $x \in (AB)_{[a]}$. 这就证明了 $A_{[a]}B_{[a]} \subseteq (AB)_{[a]}$.

(2) $(AB)^{(a)} \subseteq A^{(a)}B^{(a)}$ 可由以下蕴含得到.

$x \in (AB)^{(a)}$

$\Rightarrow (AB)(x) \nleq a$

$\Rightarrow \bigvee\limits_{x=\sum\limits_{i\in\Lambda} y_i z_i}\left(\bigwedge\limits_{i\in\Lambda}(A(y_i)\wedge B(z_i))\right) \nleq a$

$\Rightarrow$ 存在有限集 Λ 和 $y_i, z_i \in R$ 使得 $x = \sum\limits_{i\in\Lambda} y_i z_i, A(y_i) \nleq a, B(z_i) \nleq a$

$\Rightarrow$ 存在有限集 Λ 和 $y_i, z_i \in R$ 使得 $x = \sum\limits_{i\in\Lambda} y_i z_i, y_i \in A^{(a)}, z_i \in B^{(a)}$

$\Rightarrow x \in A^{(a)}B^{(a)}$.

特别地，若 $a \in P(L)$，则上述蕴含式的逆也成立. 从而 $(AB)^{(a)} = A^{(a)}B^{(a)}$.

显然有 $A^{(a)}B^{(a)} \subseteq A^{[a]}B^{[a]}$. 接下来证明 $A^{[a]}B^{[a]} \subseteq (AB)^{[a]}$. 设 $x \notin (AB)^{[a]}$. 则 $a \in \alpha((AB)(x))$. 由

$$a \in \alpha((AB)(x)) = \alpha\left(\bigvee_{x=\sum_{i\in\Lambda} y_i z_i} \left(\bigwedge_{i\in\Lambda}(A(y_i) \wedge B(z_i))\right)\right)$$

$$\subseteq \alpha\left(\bigwedge_{i\in\Lambda}(A(y_i) \wedge B(z_i))\right)$$

$$= \bigcup_{i\in\Lambda}(\alpha(A(y_i)) \cup \alpha(B(z_i)))$$

可知对任何一个 x 的一个表达 $x = \sum_{i\in\Lambda} y_i z_i$，均存在 $i \in \Lambda$ 使得 $a \in \alpha(A(y_i))$ 或 $a \in \alpha(B(z_i))$，也就是 $y_i \notin A^{[a]}$ 或 $z_i \notin B^{[a]}$，这意味着 $x \notin A^{[a]}B^{[a]}$，这样就得到了 $A^{[a]}B^{[a]} \subseteq A^{[a]}B^{[a]}$ 的证明.

由 (1), (2) 和定理 2.1.4 可以得到 (3), (4), (5) 和 (6) 成立. □

下面给出 L-模糊子环直和的表现定理.

定理 5.2.4 设 R 是一个环，$\{A_i \mid i \in \Omega\}$ 是环 R 的一族 L-模糊子环. 则下列条件成立:

(1) $\forall a \in L, \left(\sum_{i\in\Omega} A_i\right)_{(a)} \subseteq \sum_{i\in\Omega}(A_i)_{(a)} \subseteq \sum_{i\in\Omega}(A_i)_{[a]} \subseteq \left(\sum_{i\in\Omega} A_i\right)_{[a]}$.

(2) $\forall a \in L, \left(\sum_{i\in\Omega} A_i\right)^{(a)} \subseteq \sum_{i\in\Omega}(A_i)^{(a)} \subseteq \sum_{i\in\Omega}(A_i)^{[a]} \subseteq \left(\sum_{i\in\Omega} A_i\right)^{[a]}$; 特别地, 若 $a \in P(L)$, 则 $\left(\sum_{i\in\Omega} A_i\right)^{(a)} = \sum_{i\in\Omega}(A_i)^{(a)}$.

(3) $\sum_{i\in\Omega} A_i = \bigvee_{a\in L}\left\{a \wedge \left(\sum_{i\in\Omega}(A_i)_{[a]}\right)\right\} = \bigvee_{a\in M(L)}\left\{a \wedge \left(\sum_{i\in\Omega}(A_i)_{[a]}\right)\right\}$.

(4) $\sum_{i\in\Omega} A_i = \bigvee_{a\in L}\left\{a \wedge \left(\sum_{i\in\Omega}(A_i)_{(a)}\right)\right\} = \bigvee_{a\in M(L)}\left\{a \wedge \left(\sum_{i\in\Omega}(A_i)_{(a)}\right)\right\}$.

(5) $\sum_{i\in\Omega} A_i = \bigwedge_{a\in L}\left\{a \vee \left(\sum_{i\in\Omega}(A_i)^{[a]}\right)\right\} = \bigwedge_{a\in P(L)}\left\{a \vee \left(\sum_{i\in\Omega}(A_i)^{[a]}\right)\right\}$.

(6) $\sum_{i\in\Omega} A_i = \bigwedge_{a\in L}\left\{a \vee \left(\sum_{i\in\Omega}(A_i)^{(a)}\right)\right\} = \bigwedge_{a\in P(L)}\left\{a \vee \left(\sum_{i\in\Omega}(A_i)^{(a)}\right)\right\}$.

5.2 L-模糊子环的运算

证明 (1) 对于任意的 $a \in L$, 首先证明 $\left(\sum_{i \in \Omega} A_i\right)_{(a)} \subseteq \sum_{i \in \Omega} (A_i)_{(a)}$. 设 $x \in \left(\sum_{i \in \Omega} A_i\right)_{(a)}$, 则由

$$a \in \beta\left(\sum_{i \in \Omega} A_i(x)\right) = \beta\left(\bigvee_{x = \sum_{i \in \Lambda} x_i} \left(\bigwedge_{i \in \Lambda} A_i(x_i)\right)\right)$$

$$= \bigcup_{x = \sum_{i \in \Lambda} x_i} \beta\left(\bigwedge_{i \in \Lambda} A_i(x_i)\right)$$

$$\subseteq \bigcup_{x = \sum_{i \in \Lambda} x_i} \bigcap_{i \in \Lambda} (\beta(A_i(x_i)))$$

可知存在有限个 x_i 使得 $x = \sum_{i \in \Omega} x_i$, 且 $a \in \beta(A_i(x_i))$, 即 $x_i \in (A_i)_{(a)}, \forall i \in \Lambda$. 所以有 $x = \sum_{i \in \Omega} x_i \in \sum_{i \in \Omega} (A_i)_{(a)}$. 这表明 $\left(\sum_{i \in \Omega} A_i\right)_{(a)} \subseteq \sum_{i \in \Omega} (A_i)_{(a)}$.

$\sum_{i \in \Omega} (A_i)_{(a)} \subseteq \sum_{i \in \Omega} (A_i)_{[a]}$ 是显然的.

最后我们只需要证明 $\sum_{i \in \Omega} (A_i)_{[a]} \subseteq \left(\sum_{i \in \Omega} A_i\right)_{[a]}$. 假设 $x \in \sum_{i \in \Omega} (A_i)_{[a]}$. 由

$$\sum_{i \in \Omega} (A_i)_{[a]} = \left\{ x \,\middle|\, 存在 \,\Omega\, 的有限子集 \,\Lambda\, 和 \,x_i \in (A_i)_{[a]}\, 使得 \,x = \sum_{i \in \Lambda} x_i \right\}$$

可得

$$\left(\sum_{i \in \Omega} A_i\right)(x) = \bigvee_{x = \sum_{i \in \Lambda} x_i} \left(\bigwedge_{i \in \Lambda} A_i(x_i)\right) = \bigvee_{x = \sum_{i \in \Lambda} x_i} \left(\bigwedge_{i \in \Lambda} a\right) \geqslant a,$$

那就是 $x \in \left(\sum_{i \in \Omega} A_i\right)_{[a]}$, 这表明 $\sum_{i \in \Omega} (A_i)_{[a]} \subseteq \left(\sum_{i \in \Omega} A_i\right)_{[a]}$.

(2) $\left(\sum_{i \in \Omega} A_i\right)^{(a)} \subseteq \sum_{i \in \Omega} (A_i)^{(a)}$ 的证明可由下面蕴含得到.

$$x \in \left(\sum_{i \in \Omega} A_i\right)^{(a)} \Rightarrow \left(\sum_{i \in \Lambda} A_i\right)(x) \not\leq a$$

$$\Rightarrow \bigvee_{x=\sum_{i\in\Lambda} x_i} \left(\bigwedge_{i\in\Lambda} A_i(x_i)\right) \nleq a$$

$$\Rightarrow \exists x_i \in R \text{ 和有限集 } \Lambda \text{ 使得 } x = \sum_{i\in\Lambda} x_i \text{ 且 } A_i(x_i) \nleq a, \forall i \in \Lambda$$

$$\Rightarrow \exists x_i \in (A_i)^{(a)} \text{ 和有限集 } \Lambda \text{ 使得 } x = \sum_{i\in\Lambda} x_i$$

$$\Rightarrow x \in \sum_{i\in\Omega} (A_i)^{(a)}.$$

特别地, 若 $a \in P(L)$, 则上述证明的逆蕴含成立, 所以有 $\left(\sum_{i\in\Omega} A_i\right)^{(a)} = \sum_{i\in\Omega} (A_i)^{(a)}$.

显然 $\sum_{i\in\Omega} (A_i)^{(a)} \subseteq \sum_{i\in\Omega} (A_i)^{[a]}$. 现在证明 $\sum_{i\in\Omega} (A_i)^{[a]} \subseteq \left(\sum_{i\in\Omega} A_i\right)^{[a]}$.

假设 $x \notin \left(\sum_{i\in\Omega} A_i\right)^{[a]}$. 则 $a \in \alpha\left(\left(\sum_{i\in\Omega} A_i\right)(x)\right)$. 由

$$a \in \alpha\left(\left(\sum_{i\in\Omega} A_i(x)\right)(x)\right) = \alpha\left(\bigvee_{x=\sum_{i\in\Lambda} x_i} \left(\bigwedge_{i\in\Lambda} A_i(x_i)\right)\right)$$

$$\subseteq \alpha\left(\bigwedge_{i\in\Lambda} A_i(x_i)\right) \left(\text{对任何有限和 } x = \sum_{i\in\Lambda} x_i\right)$$

$$= \bigcup_{i\in\Lambda} \alpha(A_i(x_i))$$

可知对任何有限和 $x = \sum_{i\in\Lambda} x_i$, 都存在 $i \in \Lambda$ 使得 $a \in \alpha(A_i(x_i))$, 也就是 $x_i \notin (A_i)^{[a]}$. 这意味着 $x \notin \sum_{i\in\Omega} (A_i)^{[a]}$. 这表明 $\sum_{i\in\Omega} (A_i)^{[a]} \subseteq \left(\sum_{i\in\Omega} A_i\right)^{[a]}$.

从 (1), (2) 和定理 2.1.4 可以得到 (3), (4), (5) 和 (6). □

接下来讨论 L-模糊子环度的内积与和运算.

定理 5.2.5 设 A, B 是交换环 R 的两个 L-模糊子集. 则

$$\mathscr{R}(AB) \geqslant \mathscr{R}(A) \wedge \mathscr{R}(B).$$

证明 **方法 1**. 由定理 5.1.5(4) 可以得到以下结果:

5.2 L-模糊子环的运算

$\mathscr{R}(AB)$
$= \bigvee \{a \in L \mid a \leqslant (AB)(0), \forall b \in P(L), b \not\geqslant a, (AB)^{(b)} \text{ 是 } R \text{ 的子环}\}$
$= \bigvee \{a \in L \mid a \leqslant (AB)(0), \forall b \in P(L), b \not\geqslant a, A^{(b)}B^{(b)} \text{ 是 } R \text{ 的子环}\}$ (由 5.2.3(2))
$\geqslant (\bigvee \{a \in L \mid a \leqslant A(0), \forall b \in P(L), b \not\geqslant a, A^{(b)} \text{ 是 } R \text{ 的子环}\})$
$\quad \wedge (\bigvee \{a \in L \mid a \leqslant B(0), \forall b \in P(L), b \not\geqslant a, B^{(b)} \text{ 是 } R \text{ 的子环}\})$ (由分明结论)
$= \mathscr{R}(A) \wedge \mathscr{R}(B)$.

方法 2. 设 $\mathscr{R}(A) \wedge \mathscr{R}(B) \geqslant a$. 则 $\forall x, y \in R$, 有
$$a \leqslant A(0), \quad a \leqslant B(0),$$
$$A(x) \wedge A(y) \wedge a \leqslant A(x-y) \wedge A(xy),$$
$$B(x) \wedge B(y) \wedge a \leqslant B(x-y) \wedge B(xy).$$

于是

$(AB)(x) \wedge (AB)(y) \wedge a$

$= \left\{ \bigvee_{x=\sum\limits_{i\in\Lambda} u_i v_i} \left(\bigwedge_{i\in\Lambda} (A(u_i) \wedge B(v_i)) \right) \right\} \wedge \left\{ \bigvee_{y=\sum\limits_{i\in\Gamma} s_i t_i} \left(\bigwedge_{i\in\Gamma} (A(s_i) \wedge B(t_i)) \right) \right\} \wedge a$

$= \bigvee\limits_{x=\sum\limits_{i\in\Lambda} u_i v_i, y=\sum\limits_{j\in\Gamma} s_j t_j} \left\{ \left(\bigwedge_{i\in\Lambda} (A(u_i) \wedge B(v_i)) \right) \wedge \left(\bigwedge_{j\in\Gamma} (A(s_j) \wedge B(t_j)) \right) \right\} \wedge a$

$= \bigvee\limits_{x=\sum\limits_{i\in\Lambda,j\in\Gamma} u_i v_i s_j t_j} \left\{ \bigwedge_{i\in\Lambda} \bigwedge_{j\in\Gamma} (A(u_i) \wedge A(s_j) \wedge B(v_i) \wedge B(t_j) \wedge a) \right\}$

$= \bigvee\limits_{xy=\sum\limits_{i\in\Lambda,j\in\Gamma} (u_i s_j)(v_i t_j)} \left\{ \bigwedge_{i\in\Lambda} \bigwedge_{j\in\Gamma} (A(u_i) \wedge A(s_j) \wedge B(v_i) \wedge B(t_j) \wedge a) \right\}$

$\leqslant \bigvee\limits_{xy=\sum\limits_{i\in\Lambda,j\in\Gamma} (u_i s_j)(v_i t_j)} \left\{ \bigwedge_{i\in\Lambda,j\in\Gamma} (A(u_i s_j) \wedge B(v_i t_j) \wedge a) \right\}$

$\leqslant (AB)(xy)$.

再来证明
$$(AB)(x) \wedge (AB)(y) \wedge a \leqslant (AB)(x+y).$$

$(AB)(x) \wedge (AB)(y) \wedge a$

$$= \left\{ \bigvee_{x=\sum\limits_{i\in\Lambda} u_i v_i} \left(\bigwedge_{i\in\Lambda} (A(u_i) \wedge B(v_i)) \right) \right\} \wedge \left\{ \bigvee_{y=\sum\limits_{i\in\Gamma} s_i t_i} \left(\bigwedge_{i\in\Gamma} (A(s_i) \wedge B(t_i)) \right) \right\} \wedge a$$

$$= \bigvee_{x=\sum\limits_{i\in\Lambda} u_i v_i, y=\sum\limits_{j\in\Gamma} s_j t_j} \left\{ \left(\bigwedge_{i\in\Lambda} (A(u_i) \wedge B(v_i)) \right) \wedge \left(\bigwedge_{j\in\Gamma} (A(s_j) \wedge B(t_j)) \right) \right\} \wedge a$$

$$= \bigvee_{x+y=\sum\limits_{i\in\Lambda} u_i v_i + \sum\limits_{j\in\Gamma} s_j t_j} \left\{ \bigwedge_{i\in\Lambda} \bigwedge_{j\in\Gamma} (A(u_i) \wedge A(s_j) \wedge B(v_i) \wedge B(t_j) \wedge a) \right\}$$

$$= \bigvee_{x+y=\sum\limits_{i\in\Lambda\cup\Gamma} p_i q_i} \left\{ \bigwedge_{i\in\Lambda} \bigwedge_{j\in\Gamma} (A(u_i) \wedge A(s_j) \wedge B(v_i) \wedge B(t_j) \wedge a) \right\}$$

$$\leqslant \bigvee_{x+y=\sum\limits_{i\in\Lambda\cup\Gamma} p_i q_i} \left\{ \bigwedge_{i\in\Lambda\cup\Gamma} (A(p_i) \wedge B(q_i) \wedge a) \right\}$$

$$\leqslant (AB)(x+y).$$

另外, 容易证明 $a \leqslant (AB)(0), (AB)(x) \wedge a \leqslant (AB)(-x)$.

这表明 $a \leqslant \mathscr{R}(AB)$, 所以 $\mathscr{R}(AB) \geqslant \mathscr{R}(A) \wedge \mathscr{R}(B)$. □

下面定理的证明是类似的, 证明留给读者.

定理 5.2.6 设 $\{A_i \mid i \in \Omega\}$ 是环 R 的一族 L-模糊子集, 它们的和是 $\sum\limits_{i\in\Omega} A_i$, 则

$$\mathscr{R}\left(\sum_{i\in\Omega} A_i\right) \geqslant \bigwedge_{i\in\Omega} \mathscr{R}(A_i).$$

5.3 由 L-模糊子环度确定的 L-模糊凸结构

在这部分中, 首先研究 L-模糊子环度与 L-模糊凸结构之间的关系, 其次研究 L-模糊子集的同态像和原像的 L-模糊子环度, 最后证明环同态是 L-模糊凸保持映射和 L-模糊凸到凸的映射.

5.3 由 L-模糊子环度确定的 L-模糊凸结构

对于每个 $A \in L^R$, $\mathscr{R}(A)$ 可以看作由 $A \mapsto \mathscr{R}(A)$ 确定的映射 $\mathscr{R}: L^R \to L$. 下面的定理说明 $\mathscr{R}$ 恰好是 R 上的一个 L-模糊凸结构.

定理 5.3.1 设 R 是一个环. 则由 $A \mapsto \mathscr{R}(A)$ 所确定的映射 $\mathscr{R}: L^R \to L$ 是 R 上的 L-模糊凸结构, 称之为在 R 上由 L-模糊子环度确定的 L-模糊凸结构.

证明 (1) 显然有 $\mathscr{R}(\chi_\varnothing) = \mathscr{R}(\chi_R) = 1$.

(2) 设 $\{A_i\}_{i \in \Omega}$ 是环 R 的 L-模糊子集族. 由定理 5.1.7 可知

$$\mathscr{R}\left(\bigwedge_{i \in \Omega} A_i\right) \geqslant \bigwedge_{i \in \Omega} \mathscr{R}(A_i).$$

(3) 设 $\{A_i \mid i \in \Omega\} \subseteq L^R$ 是非空和全序的. 为了证明 $\bigwedge_{i \in \Omega} \mathscr{R}(A_i) \leqslant \mathscr{R}\left(\bigvee_{i \in \Omega} A_i\right)$, 需要证明对于任意 $a \leqslant \bigwedge_{i \in \Omega} \mathscr{R}(A_i)$, 有 $a \leqslant \mathscr{R}\left(\bigvee_{i \in \Omega} A_i\right)$. 由 $a \leqslant \bigwedge_{i \in \Omega} \mathscr{R}(A_i)$ 和引理 5.1.4 可知, 对于任意的 $i \in \Omega$ 与 $\forall x, y \in R$, 有

$$a \leqslant A_i(0), \quad A_i(x) \wedge A_i(y) \wedge a \leqslant A_i(xy), \quad A_i(x) \wedge A_i(y) \wedge a \leqslant A_i(x - y),$$

设 $b \in L$ 且其满足下面条件

$$b \prec \left(\bigvee_{i \in \Omega} A_i(x)\right) \wedge \left(\bigvee_{i \in \Omega} A_i(y)\right) \wedge a,$$

然后有

$$b \prec \bigvee_{i \in \Omega} A_i(x), \quad b \prec \bigvee_{i \in \Omega} A_i(y), \quad \text{且 } b \leqslant a.$$

因此存在 $i, j \in \Omega$ 使得 $b \leqslant A_i(x)$, $b \leqslant A_j(y)$, 且 $b \leqslant a$. 因为 $\{A_i \mid i \in \Omega\}$ 是全序的, 不妨假设 $A_j \leqslant A_i$, 则 $b \leqslant A_i(x) \wedge A_i(y) \wedge a$. 由

$$A_i(x) \wedge A_i(y) \wedge a \leqslant A_i(xy), \quad A_i(x) \wedge A_i(y) \wedge a \leqslant A_i(x - y),$$

我们可以得到 $b \leqslant A_i(xy)$ 且 $b \leqslant A_i(x - y)$. 进一步可以得到

$$b \leqslant \bigvee_{i \in \Omega} A_i(xy) \quad \text{且} \quad b \leqslant \bigvee_{i \in \Omega} A(x - y).$$

由 b 的任意性可得下面两个不等式:

$$\left(\bigvee_{i \in \Omega} A_i(x)\right) \wedge \left(\bigvee_{i \in \Omega} A_i(y)\right) \wedge a \leqslant \bigvee_{i \in \Omega} A_i(xy),$$

$$\left(\bigvee_{i\in\Omega}A_i(x)\right)\wedge\left(\bigvee_{i\in\Omega}A_i(y)\right)\wedge a\leqslant\bigvee_{i\in\Omega}A_i(x-y).$$

再根据引理 5.1.4, 便知有 $a\leqslant\mathscr{R}\left(\bigvee_{i\in\Omega}A_i\right)$ 成立. 从而由 a 的任意性, 可以得到

$$\bigwedge_{i\in\Omega}\mathscr{R}(A_i)\leqslant\mathscr{R}\left(\bigvee_{i\in\Omega}A_i\right).$$

综上可知, $\mathscr{R}$ 是 R 上的 L-模糊凸结构. □

现在研究环 R 的 L-模糊子集同态像和原像的 L-模糊子环度.

定理 5.3.2 设 $f:R\to R'$ 是环同态, $A\in L^R, B\in L^{R'}$. 则

(1) $\mathscr{R}(f_L^{\to}(A))\geqslant\mathscr{R}(A)$; 若 f 是单射, 则 $\mathscr{R}(f_L^{\to}(A))=\mathscr{R}(A)$.

(2) $\mathscr{R}(f_L^{\leftarrow}(B))\geqslant\mathscr{R}(B)$; 若 f 是满射, 则 $\mathscr{R}(f_L^{\leftarrow}(B))=\mathscr{R}(B)$.

证明 (1) 这能够由定理 5.1.5 和下面不等式得到.

$\mathscr{R}(f_L^{\to}(A))$

$=\bigvee\{a\in L\mid a\leqslant f_L^{\to}(A)(0),\forall b\in P(L),b\not\geqslant a,(f_L^{\to}(A))^{(b)}$ 是 R 的子环$\}$

$=\bigvee\left\{a\in L\;\middle|\;a\leqslant\bigvee_{f(r)=0}A(r),\forall b\in P(L),b\not\geqslant a,f_L^{\to}(A^{(b)})\text{ 是 }R\text{ 的子环}\right\}$

(这里用到了 $(f_L^{\to}(A))^{(b)}=f_L^{\to}(A^{(b)})$)

$\geqslant\bigvee\{a\in L\mid a\leqslant A(0),\forall b\in P(L),b\not\geqslant a,A^{(b)}$ 是 R 的子环$\}$

$=\mathscr{R}(A).$

若 f 是单射, 则上述 $\geqslant$ 可换成 $=$. 因此 $\mathscr{R}(A)=\mathscr{R}(f_L^{\to}(A))$.

(2) 由定理 5.1.5 和下面不等式同样可以得到 (2).

$\mathscr{R}(f_L^{\leftarrow}(B))=\bigvee\{a\in L\mid\forall b\leqslant a\leqslant f_L^{\leftarrow}(B)(0),(f_L^{\leftarrow}(B))_{[b]}$ 是 R 的子环$\}$

$=\bigvee\{a\in L\mid\forall b\leqslant a\leqslant B(0),f_L^{\leftarrow}(B_{[b]})$ 是 R 的子环$\}$

(这里用到了 $(f_L^{\leftarrow}(B))_{[b]}=f_L^{\leftarrow}(A_{[b]})$)

$\geqslant\bigvee\{a\in L\mid\forall b\leqslant a\leqslant B(0),B_{[b]}$ 是 R 的子环$\}$

$=\mathscr{R}(B).$

若 f 是满射, 则上述 $\geqslant$ 可以换成 $=$. 因此可以得到 $\mathscr{R}(B)=\mathscr{R}(f_L^{\leftarrow}(B))$. □

由定理 5.3.2, 自然可以得到下面定理.

定理 5.3.3 设 $f: R \to R'$ 是环同态. 令 $\mathscr{R}_R$ 和 $\mathscr{R}_{R'}$ 分别表示 R 和 R' 上由 L-模糊子环度确定的 L-模糊凸结构. 则 $f: (R, \mathscr{R}_R) \to (R', \mathscr{R}_{R'})$ 是 L-模糊凸保持映射和 L-模糊凸到凸的映射.

定理 5.3.4 设 $\{R_i \mid i \in \{1, 2, \cdots, n\}\}$ 是一族环, $\forall i \in \{1, 2, \cdots, n\}$, $A_i \in L^{R_i}$. 则

$$\mathscr{R}\left(\prod_{i=1}^{n} A_i\right) \geqslant \bigwedge_{i=1}^{n} \mathscr{R}(A_i).$$

证明 容易验证 $\prod_{i=1}^{n} A_i = \bigwedge_{i=1}^{n} P_i^{-1}(A_i)$, 其中 P_i 是 $\prod_{i=1}^{n} R_i$ 到 R_i 的投射. 由定理 5.1.7, 可以得到

$$\mathscr{R}\left(\prod_{i=1}^{n} A_i\right) = \mathscr{R}\left(\bigwedge_{i=1}^{n} P_i^{-1}(A_i)\right) \geqslant \bigwedge_{i=1}^{n} \mathscr{R}\left(P_i^{-1}(A_i)\right).$$

因为 $P_i: \prod_{i=1}^{n} R_i \to R_i$ 是环同态, 因此由定理 5.3.2 可知有 $\mathscr{R}\left(P_i^{-1}(A_i)\right) \geqslant \mathscr{R}(A_i)$. 于是便有

$$\mathscr{R}\left(\prod_{i=1}^{n} A_i\right) \geqslant \bigwedge_{i=1}^{n} \mathscr{R}(A_i). \qquad \square$$

推论 5.3.5 设 $\{R_i \mid i \in \{1, 2, \cdots, n\}\}$ 是一族环, $\forall i \in \{1, 2, \cdots, n\}$, $A_i \in L^{R_i}$. 若 $\forall i \in \{1, 2, \cdots, n\}$, A_i 是 R_i 的 L-模糊子环, 则 $\prod_{i=1}^{n} A_i$ 是 $\prod_{i=1}^{n} R_i$ 的 L-模糊子环.

由定理 5.3.2 可以得到下面推论.

推论 5.3.6 设 $\{R_i \mid i \in \{1, 2, \cdots, n\}\}$ 是一族环, $\mathscr{R}, \mathscr{R}_i$ 分别是 $\prod_{i=1}^{n} R_i, R_i$ 上由 L-模糊子环度确定的 L-模糊凸结构. 则 $P_i: \left(\prod_{i=1}^{n} R_i, \mathscr{R}\right) \to (R_i, \mathscr{R}_i)$ 是 L-模糊凸保持映射和 L-模糊凸到凸映射.

5.4 环上的 L-模糊同余关系

在第 4 章中, 我们介绍了群中的模糊同余关系. 本节考虑环上的 L-模糊同余关系. 我们将证明任意给定一个 L-模糊同余关系, 都可得到一个 L-模糊理想, 反

之, 任意给定一个 L-模糊理想, 也可导出一个 L-模糊同余关系, 从而证明了环上的 L-模糊理想和 L-模糊同余关系是一一对应的.

定义 5.4.1 设 R 是环, 一个从 R 到 R 的 L-模糊等价关系 E 叫做 R 上的一个 L-模糊左 (右) 同余关系. 如果 $\forall a,b,c,x \in R$, 下列条件 (L1)(或 (R1)) 成立:

(L1) $E(a,b) = E(c+a, c+b)$ 且 $E(a,b) \leqslant E(xa, xb)$;

(R1) $E(a,b) = E(a+c, b+c)$ 且 $E(a,b) \leqslant E(ax, bx)$.

若 R 既是 L-模糊左同余关系又是右同余关系, 则称 R 是环 R 上的 L-模糊同余关系.

定理 5.4.2 设 E 是环 R 上的 L-模糊等价关系. 那么下列条件是等价的:

(1) E 是环 R 上的 L-模糊 (左, 右) 同余关系;

(2) $\forall a \in L, E_{[a]}$ 是环 R 上的一个 (左, 右) 同余关系;

(3) $\forall a \in J(L), E_{[a]}$ 是环 R 上的一个 (左, 右) 同余关系;

(4) $\forall a \in L, E^{[a]}$ 是环 R 上的一个 (左, 右) 同余关系;

(5) $\forall a \in P(L), E^{[a]}$ 是环 R 上的一个 (左, 右) 同余关系;

(6) $\forall a \in P(L), E^{(a)}$ 是环 R 上的一个 (左, 右) 同余关系.

证明 由定理 2.4.9, 结论显然成立. □

定理 5.4.3 若 E 是环 R 上的 L-模糊同余关系. 定义 $A \in L^R$ 使得 $\forall x \in R$, $A(x) = E(0, x)$, 这里 0 为环中零元. 则 A 满足以下三个条件:

(1) $A(0) = 1$;

(2) $A(x-y) \geqslant A(x) \wedge A(y)$;

(3) $A(xy) \wedge A(yx) \geqslant A(y)$.

也就是说 A 为环 R 的一个 L-模糊理想.

证明 (1) 和 (2) 由定理 4.1.5 可以得到, 我们只需要证明 (3) 即可. $\forall x, y \in R$, 由

$$A(xy) = E(0, xy) = E(x0, xy) \geqslant E(0, y) = A(y),$$
$$A(yx) = E(0, yx) = E(0x, yx) \geqslant E(0, y) = A(y)$$

可知 (3) 成立, 因此 A 为环 R 的一个 L-模糊理想. □

定理 5.4.4 令 A 是环 R 上的 L-模糊理想, 定义 $E \in L^{R \times R}$ 使得 $\forall x, y \in R$, $E(x,y) = A(x-y)$, 则 E 是环 R 上的一个 L-模糊同余关系.

证明 $\forall x \in R$, 显然 $E(x,x) = A(0) = 1$. 另外, $\forall x, y \in R$,

$$R(x,y) = A(x-y) = A(-(y-x)) = A(y-x) = R(y,x).$$

再者, $\forall x, y, z \in R$, 由

$$E(x,z) = A(x-z) = \bigvee_{y \in R} A(x-y+y-z)$$
$$\geqslant \bigvee_{y \in R} \{A(x-y) \wedge A(y-z)\} = \bigvee_{y \in R} \{E(x,y) \wedge E(y,z)\}$$
$$= (E \circ E)(x,z)$$

可知 E 是一个 L-模糊等价关系.

下证 E 是一个 L-模糊同余关系. 显然

$$E(a,b) = A(a-b) = A((c+a)-(c+b)) = E(c+a, c+b),$$

又

$$E(a,b) = A(a-b) \leqslant A(x(a-b)) \wedge A((a-b)x)$$
$$= A(xa-xb) \wedge A(ax-bx) = E(xa,xb) \wedge E(ax,bx),$$

所以 E 是一个 L-模糊同余关系. $\square$

5.5 L-模糊理想度

我们知道环的理想是非常重要的概念, 因此对它的模糊化也是非常必要的. 实际上关于 L-模糊理想及 L-模糊素理想的概念早有引入, 这里我们对它们进一步模糊化.

首先给出 L-模糊理想度的定义如下:

定义 5.5.1 设 A 是环 R 的非零 L-模糊子集. A 的 L-模糊理想度 $\mathscr{I}(A)$ 被定义如下:

$$\mathscr{I}(A) = A(0) \wedge \bigwedge_{x,z \in R} (A(x) \mapsto A(xz) \wedge A(zx)) \wedge \bigwedge_{x,y \in R} (A(x) \wedge A(y) \mapsto A(x-y))$$
$$= \mathscr{R}(A) \wedge \bigwedge_{x,z \in R} (A(x) \mapsto A(xz) \wedge A(zx)).$$

上述定义的意义就在于给定一个环的 L-模糊子集, 它未必是一个 L-模糊理想, 但是在某种程度 $\mathscr{I}(A)$ 上, A 是一个 L-模糊理想.

显然, $\mathscr{I}(A) = 1$ 当且仅当 $\forall x, y, z \in R$, 皆有

$$A(0) = 1, \quad A(x) \leqslant A(xz) \wedge A(zx), \quad A(x) \wedge A(y) \leqslant A(x-y),$$

也就是 A 是一个 L-模糊理想. 因此 L-模糊理想度是 L-模糊理想的进一步模糊化.

下面我们来看几个例子.

例 5.5.2 令 $\mathbb{Z}_3$ 是模 3 的剩余类环, 并定义 $\mathbb{Z}_3$ 上的运算为常规的加法和乘法. 定义三个 L-模糊集 $A, B, C \in L^{\mathbb{Z}_3}$ 如下:

$$A(z) = \begin{cases} 0.8, & z = [0], \\ 0.3, & z = [1], \\ 0.4, & z = [2], \end{cases}$$

$$B(z) = \begin{cases} 0, & z = [0], \\ 1, & z = [1], \\ 0.4, & z = [2], \end{cases}$$

$$C(z) = \begin{cases} 0.2, & z = [0], \\ 0.2, & z = [1], \\ 0.2, & z = [2]. \end{cases}$$

容易验证

$$\mathscr{I}(A) = 0.3, \quad \mathscr{I}(B) = 0, \quad \mathscr{I}(C) = 0.2.$$

下面我们给出 L-模糊理想度的等价表达式. 类似于定理 5.1.3 的证明可以得到下面定理.

定理 5.5.3 设 A 是环 R 的 L-模糊子集, 则

$$\mathscr{I}(A) = A(0) \wedge \bigwedge_{x,z \in R} (A(x) \mapsto A(xz) \wedge A(zx))$$

$$\wedge \bigwedge_{x,y \in R} (A(x) \wedge A(y) \mapsto A(x+y)) \wedge \bigwedge_{y \in R} (A(y) \mapsto A(-y)).$$

由定义及蕴含运算可知下述引理是显然成立的.

引理 5.5.4 设 A 是环 R 的 L-模糊子集, 则 $\mathscr{I}(A) \geqslant a$ 当且仅当 $\forall x, y, z \in R$, $a \leqslant A(0)$, $A(x) \wedge a \leqslant A(xz) \wedge A(zx)$, $A(x) \wedge A(y) \wedge a \leqslant A(x - y)$.

结合引理 5.5.4, 易得 $\mathscr{I}(A)$ 的等价刻画如下.

定理 5.5.5 设 A 是环 R 的 L-模糊子集. 则

$$\mathscr{I}(A) = \bigvee \left\{ a \in L \,\middle|\, \begin{array}{l} a \leqslant A(0), A(x) \wedge a \leqslant A(xz) \wedge A(zx), \\ A(x) \wedge A(y) \wedge a \leqslant A(x-y), \forall x, y, z \in R \end{array} \right\}.$$

5.5 L-模糊理想度

下面利用四种截集给出 L-模糊理想度的等价刻画.

定理 5.5.6 设 A 是环 R 的 L-模糊子集. 则

(1) $\mathscr{I}(A) = \bigvee\{a \in L \mid \forall b \leqslant a \leqslant A(0), A_{[b]}$ 为环 R 的理想$\}$;

(2) $\mathscr{I}(A) = \bigvee\{a \in L \mid a \leqslant A(0), \forall b \notin \alpha(a), A^{[b]}$ 为环 R 的理想$\}$;

(3) $\mathscr{I}(A) = \bigvee\{a \in L \mid a \leqslant A(0), \forall b \in P(L), b \not\geqslant a, A^{(b)}$ 为环 R 的理想$\}$;

(4) 当 L 满足条件: $\forall a, b \in L, \beta(a \wedge b) = \beta(a) \cap \beta(b)$ 时, 有

$$\mathscr{I}(A) = \bigvee\{a \in L \mid a \leqslant A(0), \forall b \in \beta(a), A_{(b)} \text{ 为环 } R \text{ 的理想}\}.$$

证明 (1) 我们从引理 5.5.4 出发来证明 (1). 为此假设 $\forall x, y, z \in R$,

$$A(0) \geqslant a, \quad A(x) \wedge a \leqslant A(xz) \wedge A(zx), \quad A(x) \wedge A(y) \wedge a \leqslant A(x-y).$$

则对于任意的 $b \leqslant a$, $x, y \in A_{[b]}$, 有

$$A(xz) \wedge A(zx) \geqslant A(x) \wedge a \geqslant b,$$

$$A(x-y) \geqslant A(x) \wedge A(y) \wedge a \geqslant A(x) \wedge A(y) \wedge b \geqslant b,$$

也就是说 $xz \in A_{[b]}$, $zx \in A_{[b]}$, $x-y \in A_{[b]}$. 因此 $A_{[b]}$ 为 R 的理想. 即证得

$$\mathscr{I}(A) = \bigvee\left\{a \in L \;\middle|\; \begin{array}{l} A(0) \geqslant a, A(x) \wedge a \leqslant A(xz) \wedge A(zx), \\ A(x) \wedge A(y) \wedge a \leqslant A(x-y), \forall x, y, z \in R \end{array}\right\}$$

$$\leqslant \bigvee\{a \in L \mid \forall b \leqslant a \leqslant A(0), A_{[b]} \text{ 为环 } R \text{ 的理想}\}.$$

反之, 假设 $a \in L$ 且 $\forall b \leqslant a \leqslant A(0)$, $A_{[b]}$ 为 R 的理想. 对于任意的 $x, y, z \in R$, 令 $b = A(x) \wedge A(y) \wedge a$, 则 $b \leqslant a$ 且 $x, y \in A_{[b]}$, 于是有 $x - y \in A_{[b]}$, 也就是

$$A(x-y) \geqslant b = A(x) \wedge A(y) \wedge a.$$

为了证明

$$A(xy) \wedge A(yx) \geqslant A(x) \wedge a,$$

设 $A(x) \wedge a = c$, 则 $c \leqslant a$ 且 $x \in A_{[c]}$, 从而由 $A_{[c]}$ 是理想可知 $\forall z \in R$, 都有 $xz, zx \in A_{[c]}$, 也就是

$$A(xz) \wedge A(zx) \geqslant c = A(x) \wedge a,$$

这表明

$$\mathscr{I}(A) = \bigvee\left\{a \in L \;\middle|\; \begin{array}{l} a \leqslant A(0), A(x) \wedge a \leqslant A(xz) \wedge A(zx), \\ A(x) \wedge A(y) \wedge a \leqslant A(x-y), \forall x, y, z \in R \end{array}\right\}$$

$$\geqslant \bigvee \{a \in L \mid \forall b \leqslant a, A_{[b]} \text{ 为环 } R \text{ 的理想}\}.$$

这样就证明 (1) 成立.

下证 (2). 假设 $\forall x, y, z \in R$, 有

$$a \leqslant A(0), \quad A(x) \wedge a \leqslant A(xz) \wedge A(zx), \quad A(x) \wedge A(y) \wedge a \leqslant A(x-y).$$

则 $\forall b \notin \alpha(a), x, y \in A^{[b]}$, 都有 $b \notin \alpha(A(x))$ 且 $b \notin \alpha(A(y))$. 从而由

$$b \notin \alpha(A(x)) \cup \alpha(a) = \alpha(A(x) \wedge a),$$
$$b \notin \alpha(A(x)) \cup \alpha(A(y)) \cup \alpha(a) = \alpha(A(x) \wedge A(y) \wedge a)$$

和

$$A(x) \wedge a \leqslant A(xz) \wedge A(zx), \quad A(x) \wedge A(y) \wedge a \leqslant A(x-y)$$

可以知道

$$\alpha(A(xz)) \cup \alpha(A(zx)) \subseteq \alpha(A(x) \wedge a), \quad \alpha(A(x-y)) \subseteq \alpha(A(x) \wedge A(y) \wedge a),$$

这意味着 $b \notin \alpha(A(xz)), b \notin \alpha(A(zx))$ 且 $b \notin \alpha(A(x-y))$, 也就是 $xz, zx, x-y \in A^{[b]}$. 这表明 $A^{[b]}$ 是 R 的一个理想, 于是有

$$a \in \{a \in L \mid a \leqslant A(0), \forall b \notin \alpha(a), A^{[b]} \text{ 是 } R \text{ 的理想}\}.$$

进一步可得

$$\mathscr{I}(A) = \bigvee \left\{ a \in L \;\middle|\; \begin{array}{l} a \leqslant A(0), A(x) \wedge a \leqslant A(xz) \wedge A(zx), \\ A(x) \wedge A(y) \wedge a \leqslant A(x-y), \forall x, y \in R \end{array} \right\}$$
$$\leqslant \bigvee \{a \in L \mid a \leqslant A(0), \forall b \notin \alpha(a), A^{[b]} \text{ 是 } R \text{ 的理想}\}.$$

相反地, 设

$$a \in \{a \in L \mid a \leqslant A(0), \forall b \notin \alpha(a), A^{[b]} \text{ 是 } R \text{ 的理想}\}.$$

现在来证明对于任意的 $x, y, z \in R$,

$$a \leqslant A(0), \quad A(x) \wedge a \leqslant A(xz) \wedge A(zx), \quad A(x) \wedge A(y) \wedge a \leqslant A(x-y).$$

我们仅证明 $A(x) \wedge a \leqslant A(xz) \wedge A(zx)$, 另一个不等式类似可证. 为此假设 $b \notin \alpha(A(x) \wedge a)$. 由

$$\alpha(A(x) \wedge a) = \alpha(A(x)) \cup \alpha(a)$$

5.5 L-模糊理想度

可以知道 $b \notin \alpha(a)$ 且 $x \in A^{[b]}$. 因为 $A^{[b]}$ 是 R 的理想, 所以 $xz, zx \in A^{[b]}$, 也就是说
$$b \notin \alpha(A(xz)) \cup \alpha(A(zx)) = \alpha(A(xz) \wedge A(zx)),$$
这表明
$$A(x) \wedge a \leqslant A(xz) \wedge A(zx).$$
这就证明了
$$\mathscr{I}(A) = \bigvee \left\{ a \in L \;\middle|\; \begin{array}{l} a \leqslant A(0), A(x) \wedge a \leqslant A(xz) \wedge A(zx), \\ A(x) \wedge A(y) \wedge a \leqslant A(x-y), \forall x, y \in R \end{array} \right\}$$
$$\geqslant \bigvee \{a \in L \mid \forall b \notin \alpha(a), A^{[b]} \text{ 是 } R \text{ 的理想}\}.$$
所以 (2) 是成立的;

为了证明 (3), 假设 $a \in L$, 且对于任意的 $x, y, z \in R$, 有
$$a \leqslant A(0), \quad A(x) \wedge a \leqslant A(xz) \wedge A(zx), \quad A(x) \wedge A(y) \wedge a \leqslant A(x-y).$$

令 $b \in P(L)$, $b \not\geqslant a$ 且 $x, y \in \mu^{(b)}$. 现在证明 $xz, zx, x-y \in A^{(b)}$. 若 $xz \notin A^{(b)}$, 或者 $zx \notin A^{(b)}$, 或者 $x-y \notin \mu^{(b)}$, 也就是 $A(xz) \leqslant b$, 或者 $A(zx) \leqslant b$, 或者 $A(x-y) \leqslant b$, 则
$$A(x) \wedge a \leqslant A(xz) \wedge A(zx) \leqslant b, \quad \text{或者} \quad A(x) \wedge A(y) \wedge a \leqslant A(x-y) \leqslant b.$$
由 $b \in P(L)$ 和 $x, y \in A^{(b)}$, 我们得到 $a \leqslant b$, 这与 $b \not\geqslant a$ 矛盾. 因此 $xz, zx, x-y \in A^{(b)}$. 这表明 $A^{(b)}$ 是 R 的理想. 于是
$$\mathscr{I}(A) = \bigvee \left\{ a \in L \;\middle|\; \begin{array}{l} A(0) \geqslant a, A(x) \wedge a \leqslant A(xz) \wedge A(zx), \\ A(x) \wedge A(y) \leqslant A(x-y), \forall x, y, z \in R \end{array} \right\}$$
$$\leqslant \bigvee \{a \in L \mid a \leqslant A(0), \forall b \in P(L), b \not\geqslant a, A^{(b)} \text{ 是 } R \text{ 的理想}\}.$$
相反地, 假设
$$a \in \{a \in L \mid a \leqslant A(0), \forall b \in P(L), b \not\geqslant a, A^{(b)} \text{ 是 } R \text{ 的理想}\}.$$
现在来证明对于任意的 $x, y, z \in R$,
$$A(x) \wedge a \leqslant A(xz) \wedge A(zx), \quad A(x) \wedge A(y) \wedge a \leqslant A(x-y).$$
我们仅证明 $A(x) \wedge a \leqslant A(xz) \wedge A(zx)$, 另一个不等式证明是类似的. 令 $b \in P(L)$ 且 $A(x) \wedge a \not\leqslant b$, 则 $A(x) \not\leqslant b$ 且 $a \not\leqslant b$, 此时有 $x \in A^{(b)}$ 且 $A^{(b)}$ 是 R 的理想, 所以

$xz, zx \in A^{(b)}$, 即 $A(xz) \nleqslant b$ 且 $A(zx) \nleqslant b$. 由 b 是素元可知 $A(xz) \wedge A(zx) \nleqslant b$, 这表明
$$A(x) \wedge a \leqslant A(xz) \wedge A(zx).$$

因此
$$\mathscr{I}(\mu) = \bigvee \left\{ a \in L \,\middle|\, \begin{array}{l} a \leqslant A(0), A(x) \wedge a \leqslant A(xz) \wedge A(zx), \\ A(x) \wedge A(y) \wedge a \leqslant A(x-y), \forall x, y \in R \end{array} \right\}$$
$$\geqslant \bigvee \{a \in L \mid a \leqslant A(0), \forall b \in P(L), b \ngeqslant a, A^{(b)} \text{ 是 } R \text{ 的理想}\}.$$

所以 (3) 是成立的.

为了证明 (4), 假设
$$a \in \left\{ a \in L \,\middle|\, \begin{array}{l} a \leqslant A(0), A(x) \wedge a \leqslant A(xz) \wedge A(zx), \\ A(x) \wedge A(y) \wedge a \leqslant A(x-y), \forall x, y \in R \end{array} \right\}.$$

则 $\forall b \in \beta(a), x, y \in A_{(b)}$, 有
$$b \in \beta(A(x)) \cap \beta(a) = \beta(A(x) \wedge a)$$
$$\subseteq \beta(A(xz) \wedge A(zx)) = \beta(A(xz)) \cap \beta(A(zx))$$

和
$$b \in \beta(A(x)) \cap \beta(A(y)) \cap \beta(a) = \beta(A(x) \wedge A(y) \wedge a)$$
$$\subseteq \beta(A(x-y)),$$

即 $xz, zx, x-y \in A_{(b)}$. 所以 $A_{(b)}$ 是 R 的理想. 这表明
$$\mathscr{I}(A) = \bigvee \left\{ a \in L \,\middle|\, \begin{array}{l} a \leqslant A(0), A(x) \wedge a \leqslant A(xz) \wedge A(zx), \\ A(x) \wedge A(y) \wedge a \leqslant A(x-y), \forall x, y \in R \end{array} \right\}$$
$$\leqslant \bigvee \{a \in L \mid a \leqslant A(0), \forall b \in \beta(a), A_{(b)} \text{ 是 } R \text{ 的理想}\}.$$

相反地, 假设
$$a \in \{a \in L \mid a \leqslant A(0), \forall b \in \beta(a), A_{(b)} \text{ 是 } R \text{ 的理想}\}.$$

现在来证明对于任意的 $x, y, z \in R$,
$$A(x) \wedge a \leqslant A(xz) \wedge A(zx), \quad A(x) \wedge A(y) \wedge a \leqslant A(x-y).$$

我们仅证明第一个不等式, 第二个可类似地证明. 令 $b \in \beta(A(x) \wedge a)$. 由

$$\beta(A(x) \wedge a) = \beta(A(x)) \cap \beta(a)$$

可以知道 $x \in A_{(b)}$ 且 $b \in \beta(a)$. 因此 $A_{(b)}$ 是 R 的理想, 所以 $xz, zx \in A_{(b)}$, 即 $b \in \beta(A(xz))$ 且 $b \in \beta(A(zx))$, 这意味着

$$b \in \beta(A(xz)) \cap \beta(A(zx)) = \beta(A(xz) \wedge A(zx)),$$

这表明

$$A(x) \wedge a \leqslant A(xz) \wedge A(zx).$$

因此

$$\mathscr{I}(A) = \bigvee \left\{ a \in L \;\middle|\; \begin{array}{l} a \leqslant A(0), A(x) \wedge a \leqslant A(xz) \wedge A(zx), \\ A(x) \wedge A(y) \wedge a \leqslant A(x-y), \forall x, y \in R \end{array} \right\}$$

$$\geqslant \{a \in L \mid a \leqslant A(0), \forall b \in \beta(a), A_{(b)} \text{ 是 } R \text{ 的理想}\}.$$

所以 (4) 是成立的. $\square$

推论 5.5.7 令 A 为环 R 中的 L-模糊子集. 则下列条件等价:
(1) A 是环 R 的 L-模糊理想;
(2) $\forall b \in L$, $A_{[b]}$ 是环 R 的理想;
(3) $\forall b \in L$, $A^{[b]}$ 是环 R 的理想;
(4) $\forall b \in P(L)$, $A^{(b)}$ 是环 R 的理想;
(5) 当 L 满足条件: $\forall a, b \in L$, $\beta(a \wedge b) = \beta(a) \cap \beta(b)$ 时, $\forall b \in \beta(1)$, $A_{(b)}$ 是环 R 的理想.

5.6 由 L-模糊理想度确定的 L-模糊凸结构

在这一节中, 我们研究 L-模糊理想度与 L-模糊凸结构之间的关系. 对环 R 的任一 L-模糊子集 $A \in L^R$, $\mathscr{I}$ 可以看作是由 $A \longmapsto \mathscr{I}(A)$ 所确定的映射 $\mathscr{I}: L^R \to L$. 下面的定理表明 $\mathscr{I}$ 实际上是 R 上的 L-模糊凸结构.

定理 5.6.1 设 R 是一个环. 则映射 $\mathscr{I}: L^R \to L$ 是 R 上的一个 L-模糊凸结构, 称为由 L-模糊理想度确定的 L-模糊凸结构.

证明 (1) 显然可知 $\mathscr{I}(\chi_\varnothing) = \mathscr{I}(\chi_R) = 1$.
(2) 设 $\{A_i \mid i \in \Omega\} \subseteq L^R$ 是非空的, 为证明

$$\bigwedge_{i \in \Omega} \mathscr{I}(A_i) \leqslant \mathscr{I}\left(\bigwedge_{i \in \Omega} A_i\right),$$

需证明对于任意的 $a \leqslant \bigwedge_{i\in\Omega} \mathscr{I}(A_i)$, 都有 $a \leqslant \mathscr{I}\left(\bigwedge_{i\in\Omega} A_i\right)$. 由引理 5.5.4, $a \leqslant \bigwedge_{i\in\Omega} \mathscr{I}(A_i)$ 意味着 $\forall i \in \Omega, \forall x, y, z \in R$, 均有下面不等式成立:

$$a \leqslant A_i(0), \quad A_i(x) \wedge a \leqslant A_i(xz) \wedge A_i(zx), \quad A_i(x) \wedge A_i(y) \wedge a \leqslant A_i(x-y).$$

于是

$$a \leqslant \left(\bigwedge_{i\in\Omega} A_i\right)(0), \quad \left(\bigwedge_{i\in\Omega} A_i(x)\right) \wedge a \leqslant \left(\bigwedge_{i\in\Omega} A_i\right)(xz) \wedge \left(\bigwedge_{i\in\Omega} A_i\right)(zx)$$

且

$$\left(\bigwedge_{i\in\Omega} A_i(x)\right) \wedge \left(\bigwedge_{i\in\Omega} A_i(y)\right) \wedge a \leqslant \bigwedge_{i\in\Omega}(A_i(x) \wedge A_i(y) \wedge a) \leqslant \left(\bigwedge_{i\in\Omega} A_i\right)(x-y).$$

根据引理 5.5.4, 可以得到 $a \leqslant \mathscr{I}\left(\bigwedge_{i\in\Omega} A_i\right)$, 由 a 的任意性, 可得

$$\bigwedge_{i\in\Omega} \mathscr{I}(A_i) \leqslant \mathscr{I}\left(\bigwedge_{i\in\Omega} A_i\right).$$

(3) 设 $\{A_i \mid i \in \Omega\} \subseteq L^R$ 是非空全序的, 下面来证明

$$\bigwedge_{i\in\Omega} \mathscr{I}(A_i) \leqslant \mathscr{I}\left(\bigvee_{i\in\Omega} A_i\right),$$

这需要证明

$$a \leqslant \bigwedge_{i\in\Omega} \mathscr{I}(A_i) \Rightarrow a \leqslant \mathscr{I}\left(\bigvee_{i\in\Omega} A_i\right), \quad \forall a \in L.$$

设 $a \leqslant \bigwedge_{i\in\Omega} \mathscr{I}(A_i)$, 根据引理 5.5.4, 对于任意的 $i \in \Omega, x, y, z \in R$, 均有

$$a \leqslant A_i(0), \quad A_i(x) \wedge a \leqslant A_i(xz) \wedge A_i(zx), \quad A_i(x) \wedge A_i(y) \wedge a \leqslant A_i(x-y).$$

由 $A_i(x) \wedge a \leqslant A_i(xz) \wedge A_i(zx)$ 容易得到

$$\left(\bigvee_{i\in\Omega} A_i\right)(x) \wedge a \leqslant \bigvee_{i\in\Omega}(A_i(xz) \wedge A_i(zx)) \leqslant \left(\bigvee_{i\in\Omega} A_i\right)(xz) \wedge \left(\bigvee_{i\in\Omega} A_i\right)(zx).$$

为了证明

$$\left(\bigvee_{i\in\Omega}A_i\right)(x)\wedge\left(\bigvee_{i\in\Omega}A_i\right)(y)\wedge a\leqslant\left(\bigvee_{i\in\Omega}A_i\right)(x-y),$$

任取 $b\in L$ 使得

$$b\prec\left(\bigvee_{i\in\Omega}A_i\right)(x)\wedge\left(\bigvee_{i\in\Omega}A_i\right)(y)\wedge a,$$

则有

$$b\prec\bigvee_{i\in\Omega}A_i(x),\ b\prec\bigvee_{i\in\Omega}A_i(y)\ \text{且}\ b\leqslant a.$$

因此存在 $i,j\in\Omega$ 使得 $b\leqslant A_i(x), b\leqslant A_j(y)$ 且 $b\leqslant a$. 又因为 $\{A_i\mid i\in\Omega\}$ 是全序的, 不妨设 $A_j\leqslant A_i$, 则有

$$b\leqslant A_i(x)\wedge A_j(y)\wedge a\leqslant A_i(x)\wedge A_i(y)\wedge a.$$

根据 $A_i(x)\wedge A_i(y)\wedge a\leqslant A_i(x-y)$ 能够得到 $b\leqslant A_i(x-y)$, 所以有 $b\leqslant\left(\bigvee_{i\in\Omega}A_i\right)(x-y)$. 由 b 的任意性, 可以得到

$$\left(\bigvee_{i\in\Omega}A_i\right)(x)\wedge\left(\bigvee_{i\in\Omega}A_i\right)(y)\wedge a\leqslant\left(\bigvee_{i\in\Omega}A_i\right)(x-y).$$

$a\leqslant\left(\bigvee_{i\in\Omega}A_i\right)(0)$ 是显然的. 结合引理 5.5.4, 可以得到 $a\leqslant\mathscr{I}\left(\bigvee_{i\in\Omega}A_i\right)$. 再由 a 的任意性, 可知

$$\bigwedge_{i\in\Omega}\mathscr{I}(A_i)\leqslant\mathscr{I}\left(\bigwedge_{i\in\Omega}A_i\right).$$

这样我们就证明了 $\mathscr{I}$ 是 R 上的 L-模糊凸结构. □

下面我们来研究 L-模糊子集的同态像和原像的 L-模糊理想度.

定理 5.6.2 设 $f:R\to R'$ 为环同态, $A\in L^R$ 且 $B\in L^{R'}$. 则

(1) 如果 f 是满射, 那么 $\mathscr{I}(f_L^{\to}(A))\geqslant\mathscr{I}(A)$; 若 f 再是单射, 则 $\mathscr{I}(f_L^{\to}(A))=\mathscr{I}(A)$.

(2) $\mathscr{I}(f_L^{\leftarrow}(B))\geqslant\mathscr{I}(B)$; 若 f 为满射, 则 $\mathscr{I}(f_L^{\leftarrow}(B))=\mathscr{I}(B)$.

证明 (1) 由定理 5.5.6 可以得到

$$\mathscr{I}(f_L^{\rightarrow}(A)) = \bigvee \{a \in L \mid a \leqslant f_L^{\rightarrow}(A)(0), \forall b \not\geqslant a, f_L^{\rightarrow}(A)^{(b)} \text{ 为 } R \text{ 的理想}\}$$

$$= \bigvee \left\{a \in L \mid a \leqslant \bigvee_{f(r)=0} A(r), \forall b \not\geqslant a, f_L^{\rightarrow}(A^{(b)}) \text{ 为 } R \text{ 的理想}\right\}$$

$$\geqslant \bigvee \{a \in L \mid a \leqslant A(0), \forall b \not\geqslant a, A^{(b)} \text{ 为 } R \text{ 的理想}\}$$

$$= \mathscr{I}(A).$$

当 f 再为单射时, 上述 $\geqslant$ 可由 $=$ 替换, 因此 $\mathscr{I}(f_L^{\rightarrow}(A)) = \mathscr{I}(A)$.

(2) 由定理 5.5.6 可以得到

$$\mathscr{I}(f_L^{\leftarrow}(B)) = \bigvee\{a \in L \mid a \leqslant f_L^{\leftarrow}(B)(0), \forall b \notin \alpha(a), f_L^{\leftarrow}(B)^{[b]} \text{ 为 } R \text{ 的理想}\}$$

$$= \bigvee\{a \in L \mid a \leqslant B(f(0)), \forall b \notin \alpha(a), f_L^{\leftarrow}(B^{[b]}) \text{ 为 } R \text{ 的理想}\}$$

$$\geqslant \bigvee\{a \in L \mid a \leqslant B(0), \forall b \notin \alpha(a), B^{[b]} \text{ 为 } R \text{ 的理想}\}$$

$$= \mathscr{I}(B).$$

当 f 为满射时, 上述 $\geqslant$ 可由 $=$ 替换, 因此 $\mathscr{I}(f_L^{\leftarrow}(B)) = \mathscr{I}(B)$. □

根据定理 5.6.2, 可以得到下面的结论.

定理 5.6.3 设 $f: R \to R'$ 为环同态. 令 $\mathscr{I}_R$ 和 $\mathscr{I}_{R'}$ 分别为环 R 和 R' 中的 L-模糊理想度确定的 L-模糊凸结构. 则 $f: (R, \mathscr{I}_R) \to (R', \mathscr{I}_{R'})$ 是 L-模糊凸保持的和 L-模糊凸到凸的映射.

定理 5.6.4 设 $\{R_i \mid i \in \Omega\}$ 是一族环, 这里 Ω 是有限的, $\forall i \in \Omega, A_i \in L^{R_i}$. 则

$$\mathscr{I}\left(\prod_{i \in \Omega} A_i\right) \geqslant \bigwedge_{i \in \Omega} \mathscr{I}(A_i).$$

证明 容易验证 $\prod\limits_{i \in \Omega} A_i = \bigwedge\limits_{i \in \Omega} P_i^{-1}(A_i)$, 其中 P_i 是 $\prod\limits_{i \in \Omega} R_i$ 到 R_i 的投射. 由定理 5.6.1(2), 可以得到

$$\mathscr{I}\left(\prod_{i \in \Omega} A_i\right) = \mathscr{I}\left(\bigwedge_{i \in \Omega} P_i^{-1}(A_i)\right) \geqslant \bigwedge_{i \in \Omega} \mathscr{I}\left(P_i^{-1}(A_i)\right).$$

因为 $P_i: \prod\limits_{i \in \Omega} R_i \to R_i$ 是环同态, 因此由定理 5.6.2 可知有 $\mathscr{I}\left(P_i^{-1}(A_i)\right) \geqslant \mathscr{I}(A_i)$. 于是便有

$$\mathscr{I}\left(\prod_{i\in\Omega} A_i\right) \geqslant \bigwedge_{i\in\Omega} \mathscr{I}(A_i).$$

由定理 5.6.4 可以直接得到下面推论.

推论 5.6.5 设 $\{R_i \mid i \in \Omega\}$ 是有限个环, $\forall i \in \Omega$, $A_i \in L^{R_i}$. 若 $\forall i \in \Omega$, A_i 是 R_i 的 L-模糊理想. 则 $\prod_{i\in\Omega} A_i$ 是 $\prod_{i\in\Omega} R_i$ 的 L-模糊理想.

再结合定理 5.6.3 和定理 5.6.4 可得到下面推论.

推论 5.6.6 设 $\{R_i \mid i \in \Omega\}$ 是有限个环, $\mathscr{L}$, $\mathscr{L}_i$ 分别是 $\prod_{i\in\Omega} R_i$, R_i 上由 L-模糊理想度确定的 L-模糊凸结构. 则 $P_i : \left(\prod_{i\in\Omega} R_i, \mathscr{I}\right) \to (R_i, \mathscr{I}_i)$ 是 L-模糊凸保持映射和 L-模糊凸到凸映射.

5.7 L-模糊素理想度

在这一节中, 我们提出一种素理想模糊化的方法, 即给出 L-模糊素理想度的概念.

定义 5.7.1 设 A 是环 R 的一个 L-模糊子集. 则 A 的 L-模糊素理想度 $\mathscr{P}(A)$ 定义为

$$\mathscr{P}(A) = \mathscr{I}(A) \wedge \bigwedge_{x,y\in R} (A(xy) \mapsto A(x) \vee A(y)).$$

也可以说 A 是环 R 的一个 L-模糊素理想的程度为 $\mathscr{P}(A)$.

容易验证 $\mathscr{P}(A) = 1$ 当且仅当对任意的 $x, y, z \in R$, A 满足下面四个条件:

(1) $A(0) = 1$;
(2) $A(x) \leqslant A(xz) \wedge A(zx)$;
(3) $A(x) \wedge A(y) \leqslant A(x - y)$;
(4) $A(xy) \leqslant A(x) \vee A(y)$.

因此, L-模糊素理想度是 L-模糊理想的一种新的模糊化形式.

例 5.7.2 令 $\mathbb{Z}_3$ 是模 3 的剩余类环, 并定义 $\mathbb{Z}_3$ 上的运算为常规的加法和乘法. 定义三个 L-模糊集 $A, B, C \in L^{\mathbb{Z}_3}$ 如下:

$$A(z) = \begin{cases} 0.8, & z = [0], \\ 0.3, & z = [1], \\ 0.4, & z = [2], \end{cases}$$

$$B(z) = \begin{cases} 0, & z = [0], \\ 1, & z = [1], \\ 0.4, & z = [2], \end{cases}$$

$$C(z) = \begin{cases} 0.2, & z = [0], \\ 0.2, & z = [1], \\ 0.2, & z = [2]. \end{cases}$$

容易验证

$$\mathscr{P}(A) = 0.3, \quad \mathscr{P}(B) = 0, \quad \mathscr{P}(C) = 0.2.$$

类似于 L-模糊理想度的证明可证定理 5.7.3.

定理 5.7.3 设 A 是环 R 的一个 L-模糊子集. 则

$$\mathscr{P}(A) = \bigwedge_{x,z \in R} (A(x) \mapsto A(xz) \wedge A(zx)) \wedge \bigwedge_{x,y \in R} (A(x) \wedge A(y) \mapsto A(x-y))$$

$$\wedge \bigwedge_{x,y \in R} (A(xy) \mapsto A(x) \vee A(y))$$

$$= \bigwedge_{x,y \in R} (A(x) \mapsto A(xy) \wedge A(yx)) \wedge \bigwedge_{x,y \in R} (A(x) \wedge A(y) \mapsto A(x+y))$$

$$\wedge \bigwedge_{y \in R} (A(y) \mapsto A(-y)) \wedge \bigwedge_{x,y \in R} (A(xy) \mapsto A(x) \vee A(y)).$$

由以上定义及蕴含性质可以得到下面引理.

引理 5.7.4 设 A 是环 R 的一个 L-模糊子集. 则 $\mathscr{P}(A) \geqslant a$ 当且仅当 $\forall x, y, z \in R$, A 满足下面四个条件:

(1) $a \leqslant A(0)$;
(2) $A(x) \wedge a \leqslant A(xz) \wedge A(zx)$;
(3) $A(x) \wedge A(y) \wedge a \leqslant A(x-y)$;
(4) $A(xy) \wedge a \leqslant A(x) \vee A(y)$.

由上述引理易得 $\mathscr{P}(A)$ 的等价刻画如下.

定理 5.7.5 设 A 是环 R 的一个 L-模糊子集. 则

$$\mathscr{P}(A) = \bigvee \left\{ a \in L \;\middle|\; \begin{array}{c} a \leqslant A(0), A(x) \wedge a \leqslant A(xz) \wedge A(zx), \\ A(x) \wedge A(y) \wedge a \leqslant A(x-y), \\ A(xy) \wedge a \leqslant A(x) \vee A(y), \forall x, y \in R \end{array} \right\}.$$

5.7 L-模糊素理想度

下面利用 L-模糊集的四种截集给出 L-模糊素理想度的等价刻画.

定理 5.7.6 设 A 是环 R 的一个 L-模糊子集. 则

(1) $\mathscr{P}(A) = \bigvee\{a \in L \mid \forall b \in J(L), b \leqslant a \leqslant A(0), A_{[b]}$ 是 R 的素理想$\}$;

(2) $\mathscr{P}(A) = \bigvee\{a \in L \mid a \leqslant A(0), \forall b \in P(L), b \ngeqslant a, A^{(b)}$ 是 R 的素理想$\}$;

(3) 对于任意的 $c, d \in L$, 若 $\alpha(c \vee d) = \alpha(c) \cap \alpha(d)$, 则
$\mathscr{P}(A) = \bigvee\{a \in L \mid a \leqslant A(0), \forall b \notin \alpha(a), A^{[b]}$ 是 R 的素理想$\}$.

证明 (1) 对于任意的 $x, y, z \in R$, 假设

$$a \leqslant A(0), \qquad A(x) \wedge A(y) \wedge a \leqslant A(x-y),$$
$$A(x) \wedge a \leqslant A(xz) \wedge A(zx), \quad A(xy) \wedge a \leqslant A(x) \vee A(y),$$

则对于任意的 $b \in J(L)$ 且 $b \leqslant a$, 从前两个不等式和上一节的结果可知 $A_{[b]}$ 是 R 的理想, 为了证明 $A_{[b]}$ 是 R 的素理想, 我们假设 $xy \in A_{[b]}$, 则 $A(xy) \geqslant b$. 由上面第三个不等式可得

$$(A(x) \vee A(y)) \wedge a \geqslant A(xy) \wedge a \geqslant b.$$

于是 $b \leqslant A(x) \vee A(y)$, 这意味着 $A(x) \geqslant b$ 或 $A(y) \geqslant b$, 即 $x \in A_{[b]}$ 或者 $y \in A_{[b]}$. 因此 $A_{[b]}$ 是 R 的素理想. 这表明

$$\mathscr{P}(A) = \bigvee\left\{a \in L \;\middle|\; \begin{array}{l} a \leqslant A(0), A(x) \wedge a \leqslant A(xz) \wedge A(zx), \\ A(x) \wedge A(y) \wedge a \leqslant A(x-y), \\ A(xy) \wedge a \leqslant A(x) \vee A(y), \forall x, y \in R \end{array}\right\}$$

$$\leqslant \bigvee\{a \in L \mid \forall b \in J(L), b \leqslant a \leqslant A(0), A_{[b]} \text{ 是 } R \text{ 的素理想}\}.$$

相反地, 设 $a \in L$ 且对满足 $b \leqslant a \leqslant A(0)$ 的任何 $b \in J(L)$, $A_{[b]}$ 都是 R 的素理想. 则由上一节的结论可知 $\forall x, y, z \in R$,

$$A(x) \wedge a \leqslant A(xz) \wedge A(zx), \quad A(x) \wedge A(y) \wedge a \leqslant A(x-y).$$

下证 $\forall x, y \in R, A(xy) \wedge a \leqslant A(x) \vee A(y)$.

设 $b = A(xy) \wedge a$, 则对满足 $c \leqslant b$ 的任何 $c \in J(L)$, 都有 $A(xy) \geqslant c$, 所以 $xy \in A_{[c]}$. 因为 $A_{[c]}$ 是 R 的素理想, 所以有 $x \in A_{[c]}$ 或者 $y \in A_{[c]}$, 即 $A(x) \geqslant c$ 或 $A(y) \geqslant c$. 从而 $A(x) \vee A(y) \geqslant c$. 由 c 的任意性可知

$$A(x) \vee A(y) \geqslant b = A(xy) \wedge a.$$

这又证明了

$$\mathscr{P}(A) = \bigvee \left\{ a \in L \;\middle|\; \begin{array}{c} a \leqslant A(0), A(x) \wedge a \leqslant A(xz) \wedge A(zx), \\ A(x) \wedge A(y) \wedge a \leqslant A(x-y), \\ A(xy) \wedge a \leqslant A(x) \vee A(y), \forall x, y \in R \end{array} \right\}$$

$$\geqslant \bigvee \{a \in L \mid \forall b \in J(L),\ b \leqslant a \leqslant A(0),\ A_{[b]} \text{ 是 } R \text{ 的素理想}\}.$$

因此 (1) 的等式成立.

(2) 由上一节有关 L-模糊理想度的刻画可知, 为了证明 (2), 我们只需要证明下面的充要条件.

(条件一): $A(xy) \wedge a \leqslant A(x) \vee A(y)$ 当且仅当 $\forall x, y \in R, \forall b \in P(L)$, 当 $b \not\geqslant a$ 时, 由 $xy \in A^{(b)}$ 可推得 $x \in A^{(b)}$ 或者 $y \in A^{(b)}$.

为此我们假设 $A(xy) \wedge a \leqslant A(x) \vee A(y)$ 且 $\forall x, y \in R, \forall b \in P(L)$, 当 $b \not\geqslant a$ 时, 设 $xy \in A^{(b)}$. 则 $A(xy) \not\leqslant b$. 于是 $A(x) \vee A(y) \not\leqslant b$, 这意味着 $A(x) \not\leqslant b$ 或者 $A(y) \not\leqslant b$, 也就是 $x \in A^{(b)}$ 或 $y \in A^{(b)}$.

反之, 假设 $\forall x, y \in R, \forall b \in P(L)$, 当 $b \not\geqslant a$ 时, 由 $xy \in A^{(b)}$ 可推得 $x \in A^{(b)}$ 或者 $y \in A^{(b)}$. 为了证明

$$A(xy) \wedge a \leqslant A(x) \vee A(y),$$

假设 $b \in P(L)$ 且 $b \not\geqslant A(xy) \wedge a$. 则 $A(xy) \not\leqslant b$ 且 $a \not\leqslant b$, 这意味着 $xy \in A^{(b)}$. 进一步可得到 $x \in A^{(b)}$ 或者 $y \in A^{(b)}$. 从而 $A(x) \not\leqslant b$ 或者 $A(y) \not\leqslant b$, 于是 $A(x) \vee A(y) \not\leqslant b$, 这就证明了 $A(xy) \wedge a \leqslant A(x) \vee A(y)$.

综上可知 (2) 是成立的.

(3) 设对任意的 $c, d \in L$, $\alpha(c \vee d) = \alpha(c) \cap \alpha(d)$. 为了证明 (3), 我们只需要证明

(条件二): $A(xy) \wedge a \leqslant A(x) \vee A(y)$ 当且仅当 $\forall x, y \in R, \forall b \in P(L)$, 当 $b \notin \alpha(a)$ 时, 由 $xy \in A^{[b]}$ 可推得 $x \in A^{[b]}$ 或者 $y \in A^{[b]}$.

假设 $\forall x, y \in R, \forall b \in P(L)$, 当 $b \notin \alpha(a)$ 时, 由 $xy \in A^{[b]}$ 可推得 $x \in A^{[b]}$ 或者 $y \in A^{[b]}$. 下面来证明

$$A(xy) \wedge a \leqslant A(x) \vee A(y).$$

为此我们假设 $b \notin \alpha(A(xy) \wedge a)$, 则由

$$\alpha(A(xy) \wedge a) = \alpha(A(xy)) \cup \alpha(a)$$

可知 $b \notin \alpha(A(xy))$ 且 $b \notin \alpha(a)$. 于是 $xy \in A^{[b]}$, 这意味着 $x \in A^{[b]}$ 或者 $y \in A^{[b]}$, 也就是 $b \notin \alpha(A(x))$ 或者 $b \notin \alpha(A(y))$, 从而有

$$b \notin \alpha(A(x)) \cap \alpha(A(y)) = \alpha(A(x) \vee A(y)).$$

这表明 $A(xy) \wedge a \leqslant A(x) \vee A(y)$.

反之,假设 $A(xy) \wedge a \leqslant A(x) \vee A(y)$ 且 $\forall x, y \in R, \forall b \in P(L)$, 当 $b \notin \alpha(a)$ 时,有 $xy \in A^{[b]}$ 成立,则

$$b \notin \alpha(A(xy)) \cup \alpha(a) = \alpha(A(xy) \wedge a).$$

于是可得到

$$b \notin \alpha(A(x) \vee A(y)) = \alpha(A(x)) \cap \alpha(A(y)).$$

这意味着 $b \notin \alpha(A(x))$ 或者 $b \notin \alpha(A(y))$, 即 $x \in A^{[b]}$ 或者 $y \in A^{[b]}$.

综上可得 (3) 成立. □

5.8　L-模糊子域

在这一节中,我们引入 L-模糊子域度的概念,给出 L-模糊子域度的等价刻画,建立模糊子域度与模糊凸结构之间的关系,指出域同态可以看成 L-模糊凸保持映射和 L-模糊凸到凸映射,讨论域的 L-模糊子集在同态映射下的像和原像的 L-模糊子域度变化情况,并且进一步讨论 L-模糊子域乘积的表现定理.

域的模糊化工作早已有之,其定义如下:

定义 5.8.1　设 A 为域 F 的一个 L-模糊子集. 若 $\forall x, y \in F$, 有

(1) $A(1) = 1$;

(2) $A(x - y) \geqslant A(x) \wedge A(y)$;

(3) $A(xy^{-1}) \geqslant A(x) \wedge A(y)$,

则称 A 是域 F 的 L-模糊子域.

从上述定义可见,给定域 F 的一个 L-模糊子集 A, 其要么是 F 的子域,要么不是,而这只居其一,且仅居其一. 这一属性是分明的,借助于 L 的蕴含算子,我们能够将这一属性进一步模糊化.

定义 5.8.2　令 A 是域 F 中一个 L-模糊子集. A 的 L-模糊子域度 $\mathscr{F}(A)$ 定义为

$$\mathscr{F}(A) = A(1) \wedge \bigwedge_{x,y \in F} ((A(x) \wedge A(y)) \mapsto A(xy^{-1}))$$

$$\wedge \bigwedge_{x,y \in F} ((A(x) \wedge A(y)) \mapsto A(x - y)).$$

可以验证 A 是域 F 中的一个 L-模糊子域当且仅当 $\mathscr{F}(A) = 1$.

例 5.8.3　设 $\mathbb{R}$ 为一个实数域,$L = \mathbf{2}^{\{a,b\}}, a \neq b$. 通过下列式子定义 $A, B, C:$ $\mathbb{R} \to L$

$$A(x) = \begin{cases} \{a,b\}, & \text{如果 } x = 1, \\ \varnothing, & \text{如果 } x \neq 1, \end{cases}$$

$$B(x) = \begin{cases} \{a,b\}, & \text{如果 } x \text{ 是整数}, \\ \{b\}, & \text{如果 } x \text{ 不是整数}, \end{cases}$$

$$C(x) = \begin{cases} \{a\}, & \text{如果 } x \text{ 是有理数}, \\ \varnothing, & \text{如果 } x \text{ 是无理数}, \end{cases}$$

可以得到

$$\mathscr{F}(A) = \varnothing, \quad \mathscr{F}(B) = \{b\}, \quad \mathscr{F}(C) = \{a,b\}.$$

(1) 事实上, 若 $x = y = 1$, 即 $x - y = 0$, 则

$$A(x) \wedge A(y) \mapsto A(0) = \{a,b\} \mapsto \varnothing = \varnothing,$$

因此我们可以得到 $\mathscr{F}(A) = \varnothing$.

(2) 如果 $x = 2, y = 3$, 那么 xy^{-1} 不是整数, 则

$$B(x) \wedge B(y) \mapsto B(xy^{-1}) = \{a,b\} \mapsto \{b\} = \{b\},$$

因此可以得到 $\mathscr{F}(B) = \{b\}$.

(3) 如果 x 和 y 均为有理数, 那么 $x - y, xy^{-1}$ 都是有理数, 则

$$C(x) \wedge C(y) \mapsto C(x-y) \wedge C(xy^{-1}) = \{a,b\} \mapsto \{a,b\} = \{a,b\}.$$

如果 x 和 y 中有一个是无理数, 那么 $x - y$ 和 xy^{-1} 均为无理数, 则

$$C(x) \wedge C(y) \mapsto C(x-y) \wedge C(xy^{-1}) = \varnothing \mapsto \varnothing = \{a,b\}.$$

又因为 $C(0) = C(1) = \{a\}$, 所以可得 $\mathscr{F}(C) = \{a\}$.

例 5.8.4 设 $\mathbb{R}$ 为一个实数域, $L = [0,1]$. 通过下列式子定义 $A, B, C : \mathbb{R} \to [0,1]$,

$$A(x) = \begin{cases} 0, & \text{如果 } x = 0, 1, \\ 1, & \text{如果 } x \neq 0, 1, \end{cases}$$

$$B(x) = \begin{cases} 0.5, & \text{如果 } x \text{ 是有理数}, \\ 1, & \text{如果 } x \text{ 是无理数}, \end{cases}$$

$$C(x) = 0.3, \quad \forall x \in \mathbb{R},$$

5.8 L-模糊子域

可以得到
$$\mathscr{F}(A) = 0, \quad \mathscr{F}(B) = 0.5, \quad \mathscr{F}(C) = 1.$$

(1) 事实上,如果 $x = y \neq 0, 1$,那么对于 $x - y = 0$ 和 $xy^{-1} = 1$ 这种情况有
$$A(x) \wedge A(y) \mapsto A(0) \wedge A(1) = 1 \mapsto 0 = 0.$$

所以可以得到 $\mathscr{F}(A) = 0$.

(2) 如果 $x = y$ 都不是有理数,又 $x - y = 0$, $xy^{-1} = 1$ 为有理数,那么
$$B(x) \wedge B(y) \mapsto B(0) \wedge B(1) = 1 \mapsto 0.5 = 0.5.$$

因此 $\mathscr{F}(B) = 0.5$.

(3) $\mathscr{F}(C) = 0.3$ 是显然的.

定理 5.8.5 设 A 是域 F 中一个 L-模糊集. 那么

$$\mathscr{F}(A) = A(1) \wedge \bigwedge_{x,y \in F} (A(x) \wedge A(y) \mapsto A(xy)) \wedge \bigwedge_{y \in F, y \neq 0} (A(y) \mapsto A(y^{-1}))$$
$$\wedge \bigwedge_{x,y \in F} (A(x) \wedge A(y) \mapsto A(x+y)) \wedge \bigwedge_{y \in F} (A(y) \mapsto A(-y)).$$

证明 由定义 5.8.2 可得对于任意的 $x, y \in F$, ($y \neq 0$ 时考虑 y^{-1})

$$\mathscr{F}(A) \leqslant A(0) \wedge A(1),$$
$$\mathscr{F}(A) \wedge A(x) \wedge A(y) \leqslant A(xy^{-1}),$$
$$\mathscr{F}(A) \wedge A(x) \wedge A(y) \leqslant A(x - y),$$
$$\mathscr{F}(A) \wedge A(y) \leqslant A(1),$$
$$\mathscr{F}(A) \wedge A(y) \leqslant A(0).$$

特别地,有

$$\mathscr{F}(A) \wedge A(y) \leqslant \mathscr{F}(A) \wedge A(1) \wedge A(y) \leqslant A(y^{-1}),$$
$$\mathscr{F}(A) \wedge A(y) \leqslant \mathscr{F}(A) \wedge A(0) \wedge A(y) \leqslant A(-y),$$
$$\mathscr{F}(A) \wedge A(y) = \mathscr{F}(A) \wedge A(y^{-1}),$$
$$\mathscr{F}(A) \wedge A(y) = \mathscr{F}(A) \wedge A(-y),$$
$$\mathscr{F}(A) \wedge A(x) \wedge A(y) = \mathscr{F}(A) \wedge A(x) \wedge A(y^{-1}),$$

$$\mathscr{F}(A) \wedge A(x) \wedge A(y) = \mathscr{F}(A) \wedge A(x) \wedge A(-y).$$

这表明

$$\mathscr{F}(A) \leqslant \bigwedge_{x,y \in F} (A(x) \wedge A(y) \mapsto A(xy)) \wedge \bigwedge_{y \in F} (A(y) \mapsto A(y^{-1})),$$

$$\mathscr{F}(A) \leqslant \bigwedge_{x,y \in F} (A(x) \wedge A(y) \mapsto A(x+y)) \wedge \bigwedge_{y \in F} (A(y) \mapsto A(-y)).$$

因此

$$\mathscr{F}(A) \leqslant A(1) \wedge \bigwedge_{x,y \in F} (A(x) \wedge A(y) \mapsto A(xy)) \wedge \bigwedge_{y \in F} (A(y) \mapsto A(y^{-1}))$$

$$\wedge \bigwedge_{x,y \in F} (A(x) \wedge A(y) \mapsto A(x+y)) \wedge \bigwedge_{y \in F} (A(y) \mapsto A(-y)).$$

类似地, 可以证明

$$\mathscr{F}(A) \geqslant A(1) \wedge \bigwedge_{x,y \in F} (A(x) \wedge A(y) \mapsto A(xy)) \wedge \bigwedge_{y \in F} (A(y) \mapsto A(y^{-1}))$$

$$\wedge \bigwedge_{x,y \in F} (A(x) \wedge A(y) \mapsto A(x+y)) \wedge \bigwedge_{y \in F} (A(y) \mapsto A(-y)).$$

因此

$$\mathscr{F}(A) = A(1) \wedge \bigwedge_{x,y \in F} (A(x) \wedge A(y) \mapsto A(xy)) \wedge \bigwedge_{y \in F, y \neq 0} (A(y) \mapsto A(y^{-1}))$$

$$\wedge \bigwedge_{x,y \in F} (A(x) \wedge A(y) \mapsto A(x+y)) \wedge \bigwedge_{y \in F} (A(y) \mapsto A(-y)). \qquad \square$$

下面的引理明显成立.

引理 5.8.6 设 A 是域 F 中一个 L-模糊集. 那么 $\mathscr{F}(A) \geqslant a$ 当且仅当对于任意的 $x, y \in F$,

$$A(x) \wedge A(y) \wedge a \leqslant A(xy^{-1}), \quad y \neq 0,$$

$$A(x) \wedge A(y) \wedge a \leqslant A(x-y), \quad a \leqslant A(1).$$

定理 5.8.7 给出了 L-模糊子域度的一些等价刻画.

定理 5.8.7 设 A 是域 F 中一个 L-模糊集. 那么

5.8 L-模糊子域

(1) $\mathscr{F}(A) = \bigvee \left\{ a \in L \;\middle|\; \begin{array}{l} a \leqslant A(1),\ A(x) \wedge A(y) \wedge a \leqslant A(xy^{-1}), \\ A(x) \wedge A(y) \wedge a \leqslant A(x-y),\ \forall x,y \in F, y \neq 0 \end{array} \right\};$

(2) $\mathscr{F}(A) = \bigvee \{a \in L \mid \forall b \leqslant a \leqslant A(1), A_{[b]} \text{ 是域 } F \text{ 的一个子域}\};$

(3) $\mathscr{F}(A) = \bigvee \{a \in L \mid a \leqslant A(1), \forall b \notin \alpha(a), A^{[b]} \text{ 是域 } F \text{ 的一个子域}\};$

(4) $\mathscr{F}(A) = \bigvee \{a \in L \mid a \leqslant A(1), \forall b \in P(L), b \not\geqslant a, A^{(b)} \text{ 是域 } F \text{ 的一个子域}\};$

(5) 若 $\forall a, b \in L$, $\beta(a \wedge b) = \beta(a) \cap \beta(b)$, 则 $\mathscr{F}(A) = \bigvee \{a \in L \mid a \leqslant A(1), \forall b \in \beta(a), A_{(b)} \text{ 是域 } F \text{ 的一个子域}\}.$

证明 (1) 是显然的, 下证 (2). 假如 $\forall x, y \in F, y \neq 0$, 有

$$A(x) \wedge A(y) \wedge a \leqslant A(xy^{-1}), \quad A(x) \wedge A(y) \wedge a \leqslant A(x-y),$$

那么对于任意的 $b \leqslant a \leqslant A(1)$, $x, y \in A_{[b]}$, 我们可以得到

$$A(xy^{-1}) \geqslant A(x) \wedge A(y) \wedge a \geqslant A(x) \wedge A(y) \wedge b \geqslant b,$$
$$A(x-y) \geqslant A(x) \wedge A(y) \wedge a \geqslant A(x) \wedge A(y) \wedge b \geqslant b,$$

这表明 $xy^{-1} \in A_{[b]}, x - y \in A_{[b]}$. 因此 $A_{[b]}$ 是 F 的一个子域. 容易观察到 $\mathscr{F}(A) \wedge A(1) \leqslant \mathscr{F}(A) \wedge A(0)$. 故有

$$\mathscr{F}(A) = \bigvee \left\{ a \in L \;\middle|\; \begin{array}{l} a \leqslant A(1),\ A(x) \wedge A(y) \wedge a \leqslant A(xy^{-1}), \\ A(x) \wedge A(y) \wedge a \leqslant A(x-y), \forall x,y \in F, y \neq 0 \end{array} \right\}$$
$$\leqslant \bigvee \left\{ a \in L \;\middle|\; \forall b \leqslant a \leqslant A(1), A_{[b]} \text{ 是域 } F \text{ 的一个子域} \right\}.$$

反之, 若 $a \in L$ 使得 $\forall b \leqslant a \leqslant A(1)$, $A_{[b]}$ 是域 F 的一个子域. 对于任意取定的 $x, y \in F$, 令 $b = A(x) \wedge A(y) \wedge a$, 则 $b \leqslant a$, $x, y \in A_{[b]}$, 因此 $xy^{-1}, x - y \in A_{[b]}$, 即

$$A(xy^{-1}) \geqslant b = A(x) \wedge A(y) \wedge a,$$
$$A(x-y) \geqslant b = A(x) \wedge A(y) \wedge a.$$

这意味着

$$\mathscr{F}(A) = \bigvee \left\{ a \in L \;\middle|\; \begin{array}{l} a \leqslant A(1), A(x) \wedge A(y) \wedge a \leqslant A(xy^{-1}), \\ A(x) \wedge A(y) \wedge a \leqslant A(x-y), \forall x,y \in F, y \neq 0 \end{array} \right\}$$
$$\geqslant \bigvee \left\{ a \in L \;\middle|\; \forall b \leqslant a \leqslant A(1), A_{[b]} \text{ 是域 } F \text{ 的一个子域} \right\}.$$

(2) 得证.

为了证明 (3), 假设对于任意的 $x, y \in F, y \neq 0$, 有

$$a \leqslant A(1), \quad A(x) \wedge A(y) \wedge a \leqslant A(xy^{-1}), \quad A(x) \wedge A(y) \wedge a \leqslant A(x-y).$$

对于任意的 $b \notin \alpha(a), x, y \in A^{[b]}$, 显然有

$$b \notin \alpha(A(x)) \cup \alpha(A(y)) \cup \alpha(a) = \alpha(A(x) \wedge A(y) \wedge A(a)).$$

由

$$A(x) \wedge A(y) \wedge a \leqslant A(xy^{-1}),$$
$$A(x) \wedge A(y) \wedge a \leqslant A(x-y)$$

可知

$$\alpha(A(xy^{-1})) \subseteq \alpha(A(x) \wedge A(y) \wedge a),$$
$$\alpha(A(x-y)) \subseteq \alpha(A(x) \wedge A(y) \wedge a).$$

因此有 $b \notin \alpha(A(xy^{-1}))$ 且 $b \notin \alpha(A(x-y))$, 这意味着 $xy^{-1}, x-y \in A^{[b]}$, 这表明 $A^{[b]}$ 是域 F 的一个子域且

$$a \in \{a \in L \mid a \leqslant A(1), \forall b \notin \alpha(a), A^{[b]} \text{ 是域 } F \text{ 的一个子域}\}.$$

因此

$$\mathscr{F}(A) = \bigvee \left\{ a \in L \,\middle|\, \begin{array}{l} a \leqslant A(1), A(x) \wedge A(y) \wedge a \leqslant A(xy^{-1}), \\ A(x) \wedge A(y) \wedge a \leqslant A(x-y), \forall x, y \in F, y \neq 0 \end{array} \right\}$$

$$\leqslant \bigvee \left\{ a \in L \,\middle|\, a \leqslant A(1), \forall b \notin \alpha(a), A^{[b]} \text{ 是域 } F \text{ 的一个子域} \right\}.$$

相反, 若 $a \in \{a \in L \mid a \leqslant A(1), \forall b \notin \alpha(a), A^{[b]}$ 是域 F 的一个子域$\}$. 现在证明对于任意的 $x, y \in F$, 下面两公式成立:

$$A(x) \wedge A(y) \wedge a \leqslant A(xy^{-1}), \quad A(x) \wedge A(y) \wedge a \leqslant A(x-y).$$

为此设 $b \notin \alpha(A(x) \wedge A(y) \wedge a)$. 由于

$$\alpha(A(x) \wedge A(y) \wedge a) = \alpha(A(x)) \cup \alpha(A(x)) \cup \alpha(a),$$

5.8 L-模糊子域

所以可得 $b \notin \alpha(a)$ 以及 $x, y \in A^{[b]}$. 再由 $A^{[b]}$ 是域 F 的一个子域可知 $xy^{-1}, x - y \in A^{[b]}$, 进一步有 $b \notin \alpha(A(xy^{-1}))$, 且 $b \notin \alpha(A(x-y))$. 这表明

$$A(x) \wedge A(y) \wedge a \leqslant A(xy^{-1}), \quad A(x) \wedge A(y) \wedge a \leqslant A(x-y).$$

因此

$$\mathscr{F}(A) = \bigvee \left\{ a \in L \,\middle|\, \begin{array}{l} a \leqslant A(1), A(x) \wedge A(y) \wedge a \leqslant A(xy^{-1}), \\ A(x) \wedge A(y) \wedge a \leqslant A(x-y), \forall x, y \in F, y \neq 0 \end{array} \right\}$$

$$\geqslant \bigvee \left\{ a \in L \,\middle|\, a \leqslant A(1), \forall b \notin \alpha(a), A^{[b]} \text{ 是域 } F \text{ 的一个子域} \right\}.$$

这样就证明了 (3).

为了证明 (4), 假设 $a \in L$ 使得 $a \leqslant A(1)$, 且 $\forall x, y \in F, y \neq 0$, 有

$$A(x) \wedge A(y) \wedge a \leqslant A(xy^{-1}), \quad A(x) \wedge A(y) \wedge a \leqslant A(x-y).$$

设 $b \in P(L), b \not\geqslant a, x, y \in A^{(b)}$. 以下证明 $xy^{-1}, x - y \in A^{(b)}$. 若 $xy^{-1} \notin A^{(b)}$ 或者 $x - y \notin A^{(b)}$, 则有 $A(xy^{-1}) \leqslant b$ 或者 $A(x-y) \leqslant b$, 于是可得

$$A(x) \wedge A(y) \wedge a \leqslant A(xy^{-1}) \leqslant b, \quad \text{或者} \quad A(x) \wedge A(y) \wedge a \leqslant A(x-y) \leqslant b.$$

由于 $b \in P(L), x, y \in A^{(b)}$, 所以必有 $a \leqslant b$, 这与 $b \not\geqslant a$ 矛盾. 因此必有 $xy^{-1}, x - y \in A^{(b)}$. 这意味着 $A^{(b)}$ 是域 F 的一个子域. 因此

$$\mathscr{F}(A) = \bigvee \left\{ a \in L \,\middle|\, \begin{array}{l} a \leqslant A(1), A(x) \wedge A(y) \wedge a \leqslant A(xy^{-1}), \\ A(x) \wedge A(y) \wedge a \leqslant A(x-y), \forall x, y \in F, y \neq 0 \end{array} \right\}$$

$$\leqslant \bigvee \left\{ a \in L \,\middle|\, a \leqslant A(1), \forall b \in P(L), b \not\geqslant a, A^{(b)} \text{ 是域 } F \text{ 的一个子域} \right\}.$$

相反, 若 $a \leqslant A(1)$ 且 $a \in \{a \in L \mid \forall b \in P(L), b \not\geqslant a, A^{(b)} \text{ 是域 } F \text{ 的一个子域}\}$. 下面要证明对于任意的 $x, y \in F, y \neq 0$,

$$A(x) \wedge A(y) \wedge a \leqslant A(xy^{-1}), \quad A(x) \wedge A(y) \wedge a \leqslant A(x-y).$$

为此我们令 $b \in P(L)$ 使得 $A(x) \wedge A(y) \wedge a \not\leqslant b$, 则有 $A(x) \not\leqslant b, A(y) \not\leqslant b, a \not\leqslant b$, 这意味着 $x, y \in A^{(b)}$. 因为 $A^{(b)}$ 是域 F 的一个子域, 所以可得 $xy^{-1} \in A^{(b)}, x - y \in A^{(b)}$, 即 $A(xy^{-1}) \not\leqslant b$ 且 $A(x-y) \not\leqslant b$, 也就是说下面两式成立:

$$A(x) \wedge A(y) \wedge a \leqslant A(xy^{-1}), \quad A(x) \wedge A(y) \wedge a \leqslant A(x-y).$$

因此我们有

$$\mathscr{F}(A) = \bigvee \left\{ a \in L \,\middle|\, \begin{array}{l} a \leqslant A(1), A(x) \wedge A(y) \wedge a \leqslant A(xy^{-1}), \\ A(x) \wedge A(y) \wedge a \leqslant A(x-y), \forall x, y \in F, y \neq 0 \end{array} \right\}$$

$$\geqslant \bigvee \left\{ a \in L \,\middle|\, a \leqslant A(1), \forall b \in P(L), b \not\geqslant a, A^{(b)} \text{ 是域 } F \text{ 的一个子域} \right\}.$$

这样就证明 (4).

为了证明 (5), 假设

$$a \in \left\{ a \in L \,\middle|\, \begin{array}{l} a \leqslant A(1), A(x) \wedge A(y) \wedge a \leqslant A(xy^{-1}), \\ A(x) \wedge A(y) \wedge a \leqslant A(x-y), \forall x, y \in F, y \neq 0 \end{array} \right\}.$$

$\forall b \in \beta(a)$, 令 $x, y \in A_{(b)}$, 则容易验证

$$b \in \beta(A(x)) \cap \beta(A(y)) \cap \beta(a) = \beta(A(x) \wedge A(y) \wedge a) \subseteq \beta(A(xy^{-1}))$$

且

$$b \in \beta(A(x)) \cap \beta(A(y)) \cap \beta(a) = \beta(A(x) \wedge A(y) \wedge a) \subseteq \beta(A(x-y)).$$

这意味着 $xy^{-1}, x-y \in A_{(b)}$. 因此 $A_{(b)}$ 是域 F 的一个子域, 也就是说

$$\mathscr{F}(A) = \bigvee \left\{ a \in L \,\middle|\, \begin{array}{l} a \leqslant A(1), A(x) \wedge A(y) \wedge a \leqslant A(xy^{-1}), \\ A(x) \wedge A(y) \wedge a \leqslant A(x-y), \forall x, y \in F, y \neq 0 \end{array} \right\}$$

$$\leqslant \bigvee \left\{ a \in L \,\middle|\, a \leqslant A(1), \forall b \in \beta(a), A_{(b)} \text{ 是域 } F \text{ 的一个子域} \right\}.$$

相反, 假设 $a \in \{a \in L \mid a \leqslant A(1), \forall b \in \beta(a), A_{(b)} \text{ 是域 } F \text{ 的一个子域}\}$. 下面我们要证明 $\forall x, y \in F$,

$$A(x) \wedge A(y) \wedge a \leqslant A(xy^{-1}), \quad \text{这里 } y \neq 0, \quad A(x) \wedge A(y) \wedge a \leqslant A(x-y).$$

为此令 $b \in \beta(A(x) \wedge A(y) \wedge a)$. 由

$$\beta(A(x) \wedge A(y) \wedge a) = \beta(A(x)) \cap \beta(A(y)) \cap \beta(a),$$

可得 $x, y \in A_{(b)}, b \in \beta(a)$. 因为 $A_{(b)}$ 是域 F 的一个子域, 所以有 $xy^{-1}, x-y \in A_{(b)}$, 也就是 $b \in \beta(A(xy^{-1}))$ 且 $b \in \beta(A(x-y))$, 这意味着

$$A(x) \wedge A(y) \wedge a \leqslant A(xy^{-1}), \quad A(x) \wedge A(y) \wedge a \leqslant A(x-y).$$

因此我们可得

$$\mathscr{F}(A) = \bigvee \left\{ a \in L \,\middle|\, \begin{array}{l} a \leqslant A(1), A(x) \wedge A(y) \wedge a \leqslant A(xy^{-1}), \\ A(x) \wedge A(y) \wedge a \leqslant A(x-y), \forall x, y \in F, y \neq 0 \end{array} \right\}$$

$$\geqslant \{ a \in L \mid a \leqslant A(1), \forall b \in \beta(a), A_{(b)} \text{ 是域 } F \text{ 的一个子域}\}.$$

这样就证明了 (5). □

5.9 由 L-模糊子域度确定的 L-模糊凸结构

在这一节中,我们研究 L-模糊子域度和 L-模糊凸结构之间的关系,并且研究 L-模糊子集的同态像和原像的子域度,证明一个域同态恰好是一个 L-模糊凸保持映射和一个 L-模糊凸到凸映射.

对任意 $A \in L^F$,$\mathscr{F}(A)$ 自然可以被认为是一种由 $A \mapsto \mathscr{F}(A)$ 确定的映射 $\mathscr{F} : L^F \to L$ 的像. 下面定理表明 $\mathscr{F}$ 是域 F 上一个 L-模糊凸结构.

定理 5.9.1 设 F 是一个域. 由 $A \mapsto \mathscr{F}(A)$ 确定的映射 $\mathscr{F} : L^F \to L$ 是 F 上一个 L-模糊凸结构, 它叫做 F 上由 L 模糊子域度确定的 L-模糊凸结构.

证明 (1) 容易得到

$$\mathscr{F}(\chi_\varnothing) = \mathscr{F}(\chi_F) = 1.$$

(2) 设 $\{A_i \mid i \in \Omega\} \subseteq L^F$ 非空. 下面要证明

$$\bigwedge_{i \in \Omega} \mathscr{F}(A_i) \leqslant \mathscr{F}\left(\bigwedge_{i \in \Omega} A_i\right).$$

对于任意 $a \leqslant \bigwedge_{i \in \Omega} \mathscr{F}(A_i)$,由定理 5.8.7,对于任意的 $i \in \Omega$ 和任意的 $x, y \in F, y \neq 0$, 都有

$$a \leqslant \bigwedge_{i \in \Omega} A_i(1),$$

$$A_i(x) \wedge A_i(y) \wedge a \leqslant A_i(xy^{-1}),$$

$$A_i(x) \wedge A_i(y) \wedge a \leqslant A_i(x-y).$$

这意味着

$$\left(\bigwedge_{i \in \Omega} A_i(x)\right) \wedge \left(\bigwedge_{j \in \Omega} A_j(y)\right) \wedge a \leqslant \bigwedge_{i \in \Omega} (A_i(x) \wedge A_i(y) \wedge a) \leqslant \bigwedge_{i \in \Omega} A_i(xy^{-1}),$$

$$\left(\bigwedge_{i\in\Omega}A_i(x)\right)\wedge\left(\bigwedge_{j\in\Omega}A_j(y)\right)\wedge a\leqslant\bigwedge_{i\in\Omega}(A_i(x)\wedge A_i(y)\wedge a)\leqslant\bigwedge_{i\in\Omega}A_i(x-y).$$

由引理 5.8.6, 我们知道 $a\leqslant\mathscr{F}\left(\bigwedge_{i\in\Omega}A_i\right)$. 由 a 的任意性, 可以得到

$$\bigwedge_{i\in\Omega}\mathscr{F}(A_i)\leqslant\mathscr{F}\left(\bigwedge_{i\in\Omega}A_i\right).$$

(3) 设 $\{A_i \mid i\in\Omega\}\subseteq L^F$ 是非空且全序的. 现在来证明

$$\bigwedge_{i\in\Omega}\mathscr{F}(A_i)\leqslant\mathscr{F}\left(\bigvee_{i\in\Omega}A_i\right).$$

对于任意的 $a\leqslant\bigwedge_{i\in\Omega}\mathscr{F}(A_i)$. 由引理 5.8.6 可知, $\forall i\in\Omega$ 和对于任意的 $x,y\in F, y\neq 0$, 都有

$$A_i(x)\wedge A_i(y)\wedge a\leqslant A_i(xy^{-1}),$$
$$A_i(x)\wedge A_i(y)\wedge a\leqslant A_i(x-y).$$

取 $b\in L$ 使得

$$b\prec\left(\bigvee_{i\in\Omega}A_i(x)\right)\wedge\left(\bigvee_{i\in\Omega}A_i(y)\right)\wedge a,$$

则可以得到

$$b\prec\bigvee_{i\in\Omega}A_i(x),\ \ b\prec\bigvee_{i\in\Omega}A_i(y),\ \text{且}\ b\leqslant a.$$

因此存在 $i,j\in\Omega$ 使得 $b\leqslant A_i(x), b\leqslant A_j(y)$, 且 $b\leqslant a$. 由于 $\{A_i\mid i\in\Omega\}$ 是全序的, 所以要么 $A_j\leqslant A_i$, 要么 $A_i\leqslant A_j$, 不妨设 $A_j\leqslant A_i$, 则

$$b\leqslant A_i(x)\wedge A_i(y)\wedge a.$$

因为

$$A_i(x)\wedge A_i(y)\wedge a\leqslant A_i(xy^{-1}),$$
$$A_i(x)\wedge A_i(y)\wedge a\leqslant A_i(x-y),$$

5.9 由 L-模糊子域度确定的 L-模糊凸结构

所以可以得到 $b \leqslant A_i(xy^{-1})$ 和 $b \leqslant A_i(x-y)$. 因此 $b \leqslant \bigvee_{i\in\Omega} A_i(xy^{-1})$, $b \leqslant \bigvee_{i\in\Omega} A(x-y)$. 由 b 的任意性, 可以得到

$$\left(\bigvee_{i\in\Omega} A_i(x)\right) \wedge \left(\bigvee_{i\in\Omega} A_i(y)\right) \wedge a \leqslant \bigvee_{i\in\Omega} A_i(xy^{-1}),$$

$$\left(\bigvee_{i\in\Omega} A_i(x)\right) \wedge \left(\bigvee_{i\in\Omega} A_i(y)\right) \wedge a \leqslant \bigvee_{i\in\Omega} A_i(x-y).$$

结合引理 5.8.6, 可以得到 $a \leqslant \mathscr{F}\left(\bigvee_{i\in\Omega} A_i\right)$. 由 a 的任意性可得 $\bigwedge_{i\in\Omega} \mathscr{F}(A_i) \leqslant \mathscr{F}\left(\bigvee_{i\in\Omega} A_i\right)$.

综上可知 $\mathscr{F}$ 是 F 上一个 L-模糊凸结构. □

现在讨论 L-模糊子集同态映射的像和原像的 L-模糊子域度.

定理 5.9.2 设 $f: F \to F'$ 是一个域同态, $A \in L^F$, $B \in L^{F'}$. 则有
(1) $\mathscr{F}(f_L^{\to}(A)) \geqslant \mathscr{F}(A)$; 如果 f 是单射, 那么 $\mathscr{F}(f_L^{\to}(A)) = \mathscr{F}(A)$.
(2) $\mathscr{F}(f_L^{\leftarrow}(B)) \geqslant \mathscr{F}(B)$; 如果 f 是满射, 那么 $\mathscr{F}(f_L^{\leftarrow}(B)) = \mathscr{F}(B)$.

证明 **方法 1**. (1) 从定理 5.8.7 出发可得证明如下:

$\mathscr{F}(f_L^{\to}(A))$

$= \bigvee \left\{ a \in L \left| \begin{array}{l} a \leqslant f_L^{\to}(A)(1),\ f_L^{\to}(A)(x') \wedge f_L^{\to}(A)(y') \wedge a \leqslant f_L'(A)(x'y'^{-1}), \\ f_L^{\to}(A)(x') \wedge f_L^{\to}(A)(y') \wedge a \leqslant f_L'(A)(x'-y'),\ x',y' \in F', y' \neq 0 \end{array} \right.\right\}$

$= \bigvee \left\{ a \in L \left| \begin{array}{l} a \leqslant \bigvee_{f(r)=1} A(r),\ \bigvee_{f(x)=x'} A(x) \wedge \bigvee_{f(y)=y'} A(y) \wedge a \leqslant \bigvee_{f(z)=x'y'^{-1}} A(z), \\ \bigvee_{f(x)=x'} A(x) \wedge \bigvee_{f(y)=y'} A(y) \wedge a \leqslant \bigvee_{f(z)=x'-y'} A(z),\ x',y' \in F', y' \neq 0 \end{array} \right.\right\}$

$\geqslant \bigvee \left\{ a \in L \left| \begin{array}{l} a \leqslant A(1),\ A(x) \wedge A(y) \wedge a \leqslant A(xy^{-1}), \\ A(x) \wedge A(y) \wedge a \leqslant A(x-y),\ x,y \in F, y \neq 0 \end{array} \right.\right\}$

$= \mathscr{F}(A).$

如果 f 是单射, 以上 $\geqslant$ 可以替换为 $=$, 即 $\mathscr{F}(A) = \mathscr{F}(f_L^{\to}(A))$.
(2) 的前半部分由以下事实可以证明:

$\mathscr{F}(f_L^{\leftarrow}(B))$

$$= f_L^{\leftarrow}(B)(1) \wedge \bigwedge_{x,y \in F} ((f_L^{\leftarrow}(B)(x) \wedge f_L^{\leftarrow}(B)(y)) \mapsto f_L^{\leftarrow}(B)(xy^{-1}))$$

$$\wedge \bigwedge_{x,y \in F} ((f_L^{\leftarrow}(B)(x) \wedge f_L^{\leftarrow}(B)(y)) \mapsto f_L^{\leftarrow}(B)(x-y))$$

$$= B(f(1)) \wedge \bigwedge_{x,y \in F} ((B(f(x)) \wedge B(f(y))) \mapsto B(f(x)f(y^{-1})))$$

$$\wedge \bigwedge_{x,y \in F} ((B(f(x)) \wedge B(f(y))) \mapsto B(f(x) - f(y)))$$

$$\geqslant B(1) \wedge \bigwedge_{x',y' \in F'} ((B(x') \wedge B(y')) \mapsto B(x'y'^{-1}))$$

$$\wedge \bigwedge_{x',y' \in F'} ((B(x') \wedge B(y')) \mapsto B(x' - y'))$$

$$= \mathscr{F}(B).$$

如果 f 是满射, 以上 $\geqslant$ 可以替换为 $=$, 即可得到 $\mathscr{F}(B) = \mathscr{F}(f_L^{\leftarrow}(B))$.

方法 2. (1) $\mathscr{F}(f_L^{\rightarrow}(A))$

$$= \bigvee \{a \in L \mid a \leqslant f_L^{\rightarrow}(A)(1), \forall b \in P(L), b \not\geqslant a,$$
$$(f_L^{\rightarrow}(A)^{(b)}) \text{ 是域 } F \text{ 的一个子域}\}$$

$$= \bigvee \{a \in L \mid a \leqslant f_L^{\rightarrow}(A)(1), \forall b \in P(L), b \not\geqslant a,$$
$$f_L^{\rightarrow}(A^{(b)}) \text{ 是域 } F \text{ 的一个子域}\}$$

$$\geqslant \bigvee \{a \in L \mid a \leqslant A(1), \forall b \in P(L), b \not\geqslant a,$$
$$A^{(b)} \text{ 是域 } F \text{ 的一个子域}\}$$

$$= \mathscr{F}(A),$$

如果 f 是单射, 以上 $\geqslant$ 可以替换为 $=$, 即 $\mathscr{F}(A) = \mathscr{F}(f_L^{\rightarrow}(A))$.

(2) $\mathscr{F}(f_L^{\leftarrow}(B))$

$$= \bigvee \{a \in L \mid \forall b \leqslant a \leqslant f_L^{\leftarrow}(B)(1), (f_L^{\leftarrow}(B)_{[b]}) \text{ 是域 } F \text{ 的一个子域}\}$$

$$= \bigvee \{a \in L \mid \forall b \leqslant a, f_L^{\leftarrow}(B_{[b]}) \text{ 是域 } F \text{ 的一个子域}\}$$

$$\geqslant \bigvee \{a \in L \mid \forall b \leqslant a \leqslant B(1), B_{[b]} \text{ 是域 } F \text{ 的一个子域}\}$$

$$= \mathscr{F}(B),$$

如果 f 是满射, 以上 $\geqslant$ 可以替换为 $=$, 即可得到 $\mathscr{F}(B) = \mathscr{F}(f_L^{\leftarrow}(B))$. □

5.9 由 L-模糊子域度确定的 L-模糊凸结构

由定理 5.9.2 的 (1) 和 (2), 可以得到下面定理.

定理 5.9.3 设 $\mathscr{F}$ 和 $\mathscr{F}'$ 分别是域 F 和 F' 上由 L-模糊子域度所确定的 L-模糊凸结构. 若 $f: F \to F'$ 是一个域同态, 则 $f: (F, \mathscr{F}) \to (F', \mathscr{F}')$ 是 L-模糊凸保持映射和 L-模糊凸到凸映射.

下面推论显而易见.

推论 5.9.4 设 $\mathscr{F}$ 和 $\mathscr{F}'$ 是 F 和 F' 上由 L-模糊子域度确定的 L-模糊凸结构. 若 $f: F \to F'$ 是一个域同构, 那么 $f: (F, \mathscr{F}) \to (F', \mathscr{F}')$ 是 L-模糊凸空间之间的 L-模糊同构.

定理 5.9.5 设 $\{F_i \mid i \in \Omega\}$ 是有限个域, $\forall i \in \Omega, A_i \in L^{F_i}$. 则

$$\mathscr{F}\left(\prod_{i \in \Omega} A_i\right) \geqslant \bigwedge_{i \in \Omega} \mathscr{F}(A_i).$$

证明 容易验证 $\prod_{i \in \Omega} A_i = \bigwedge_{i \in \Omega} P_i^{-1}(A_i)$, 其中 P_i 是 $\prod_{i \in \Omega} F_i$ 到 F_i 的投射. 由定理 5.9.1 证明中的 (2), 可以得到

$$\mathscr{F}\left(\prod_{i \in \Omega} A_i\right) = \mathscr{F}\left(\bigwedge_{i \in \Omega} P_i^{-1}(A_i)\right) \geqslant \bigwedge_{i \in \Omega} \mathscr{F}\left(P_i^{-1}(A_i)\right).$$

因为 $P_i: \prod_{i \in \Omega} F_i \to F_i$ 是域同态, 因此由定理 5.9.2 可知有 $\mathscr{F}(P_i^{-1}(A_i)) \geqslant \mathscr{F}(A_i)$. 于是便有

$$\mathscr{F}\left(\prod_{i \in \Omega} A_i\right) \geqslant \bigwedge_{i \in \Omega} \mathscr{F}(A_i). \qquad \square$$

由定理 5.9.5 可以直接得到下面推论.

推论 5.9.6 设 $\{F_i \mid i \in \Omega\}$ 是有限个域, $\forall i \in \Omega, A_i \in L^{F_i}$. 若 $\forall i \in \Omega, A_i$ 是 F_i 的 L-模糊子域. 则 $\prod_{i \in \Omega} A_i$ 是 $\prod_{i \in \Omega} F_i$ 的 L-模糊子域.

再结合定理 5.6.3 和定理 5.9.5 可得到下面推论.

推论 5.9.7 设 $\{F_i \mid i \in \Omega\}$ 是有限个域, $\mathscr{F}, \mathscr{F}_i$ 分别是 $\prod_{i \in \Omega} F_i, F_i$ 上由 L-模糊子域度确定的 L-模糊凸结构, 则 $P_i: \left(\prod_{i \in \Omega} F_i, \mathscr{F}\right) \to (F_i, \mathscr{F}_i)$ 是 L-模糊凸保持映射和 L-模糊凸到凸映射.

习 题 5

1. 令 $\mathbb{Z}_5$ 是模 5 的剩余类环, $\mathbb{Z}_5$ 上的运算为常规的加法和乘法. 定义三个 L-模糊集 $A, B, C \in L^{\mathbb{Z}_5}$ 如下:

$$A(z) = \begin{cases} 0.8, & z = [0], \\ 0.3, & z = [1], \\ 0.4, & z = [2], \\ 0.5, & z = [3], \\ 0.6, & z = [4], \end{cases} \quad B(z) = \begin{cases} 0, & z = [0], \\ 1, & z = [1], \\ 0.4, & z = [2], \\ 0.5, & z = [3], \\ 0.6, & z = [4], \end{cases}$$

$$C(z) = \begin{cases} 0.2, & z = [0], \\ 0.2, & z = [1], \\ 0.2, & z = [2], \\ 0.2, & z = [3], \\ 0.2, & z = [4], \end{cases}$$

试计算 A, B, C 的环度.

2. 试给出定理 5.2.6 的证明.

3. 令 $\mathbb{Z}_5$ 是模 5 的剩余类环, $\mathbb{Z}_5$ 上的运算为常规的加法和乘法. 定义三个 L-模糊集 $A_1, A_2, A_3 \in L^{\mathbb{Z}_5}$ 如下:

$$A_1(z) = \begin{cases} 0.8, & z = [0], \\ 0.3, & z = [1], \\ 0.4, & z = [2], \\ 0.5, & z = [3], \\ 0.6, & z = [4], \end{cases} \quad A_2(z) = \begin{cases} 0, & z = [0], \\ 1, & z = [1], \\ 0.4, & z = [2], \\ 0.5, & z = [3], \\ 0.6, & z = [4], \end{cases}$$

$$A_3(z) = \begin{cases} 0.2, & z = [0], \\ 0.2, & z = [1], \\ 0.2, & z = [2], \\ 0.2, & z = [3], \\ 0.2, & z = [4], \end{cases}$$

试计算 $A_1 A_2$ 和 $\sum\limits_{i=1}^{3} A_i$ 的环度.

4. 试用截集刻画定理的结论证明定理 5.3.4.

习 题 5

5. 令 $\mathbb{Z}_5$ 是模 5 的剩余类环, 定义三个 L-模糊集 $A, B, C \in L^{\mathbb{Z}_5}$ 如下:

$$A(z) = \begin{cases} 0.8, & z = [0], \\ 0.3, & z = [1], \\ 0.4, & z = [2], \\ 0.5, & z = [3], \\ 0.6, & z = [4], \end{cases} \quad B(z) = \begin{cases} 0, & z = [0], \\ 1, & z = [1], \\ 0.4, & z = [2], \\ 0.5, & z = [3], \\ 0.6, & z = [4], \end{cases}$$

$$C(z) = \begin{cases} 0.2, & z = [0], \\ 0.2, & z = [1], \\ 0.2, & z = [2], \\ 0.2, & z = [3], \\ 0.2, & z = [4], \end{cases}$$

试计算 A, B, C 以及 $A \wedge B \wedge C$ 的环度.

6. 证明定理 5.4.2.

7. 请写出例 5.5.2 的计算过程.

8. 证明推论 5.5.7.

9. 试用截集刻画定理的结论证明定理 5.6.4.

10. 请计算例 5.5.2 中 $A \wedge B \wedge C$ 的理想度.

11. 请仔细验证例 5.7.2.

12. 在 L-模糊理想度的定义中, 映射 $\mathscr{P} : L^R \to L$ 使得 $A \to \mathscr{P}(A)$ 是否构成一个 L-模糊凸结构? 试证明你的结论.

13. 一个 L-模糊素理想的同态像是不是一个 L-模糊素理想?

14. 一个 L-模糊素理想在同态映射下的原像是不是一个 L-模糊素理想?

15. 请详细验证例 5.8.3 和例 5.8.4.

16. 请计算例 5.8.3 中 $A \wedge B \wedge C$ 的模糊子域度.

17. 试用模糊子域的乘法群度和加法群度来表示子域度.

18. 试用截集刻画定理的结论证明定理 5.9.5.

第 6 章 L-模糊向量子空间

在模糊向量空间概念被提出后，其维数的概念也被提出了[99]，但是从这些维数定义出发得到的维数性质并不理想. 于是我与我的毕业生黄春娥共同提出了一种新模糊维数的定义[136]，取得了相当理想的结果. 在这一章中，我们首先引入模糊数的概念，其次借助模糊数给出模糊子空间维数的概念，并研究其性质和刻画. 在此之后，又提出 L-模糊向量子空间度的概念，并借助 L-模糊集的四种截集给出其等价刻画，然后研究 L-模糊向量子空间度与 L-模糊凸结构的关系，并进一步研究 L-模糊集在线性映射下的像和原像的子空间度，最后证明线性映射是 L-模糊凸保持映射和 L-模糊凸到凸映射.

6.1 模糊数与模糊自然数

模糊集的势是由 Zadeh L A 在 [166] 中首次引入的，可以很容易地看出这种势实际上就是模糊自然数 [82,133,134]，而模糊自然数又是模糊数的特殊情形. 本节我们将介绍什么是模糊数和模糊自然数，它们在模糊数学的各分支中的作用是非常重要的.

定义 6.1.1 设 L 是一个完备格. 一个 L-模糊数就是一个单调递减映射 $\lambda : \mathbb{R} \to L$ 的等价类 $[\lambda]$ 使得它满足

$$\lambda(-\infty) = \bigvee_{t \in \mathbb{R}} \lambda(t) = 1 \quad \text{且} \quad \lambda(+\infty) = \bigwedge_{t \in \mathbb{R}} \lambda(t) = 0,$$

这里 1 和 0 分别表示 L 中的最大元和最小元，两个映射 λ, μ 等价当且仅当 $\forall t \in (-\infty, +\infty)$, $\lambda(t-) = \mu(t-)$，或者等价地 $\lambda(t+) = \mu(t+)$.

所有 L-模糊数的集合记为 $\mathbb{R}(L)$.

定义 6.1.2 让 $\mathbb{N}$ 表示自然数的集合. 一个 L-模糊自然数就是一个单调递减的映射 $\lambda : \mathbb{N} \to L$ 使得它满足

$$\lambda(0) = 1, \quad \bigwedge_{n \in \mathbb{N}} \lambda(n) = 0.$$

如果存在 $n \in \mathbb{N}$ 使得 $\lambda(n) = 0$，那么就称 λ 为有限的 L-模糊自然数. 所有 L-模糊自然数的集合记为 $\mathbb{N}(L)$.

6.1 模糊数与模糊自然数

从上述定义可看出, 一个 L-模糊自然数可看成一个 L-模糊数.

定义 6.1.3 让 A 是一个模糊集, 定义一个映射 $|A|: \mathbb{N} \to L$ 使得 $\forall n \in \mathbb{N}$, $|A|(n) = \bigvee \{a \in L \mid |A_{[a]}| \geqslant n\}$. 那么 $|A| \in \mathbb{N}(L)$, 叫做 A 的势. 当 $L = [0,1]$ 时, 这就是 Zadeh L A [166] 所定义的模糊集的势.

容易看出模糊集合的势是一个模糊自然数.

定义 6.1.4 对任何 $\lambda, \mu \in \mathbb{N}(L)$, 定义 λ 与 μ 的和 $\lambda + \mu$ 如下: 对任何 $n \in \mathbb{N}$,
$$(\lambda + \mu)(n) = \bigvee_{k+l=n} (\lambda(k) \wedge \mu(l)).$$

注 6.1.5 设 λ 是一个 L-模糊自然数, 对任意的 $a \in L$, λ 的截集 $\lambda_{[a]}$ 是一个自然数的集合, 我们将不区别 $\lambda_{[a]}$ 与 $\max\{n \mid n \in \lambda_{[a]}\}$. 也就是说在下面的叙述中, $\lambda_{[a]}$ 都可以看成 $\max\{n \mid n \in \lambda_{[a]}\}$. 例如, 给定两个 L-模糊自然数 λ 和 μ, 为了方便, 我们常把 $\lambda_{[a]} \subseteq \mu_{[a]}$ 写成 $\lambda_{[a]} \leqslant \mu_{[a]}$. 另外, $n \in \lambda_{[a]}$ 也常常记为 $n \leqslant \lambda_{[a]}$.

定理 6.1.6 对任意的 $\lambda, \mu \in \mathbb{N}(L)$ 和对任意的 $a \in L$, 都有
(1) $\forall a \in J(L), (\lambda + \mu)_{[a]} = \lambda_{[a]} + \mu_{[a]}$;
(2) $\forall a \in P(L), (\lambda + \mu)^{(a)} = \lambda^{(a)} + \mu^{(a)}$.

证明 (1) 我们先来证明 $(\lambda + \mu)_{[a]} \leqslant \lambda_{[a]} + \mu_{[a]}$. 假设 $n \leqslant (\lambda + \mu)_{[a]}$, 那么
$$(\lambda + \mu)(n) = \bigvee_{k+l=n} \lambda(k) \wedge \mu(l) \geqslant a.$$

因为 a 是余素元, 所以存在 $k, l \in \mathbb{N}$ 使得 $n = k + l$ 且 $\lambda(k) \wedge \mu(l) \geqslant a$. 这意味着 $k \leqslant \lambda_{[a]}$ 且 $l \leqslant \mu_{[a]}$, 也就是 $n \leqslant \lambda_{[a]} + \mu_{[a]}$. 从而 $(\lambda + \mu)_{[a]} \leqslant \lambda_{[a]} + \mu_{[a]}$.

反之, 设 $n \leqslant \lambda_{[a]} + \mu_{[a]}$, 则存在 $k, l \in \mathbb{N}$ 使得 $n = k+l$, $\lambda(k) \geqslant a$ 且 $\mu(l) \geqslant a$. 于是
$$(\lambda + \mu)(n) = \bigvee_{k+l=n} \lambda(k) \wedge \mu(l) \geqslant a,$$

这意味着 $n \leqslant (\lambda + \mu)_{[a]}$. 因此 $(\lambda + \mu)_{[a]} \geqslant \lambda_{[a]} + \mu_{[a]}$.

这样我们就证明了 $(\lambda + \mu)_{[a]} = \lambda_{[a]} + \mu_{[a]}$.

(2) 我们先来证明 $(\lambda + \mu)^{(a)} \leqslant \lambda^{(a)} + \mu^{(a)}$. 假设 $n \leqslant (\lambda + \mu)^{(a)}$, 那么
$$(\lambda + \mu)(n) = \bigvee_{k+l=n} \lambda(k) \wedge \mu(l) \nleqslant a.$$

因为 a 是素元, 所以存在 $k, l \in \mathbb{N}$ 使得 $n = k + l$, $\lambda(k) \nleqslant a$ 且 $\mu(l) \nleqslant a$. 这意味着 $k \leqslant \lambda^{(a)}$ 且 $l \leqslant \mu^{(a)}$, 也就是 $n \leqslant \lambda^{(a)} + \mu^{(a)}$. 从而 $(\lambda + \mu)^{(a)} \leqslant \lambda^{(a)} + \mu^{(a)}$.

反之，设 $n \leqslant \lambda^{(a)} + \mu^{(a)}$，则存在 $k, l \in \mathbb{N}$ 使得 $n = k + l$，$\lambda(k) \not\leqslant a$ 且 $\mu(l) \not\leqslant a$. 于是 $(\lambda + \mu)(n) = \bigvee_{k+l=n} \lambda(k) \wedge \mu(l) \not\leqslant a$，这意味着 $n \leqslant (\lambda + \mu)^{(a)}$. 因此 $(\lambda + \mu)^{(a)} \geqslant \lambda^{(a)} + \mu^{(a)}$.

这样我们就证明了 $(\lambda + \mu)^{(a)} = \lambda^{(a)} + \mu^{(a)}$. □

定义 6.1.7 让 V 是域 F 上的一个向量空间且 $A \in L^V$. A 叫做 V 的一个 L-模糊子空间，如果它满足

$$A(kx + ly) \geqslant A(x) \wedge A(y), \quad \forall x, y \in V, \quad \forall k, l \in F.$$

类似于模糊子群的截集是子群的证明可以得到下面结果.

定理 6.1.8 让 V 是域 F 上的一个向量空间且 $A \in L^V$. 则下列条件等价：
(1) A 是 V 的 L-模糊子空间；
(2) $\forall a \in L$，$A_{[a]}$ 是 V 的子空间；
(3) $\forall a \in J(L)$，$A_{[a]}$ 是 V 的子空间；
(4) $\forall a \in L$，$A^{[a]}$ 是 V 的子空间；
(5) $\forall a \in P(L)$，$A^{[a]}$ 是 V 的子空间；
(6) $\forall a \in P(L)$，$A^{(a)}$ 是 V 的子空间.

定义 6.1.9 设 A, B 是向量空间 V 的两个 L-模糊子空间. 定义 A 与 B 的交 $A \wedge B$ 与和 $A + B$ 分别为 $\forall x \in V$,

$$(A \wedge B)(x) = A(x) \wedge B(x);$$

$$(A + B)(x) = \bigvee_{x = x_1 + x_2} (A(x_1) \wedge B(x_2)) = \bigvee_{x_1 \in V} (A(x_1) \wedge B(x - x_1)).$$

容易验证 $A \wedge B$ 和 $A + B$ 都是 V 的 L-模糊子空间.

定理 6.1.10 对于向量空间 V 的两个 L-模糊子空间 A 和 B 而言，下面等式成立：
(1) 对任意的 $a \in L$，$(A \cap B)_{[a]} = A_{[a]} \cap B_{[a]}$；
(2) 对任意的 $a \in P(L)$，$(A \cap B)^{(a)} = A^{(a)} \cap B^{(a)}$；
(3) 对任意的 $a \in P(L)$，$(A + B)^{(a)} = A^{(a)} + B^{(a)}$.

证明 由定义 6.1.9，我们容易得到 (1) 和 (2). 下面等式表明 (3) 成立. 对任意的 $a \in P(L)$,

$$x \in (A + B)^{(a)}$$
$$\Leftrightarrow \bigvee_{x_1 + x_2 = x} (A(x_1) \wedge B(x_2)) \not\leqslant a$$

$\Leftrightarrow \exists x_1, x_2$ 使得 $x_1 + x_2 = x$ 且 $A(x_1) \wedge B(x_2) \not\leq a$

$\Leftrightarrow \exists x_1, x_2$ 使得 $x_1 + x_2 = x$, $x_1 \in A^{(a)}$ 且 $x_2 \in B^{(a)}$

$\Leftrightarrow x \in A^{(a)} + B^{(a)}$. □

6.2 模糊向量空间的模糊基和模糊维数

在下面的结果中,我们总假定 $L = [0,1]$,而且 V 是一个有限维的向量空间. 在这一节中,我们给出模糊基的一个新定义.

让 V 是一个 n 维向量空间,A 是 V 的一个模糊子空间. Lwen [98] 证明了 $A(E)$ 是 $[0,1]$ 的有限子集. 容易证明下面引理.

引理 6.2.1 如果 A 是向量空间 V 的模糊子空间,那么存在一个有限系列 $1 = \alpha_0 \geqslant \alpha_1 > \alpha_2 > \cdots > \alpha_r \geqslant 0$ 使得

(1) 如果 $a, b \in (\alpha_{i+1}, \alpha_i]$,那么 $A_{[a]} = A_{[b]}$;

(2) 如果 $a \in (\alpha_{i+1}, \alpha_i]$ 且 $b \in (\alpha_i, \alpha_{i-1}]$,那么 $A_{[a]} \supsetneq A_{[b]}$;

(3) 如果 $a, b \in [\alpha_{i+1}, \alpha_i)$,那么 $A_{(a)} = A_{(b)}$;

(4) 如果 $a \in [\alpha_{i+1}, \alpha_i)$ 且 $b \in [\alpha_i, \alpha_{i-1})$,那么 $A_{(a)} \supsetneq A_{(b)}$.

对于一个模糊子空间 A,借助于引理 6.2.1,我们能够得到一族子空间如下:

$$\{0\} \subseteq A_{[\alpha_1]} \subsetneq A_{[\alpha_2]} \subsetneq \cdots \subsetneq A_{[\alpha_r]} \subseteq V. \qquad (6.2.1)$$

我们称 (6.2.1) 为 A 的不可约水平子空间族. 假设 $A_{[\alpha_1]} \neq \{0\}$,否则我们就选择 $A_{[\alpha_2]}$. 让 B_{α_1} 是 $A_{[\alpha_1]}$ 的一个基. 靠扩张 B_{α_1},我们能够得到 $A_{[\alpha_2]}$ 的一个基 B_{α_2}. 进一步靠扩张 B_{α_2} 我们能够得到 $A_{[\alpha_3]}$ 的基 B_{α_3}. 类似地,继续下去,靠扩张 B_{r-1},我们能够得到 $A_{[\alpha_r]}$ 的基 B_{α_r}. 这样我们就得到了一个系列

$$B_{\alpha_1} \subsetneq B_{\alpha_2} \subsetneq B_{\alpha_3} \subsetneq \cdots \subsetneq B_{\alpha_r}, \qquad (6.2.2)$$

这里 B_{α_i} 是 $A_{[\alpha_i]}(1 \leqslant i \leqslant r)$ 的一个基. 于是我们能够定义 V 的一个模糊子集 β 如下:

$$\beta(x) = \bigvee \{\alpha_i \mid x \in B_{\alpha_i}\}, \qquad (6.2.3)$$

其中 β 称为 A 对应于 (6.2.3) 的模糊基.

下面定理是平凡的.

定理 6.2.2 令 β 是借助 (6.2.3) 得到的 A 的模糊基. 那么下面条件成立:
(1) 若 $a, b \in (\alpha_{i+1}, \alpha_i)$, 则 $\beta_{[a]} = \beta_{[b]} = B_{\alpha_i}$;
(2) 若 $a \in (\alpha_{i+1}, \alpha_i]$ 且 $b \in (\alpha_i, \alpha_{i-1}]$, 则 $\beta_{[a]} \supsetneq \beta_{[b]}$;
(3) 若 $a, b \in [\alpha_{i+1}, \alpha_i)$, 则 $\beta_{(a)} = \beta_{(b)} = B_{\alpha_{i+1}}$;
(4) 若 $a \in [\alpha_{i+1}, \alpha_i)$ 且 $b \in [\alpha_i, \alpha_{i-1})$, 则 $\beta_{(a)} \supsetneq \beta_{(b)}$.

推论 6.2.3 设 β 是由 (6.2.3) 得到的 A 的模糊基. 则下面条件成立:
(1) $a \in (0, 1]$, $\beta_{[a]}$ 是 $A_{[a]}$ 的基;
(2) $a \in [0, 1)$, $\beta_{(a)}$ 是 $A_{(a)}$ 的基.

从上面推论我们很容易得到下面推论.

推论 6.2.4 设 A 是 V 的模糊子空间, 并设 β_1 和 β_2 是 A 的两个基. 则下列条件等价:
(1) 对任意的 $a \in (0, 1]$, $|(\beta_1)_{[a]}| = |(\beta_2)_{[a]}|$;
(2) 对任意的 $a \in [0, 1)$, $|(\beta_1)_{(a)}| = |(\beta_2)_{(a)}|$;
(3) $|\beta_1| = |\beta_2|$.

接下来, 我们重新定义模糊向量空间的模糊维数. 一个分明向量空间的维数能够看作它的基的势. 类似地, 我们能够定义模糊向量空间的维数就是它的模糊基的势.

定义 6.2.5 设 A 是带有模糊基 β 的模糊向量空间. 定义 $\dim(A) = |\beta|$, 则称 $\dim(A)$ 为 A 的模糊维数.

定理 6.2.6 设 A 是带有模糊基 β 的模糊向量空间. 则
$$\dim(A)(n) = \bigvee \{a \in [0,1) \mid |\beta_{(a)}| \geqslant n\}$$
$$= \bigvee \{a \in [0,1) \mid \dim(A_{(a)}) \geqslant n\}.$$

证明 由推论 6.2.3 可知 $\dim(A_{(a)}) = |\beta_{(a)}|$. 对任意的 $n \in \mathbb{N}$, 令
$$\lambda = \bigvee \{a \in [0,1) \mid \dim(A_{(a)}) \geqslant n\}.$$
显然有
$$\lambda \leqslant \dim(A)(n) = \bigvee \{a \in (0,1] \mid |\beta_{[a]}| \geqslant n\}.$$
为了证明 $\lambda \geqslant \dim(A)(n)$, 假设 $\dim(A)(n) \neq 0$ 且 $\dim(A)(n) > b$. 则存在 $a > b$ 使得 $|\beta_{[a]}| \geqslant n$. 此时有 $n \leqslant |\beta_{[a]}| \leqslant |\beta_{(b)}| \leqslant |\beta_{[b]}|$. 这意味着 $\lambda = \bigvee \{a \in [0,1) : \dim(A)_{(a)} \geqslant n\} \geqslant b$. 这样我们就得到
$$\lambda \geqslant \bigvee \{b : 0 \leqslant b < \dim(A)(n)\} = \dim(A)(n). \quad \square$$

定理 6.2.7 设 A 是 V 的模糊子空间. 则
(1) 对任意的 $a \in (0,1]$, $(\dim(A))_{[a]} = \dim(A_{[a]})$;
(2) 对任意的 $a \in [0,1)$, $(\dim(A))_{(a)} = \dim(A_{(a)})$.

证明 设
$$\{0\} \subseteq A_{[\alpha_1]} \subsetneq A_{[\alpha_2]} \subsetneq \cdots \subsetneq A_{[\alpha_r]} \subseteq V$$
是 A 的不可约水平子空间族.

(1) 从 $\dim(A)$ 的定义我们知道对任意的 $a \in (0,1]$, 有 $\dim(A_{[a]}) \leqslant (\dim(A))_{[a]}$. 现在需要证明 $(\dim(A))_{[a]} \leqslant \dim(A_{[a]})$. 从模糊维数的定义, 我们能够得到

$$n \leqslant (\dim(A))_{[a]} \Rightarrow \dim(A)(n) \geqslant a$$
$$\Rightarrow \bigvee\{\alpha_i \mid \dim(A_{[\alpha_i]}) \geqslant n\} \geqslant a$$
$$\Rightarrow \exists \alpha_i \geqslant a \text{ 使得 } n \leqslant \dim(A_{[\alpha_i]})$$
$$\Rightarrow n \leqslant \dim(A_{[\alpha_i]}) \leqslant \dim(A_{[a]}).$$

因此对任意的 $a \in (0,1]$, 有 $(\dim(A))_{[a]} = \dim(A_{[a]})$ 成立.

(2) 为了证明 $(\dim(A))_{(a)} \leqslant \dim(A_{(a)})$, 我们假设 $n = (\dim(A))_{(a)}$. 则 $\dim(A)(n) > a$, 即, $\bigvee\{b \in (0,1] \mid \dim(A_{[b]}) \geqslant n\} > a$. 从而存在 $b \in (0,1]$ 使得 $a < b$ 且 $n \leqslant \dim(A_{[b]})$. 因为 $A_{[b]} \subseteq A_{(a)}$, 所以 $n \leqslant \dim(A_{(a)})$. 因此有 $(\dim(A))_{(a)} \leqslant \dim(A_{(a)})$.

为了证明 $\dim(A_{(a)}) \leqslant (\dim(A))_{(a)}$, 取 $\alpha_i > a$ 使得 $A_{(a)} = A_{[\alpha_i]}$. 那么容易得到

$$\dim(A_{(a)}) = \dim(A_{[\alpha_i]}) \leqslant (\dim(A))_{[\alpha_i]} \leqslant (\dim(A))_{(a)}. \qquad \square$$

定理 6.2.8 设 A 和 B 是 V 的两个模糊子空间, 则
$$\dim(A+B) + \dim(A \cap B) = \dim(A) + \dim(B).$$

证明 对任意的 $a \in [0,1)$, 由定理 6.1.6、定理 6.1.10 和定理 6.2.7 我们能够得到

$$(\dim(A+B) + \dim(A \cap B))_{(a)}$$
$$= (\dim(A+B))_{(a)} + (\dim(A \cap B))_{(a)}$$
$$= \dim((A+B)_{(a)}) + \dim((A \cap B)_{(a)})$$

$$= \dim\left((A)_{(a)} + (B)_{(a)}\right) + \dim\left((A)_{(a)} \cap (B)_{(a)}\right)$$
$$= \dim\left((A)_{(a)}\right) + \dim\left((B)_{(a)}\right)$$
$$= (\dim(A))_{(a)} + (\dim(B))_{(a)}$$
$$= (\dim(A) + \dim(B))_{(a)}$$

因此 $\dim(A+B) + \dim(A \cap B) = \dim(A) + \dim(B)$. □

定义 6.2.9 设 A 是 V 的模糊子空间且 $f: V \to V$ 是一个线性变换. 如果对任意的 $x \in V$, $A(f(x)) \geqslant A(x)$, 那么就称 f 是 A 的一个线性变换.

引理 6.2.10 设 A 是 V 的模糊子空间且 f 是 A 的一个线性变换, 则 $\widetilde{\ker f} = A|_{\ker f}$ 与 $\widetilde{\mathrm{im} f} = A|_{\mathrm{im} f}$ 是两个模糊子空间, 这里 $\ker f$ 与 $\mathrm{im} f$ 分别表示 f 关于 V 的核与像, $A|_{\ker f}$ 与 $A|_{\mathrm{im} f}$ 分别表示 A 在 $\ker f$ 与 $\mathrm{im} f$ 上的限制.

证明是简单的, 故略去.

定理 6.2.11 设 A 是 V 的模糊子空间且 $f: V \to V$ 是 A 上的线性变换, 则
$$\dim(\widetilde{\ker f}) + \dim(\widetilde{\mathrm{im} f}) = \dim(A).$$

证明 由定理 6.1.6 和定理 6.2.7 可知, 对任意的 $a \in [0, 1)$,
$$\left(\dim(\widetilde{\mathrm{im} f}) + \dim(\widetilde{\ker f})\right)_{(a)} = \left(\dim(\widetilde{\mathrm{im} f})\right)_{(a)} + \left(\dim(\widetilde{\ker f})\right)_{(a)}$$
$$= \dim\left((\widetilde{\mathrm{im} f})_{(a)}\right) + \dim\left((\widetilde{\ker f})_{(a)}\right)$$
$$= \dim\left(A_{(a)} \cap \mathrm{im} f\right) + \dim\left(A_{(a)} \cap \ker f\right).$$

容易验证 $f|_{A_{(a)}}$ 是 $A_{(a)}$ 上的一个线性变换. 于是我们有
$$\left(\dim(\widetilde{\mathrm{im} f}) + \dim(\widetilde{\ker f})\right)_{(a)} = \dim(\mathrm{im} f|_{A_{(a)}}) + \dim(\ker f|_{A_{(a)}})$$
$$= \dim\left(A_{(a)}\right) = (\dim(A))_{(a)}.$$

因此 $\dim(\widetilde{\ker f}) + \dim(\widetilde{\mathrm{im} f}) = \dim(A)$. □

6.3 L-模糊向量子空间度

设 V 是域 F 上的一个向量空间, A 是 V 的一个 L-模糊子集, A 称为 V 的一个 L-模糊子空间, 如果它满足下面条件:
$$A(kx + ly) \geqslant A(x) \wedge A(y), \quad \forall x, y \in V, \quad \forall k, l \in F.$$

6.3 L-模糊向量子空间度

给定 V 的一个 L-模糊子集 A, 从上述定义可知, A 要么是 V 的一个 L-模糊子空间, 要么不是, 二者必居其一, 且仅居其一, 没有模糊性可言. 下面我们给出其模糊化的一种新的方法.

定义 6.3.1 设 A 是域 F 上的向量空间 V 的一个 L-模糊子集. 定义

$$\mathscr{V}(A) = \bigwedge_{x,y\in V, k,l\in F} (A(x) \wedge A(y) \mapsto A(kx+ly)).$$

称 $\mathscr{V}(A)$ 为 A 的 L-模糊子空间度, 容易验证 A 是向量空间 V 的一个 L-模糊子空间当且仅当 $\mathscr{V}(A) = 1$.

由以上定义及蕴含性质可以得到下面引理.

引理 6.3.2 设 A 是向量空间 V 的一个 L-模糊子集. 则 $\mathscr{V}(A) \geqslant a$ 当且仅当对于任意的 $x, y \in E, k, l \in F$,

$$A(x) \wedge A(y) \wedge a \leqslant A(kx + ly).$$

由上述引理易得 $\mathscr{V}(A)$ 的等价刻画如下.

定理 6.3.3 设 A 是域 F 上的向量空间 V 的一个 L-模糊子集. 则

$$\mathscr{V}(A) = \bigvee \{a \in L \mid A(x) \wedge A(y) \wedge a \leqslant A(kx+ly), \forall x, y \in V, \forall k, l \in F\}.$$

下面利用 L-模糊集的四种截集, 给出 L-模糊向量子空间度的等价刻画.

定理 6.3.4 设 A 是向量空间 V 的一个 L-模糊子集. 则
(1) $\mathscr{V}(A) = \bigvee \{a \in L \mid \forall b \leqslant a, \mu_{[b]}$ 是 V 的子空间$\}$;
(2) $\mathscr{V}(A) = \bigvee \{a \in L \mid \forall b \notin \alpha(a), \mu^{[b]}$ 是 V 的子空间$\}$;
(3) $\mathscr{V}(A) = \bigvee \{a \in L \mid \forall b \in P(L), b \not\geqslant a, \mu^{(b)}$ 是 V 的子空间$\}$;
(4) 若 $\forall a, b \in L$, 有 $\beta(a \wedge b) = \beta(a) \cap \beta(b)$, 则
$$\mathscr{V}(A) = \bigvee \{a \in L \mid \forall b \in \beta(a), \mu_{(b)} \text{ 是 } V \text{ 的子空间}\}.$$

证明 仅给出 (2), (4) 的证明过程, (1), (3) 类似可证.

(2) 假设 $A(x) \wedge A(y) \wedge a \leqslant A(kx+ly), \forall x, y \in E, \forall k, l \in F$. 那么 $\forall b \notin \alpha(a)$ 且 $\forall x, y \in \mu^{[b]}$, 我们有

$$b \notin \alpha(A(x)) \cup \alpha(A(y)) \cup \alpha(a) = \alpha(A(x) \wedge A(y) \wedge a).$$

由 $A(x) \wedge A(y) \wedge a \leqslant A(kx + ly)$ 可知

$$\alpha(A(kx+ly)) \subseteq \alpha(A(x) \wedge A(y) \wedge a).$$

因此 $b \notin \alpha(A(kx+ly))$, 也就是 $kx + ly \in \mu^{[b]}$. 这意味着 $\mu^{[b]}$ 是 V 的子空间, 并且

$$a \in \{a \in L \mid \forall b \notin \alpha(a), \mu^{[b]} \text{ 是 } V \text{ 的子空间}\},$$

这表明

$$\mathscr{V}(A) = \bigvee \{a \in L \mid A(x) \wedge A(y) \wedge a \leqslant A(kx+ly), x,y \in V, k,l \in F\}$$
$$\leqslant \bigvee \{a \in L \mid \forall b \notin \alpha(a), \mu^{[b]} \text{ 是 } V \text{ 的子空间}\}.$$

相反地，假设 $a \in \{a \in L \mid \forall b \notin \alpha(a), \mu^{[b]} \text{ 是 } V \text{ 的子空间}\}$，接下来我们来证明

$$A(x) \wedge A(y) \wedge a \leqslant A(kx+ly), \quad \forall x,y \in V, \quad \forall k,l \in F.$$

设 $b \notin \alpha(A(x) \wedge A(y) \wedge a)$. 由

$$\alpha(A(x) \wedge A(y) \wedge a) = \alpha(A(x)) \cup \alpha(A(y)) \cup \alpha(a)$$

可得 $b \notin \alpha(a)$ 且 $x,y \in \mu^{[b]}$. 因为 $A^{[b]}$ 是 V 的子空间，所以 $kx+ly \in A^{[b]}$，即 $b \notin \alpha(A(kx+ly))$，从而我们有

$$A(x) \wedge A(y) \wedge a \leqslant A(kx+ly),$$

这意味着

$$\mathscr{V}(A) = \bigvee \{a \in L \mid A(x) \wedge A(y) \wedge a \leqslant A(kx+ly), x,y \in V, k,l \in F\}$$
$$\geqslant \bigvee \{a \in L \mid \forall b \notin \alpha(a), A^{[b]} \text{ 是 } V \text{ 的子空间}\}.$$

这样就证明了 (2).

(4) 假设 $a \in \{a \in L \mid A(x) \wedge A(y) \wedge a \leqslant A(kx+ly), x,y \in V, k,l \in F\}$. 那么 $\forall b \in \beta(a)$ 和 $\forall x,y \in A_{(b)}$，我们得到

$$b \in \beta(A(x)) \cap \beta(A(y)) \cap \beta(a) = \beta(A(x) \wedge A(y) \wedge a) \subseteq \beta(A(kx+ly)),$$

即 $kx+ly \in A_{(b)}$，因此 $A_{(b)}$ 是 V 的一个子空间. 这意味着

$$\mathscr{V}(A) = \bigvee \{a \in L \mid A(x) \wedge A(y) \wedge a \leqslant A(kx+ly), x,y \in V, l,k \in F\}$$
$$\leqslant \bigvee \{a \in L \mid \forall b \in \beta(a), A_{(b)} \text{ 是 } V \text{ 的子空间}\}.$$

相反地，假设 $a \in \{a \in L \mid \forall b \in \beta(a), A_{(b)} \text{ 是 } V \text{ 的一个子空间}\}$. 接下来我们来证明

$$A(x) \wedge A(y) \wedge a \leqslant A(kx+ly), \quad \forall x,y \in V, \quad k,l \in F.$$

设 $b \in \beta(A(x) \wedge A(y) \wedge a)$，由

$$\beta(A(x) \wedge A(y) \wedge a) = \beta(A(x)) \cap \beta(A(y)) \cap \beta(a)$$

可知 $x,y \in A_{(b)}$ 且 $b \in \beta(a)$. 因为 $A_{(b)}$ 是 V 的一个子空间, 所以 $kx+ly \in A_{(b)}$, 即 $b \in \beta(A(kx+ly))$, 这表明 $A(x) \wedge A(y) \wedge a \leqslant A(kx+ly)$. 因此

$$\mathscr{V}(A) = \bigvee \{a \in L \mid A(x) \wedge A(y) \wedge a \leqslant A(kx+ly), x,y \in V, l,k \in F\}$$

$$\geqslant \bigvee \{a \in L \mid \forall b \in \beta(a), A_{(b)} \text{ 是 } V \text{ 的子空间}\}.$$

这样就证明了 (4). □

6.4 由 L-模糊子空间度确定的 L-模糊凸结构

在这一节中, 我们将研究 L-模糊子空间度与 L-模糊凸结构之间的关系.

对于每个 $A \in L^V$, 都有一个 $\mathscr{V}(A)$ 与之对应, 因此 $A \mapsto \mathscr{V}(A)$ 就确定了一个映射 $\mathscr{V}: L^V \to L$. 下面定理说明 $\mathscr{V}$ 恰好是 V 上的一个 L-模糊凸结构.

定理 6.4.1 设 V 是一个向量空间. 则由 $A \mapsto \mathscr{V}(A)$ 所确定的映射 $\mathscr{V}: L^V \to L$ 是 V 上的一个 L-模糊凸结构, 称为由 L-模糊子空间度确定的 L-模糊凸结构.

证明 (1) 显然我们有 $\mathscr{V}(\chi_\varnothing) = \mathscr{V}(\chi_X) = 1$.

(2) 设 $\{A_i \mid i \in \Omega\}$ 是 L^V 中的一族 L-模糊集, 我们来证明

$$\mathscr{V}\left(\bigwedge_{i \in \Omega} A_i\right) \geqslant \bigwedge_{i \in \Omega} \mathscr{V}(A_i).$$

为此假设 $a \leqslant \bigwedge_{i \in \Omega} \mathscr{V}(A_i)$, 由引理 6.3.2 可知, 对于任意的 $x,y \in V, k,l \in F$, 都有

$$A_i(x) \wedge A_i(y) \wedge a \leqslant A_i(kx+ly).$$

因此

$$\bigwedge_{i \in \Omega} A_i(x) \wedge \bigwedge_{i \in \Omega} A_i(y) \wedge a \leqslant \bigwedge_{i \in \Omega} A_i(kx+ly),$$

这表明, $a \leqslant \mathscr{V}\left(\bigwedge_{i \in \Omega} A_i\right)$. 于是可得

$$\mathscr{V}\left(\bigwedge_{i \in \Omega} A_i\right) \geqslant \bigwedge_{i \in \Omega} \mathscr{V}(A_i).$$

(3) 设 $\{A_i \mid i \in \Omega\} \subseteq L^V$ 是非空和全序的. 为证 $\bigwedge_{i \in \Omega} \mathscr{V}(A_i) \leqslant \mathscr{V}\left(\bigvee_{i \in \Omega} A_i\right)$, 需要证明对于任意 $a \leqslant \bigwedge_{i \in \Omega} \mathscr{V}(A_i)$, 有 $a \leqslant \mathscr{V}\left(\bigvee_{i \in \Omega} A_i\right)$ 成立. 为此设 $a \leqslant$

$\bigwedge_{i\in\Omega} \mathscr{V}(A_i)$. 由引理 6.3.2, 对于任意的 $i \in \Omega$ 和对于任意的 $x, y \in V, k, l \in F$, 都有

$$A_i(x) \wedge A_i(y) \wedge a \leqslant A_i(kx + ly).$$

设 $b \in L$ 使得

$$b \prec \left(\bigvee_{i\in\Omega} A_i(x)\right) \wedge \left(\bigvee_{i\in\Omega} A_i(y)\right) \wedge a,$$

进一步可得

$$b \prec \bigvee_{i\in\Omega} \mu_i(x), \quad b \prec \bigvee_{i\in\Omega} \mu_i(y), \quad b \leqslant a,$$

于是存在 $i, j \in \Omega$ 使得 $b \leqslant A_i(x), b \leqslant A_j(y)$ 且 $b \leqslant a$. 因为 $\{A_i | i \in \Omega\}$ 是全序的, 所以不妨假设 $A_j \leqslant A_i$, 那么进一步有 $b \leqslant A_i(x) \wedge A_i(y) \wedge a$. 由

$$A_i(x) \wedge A_i(y) \wedge a \leqslant A_i(kx + ly),$$

我们可以得到 $b \leqslant A_i(kx + ly)$, 这意味着 $b \leqslant \left(\bigvee_{i\in\Omega} A_i\right)(kx + ly)$. 这样, 由 b 的任意性可得

$$\left(\bigvee_{i\in\Omega} A_i\right)(x) \wedge \left(\bigvee_{i\in\Omega} A_i\right)(y) \wedge a \leqslant \left(\bigvee_{i\in\Omega} A_i\right)(kx + ly).$$

结合引理 6.3.2, 可得 $a \leqslant \mathscr{V}\left(\bigvee_{i\in\Omega} A_i\right)$. 由 a 的任意性, 进一步有

$$\bigwedge_{i\in\Omega} \mathscr{V}(A_i) \leqslant \mathscr{V}\left(\bigvee_{i\in\Omega} A_i\right).$$

因此 $\mathscr{V}$ 是 V 上的 L-模糊凸结构. □

接下来我们研究 L-模糊子集的同态像和原像的 L-模糊子空间度.

定理 6.4.2 设 $f : V \to V'$ 是向量空间之间的线性映射, $A \in L^V, B \in L^{V'}$, 则

(1) $\mathscr{V}(f_L^{\to}(A)) \geqslant \mathscr{V}(A)$; 如果 f 是单射, $\mathscr{V}(f_L^{\to}(A)) = \mathscr{V}(A)$.

(2) $\mathscr{V}(f_L^{\leftarrow}(B)) \geqslant \mathscr{V}(B)$; 如果 f 是满射, $\mathscr{V}(f_L^{\leftarrow}(B)) = \mathscr{V}(B)$.

证明 (1) 由定理 6.3.4 可得

$$\mathscr{V}(f_L^{\rightarrow}(A)) = \bigvee\{a \in L \mid \forall b \in P(L), b \not\geq a, (f_L^{\rightarrow}(A))^{(b)} \text{ 是 } V \text{ 的子空间}\}$$

$$= \bigvee\{a \in L \mid \forall b \in P(L), b \not\geq a, f_L^{\rightarrow}(A^{(b)}) \text{ 是 } V \text{ 的子空间}\}$$

$$\geq \bigvee\{a \in L \mid \forall b \in P(L), b \not\geq a, A^{(b)} \text{ 是 } V \text{ 的子空间}\}$$

$$= \mathscr{V}(A).$$

若 f 是单射, 则上述 $\geq$ 可换成 $=$, 因此 $\mathscr{V}(A) = \mathscr{V}(f_L^{\rightarrow}(A))$.

(2) 同样由定理 6.3.4可得

$$\mathscr{V}(f_L^{\leftarrow}(B)) = \bigvee\{a \in L \mid \forall b \leq a, (f_L^{\leftarrow}(B))_{[b]} \text{ 是 } V \text{ 的子空间}\}$$

$$= \bigvee\{a \in L \mid \forall b \leq a, f_L^{\leftarrow}(B_{[b]}) \text{ 是 } V \text{ 的子空间}\}$$

$$\geq \bigvee\{a \in L \mid \forall b \leq a, B_{[b]} \text{ 是 } V \text{ 的子空间}\}$$

$$= \mathscr{V}(B).$$

若 f 是满射, 则上述 $\geq$ 可换成 $=$, 因此我们可以得到 $\mathscr{V}(B) = \mathscr{V}(f_L^{\leftarrow}(B))$. □

根据定理 6.4.2, 显然有下面定理成立.

推论 6.4.3 设 $f: V \to V'$ 是线性映射, 令 $\mathscr{V}$ 和 $\mathscr{V}'$ 分别是 V 和 V' 上由 L-模糊子空间度确定的 L-模糊凸结构. 则 $f: (V, \mathscr{V}) \to (V', \mathscr{V}')$ 是 L-模糊凸保持映射和 L-模糊凸到凸映射.

习 题 6

1. 证明定理 6.1.8.

2. 设 A, B 是向量空间 V 的两个 L-模糊子空间. 试验证 $A \wedge B$ 和 $A + B$ 都是 V 的 L-模糊子空间.

3. 在定理 6.1.10 中, 条件 (2) 能否替换为 $(A \cap B)_{(a)} = A_{(a)} \cap B_{(a)}$ 和 $(A \cap B)^{[a]} = A^{[a]} \cap B^{[a]}$? 条件 (3) 能否替换为 $(A + B)^{[a]} = A^{[a]} + B^{[a]}$? 如果不能, 那么在什么条件下可以?

4. 证明定理 6.2.2、推论 6.2.3 和推论 6.2.4.

5. 请举出一个向量空间中三个模糊集, 使得它们的子空间度分别是 1, 0 和 0.5.

6. 本节我们取的模糊集都是格值为 $[0,1]$ 的情形, 假如格值不是 $[0,1]$, 那么如何定义模糊基和模糊维数呢? 我们把它留作公开问题.

7. 试用截集形式给出定理 6.4.1 证明中 (2) 的证明.

第 7 章　L-模糊子格

模糊格的概念最早源于文献 [165]，后被推广到一般格值情形[146]．实际上关于模糊格的定义有很多种，例如，从模糊偏序出发可以定义一种模糊格，这方面已有很多漂亮的工作，见 [94,162] 等．本章从一个给定格出发，考虑其上的格运算在模糊集上的遗传形式，这样的模糊子格自然构成一个凸结构，类似地，它的所有 L-模糊子格自然形成一个 L-凸结构．

7.1　L-模糊子格度

下面我们给出 L-模糊子格的定义．

定义 7.1.1　设 $(M,\vee,\wedge)$ 是一个格，$A\in L^M$ 称为 M 的一个 L-模糊子格，如果它满足下面条件：

(1) $A(x\vee y)\geqslant A(x)\wedge A(y),\forall x,y\in M$;

(2) $A(x\wedge y)\geqslant A(x)\wedge A(y),\forall x,y\in M$.

一个 L-模糊子格 A 叫做 L-模糊凸子格，如果它再满足下面条件：

(3) $\forall x,y,z\in M$，当 $x\wedge y\leqslant z\leqslant x\vee y$ 时，有 $A(x\wedge y)\wedge A(x\vee y)\leqslant A(z)$.

注 7.1.2　显然，对于一个格 M 及其运算 $\vee,\wedge$ 来说，A 是 M 的一个 L-模糊子格当且仅当 A 分别是 M 的 L-模糊 $\vee$-子半群和 $\wedge$-子半群．

对于 M 的一个 L-模糊子集 A 而言，它可能是 M 的 L-模糊子格，也可能不是，二者必居其一且仅居其一．在这一章中，我们将以蕴含算子为工具，引入 L-模糊子格度的概念，从而使得 M 的任意一个 L-模糊子集在一定程度上都是子格，这个 L-模糊子格度恰好确定了一个 L-模糊凸结构．

定义 7.1.3　设 $(M,\vee,\wedge)$ 是一个格，$A\in L^M$．令

$$\mathscr{L}(A)=\bigwedge_{x,y\in X}(A(x)\wedge A(y)\mapsto A(x\wedge y)\wedge A(x\vee y)).$$

则称 $\mathscr{L}(A)$ 为 A 的 L-模糊子格度．

注 7.1.4　显然 A 是一个 L-模糊子格当且仅当 $\mathscr{L}(A)=1$，所以 L-模糊子格度是 L-模糊子格的一种广义的定义．

7.1 L-模糊子格度

例 7.1.5 设 $M = \{\varnothing, \{a\}, \{b\}, \{c\}, \{a,b,c\}\}$ 是幂集格 $2^{\{a,b,c\}}$ 的子格, $L = [0,1]$. 定义 $A \in L^M$ 为

$$A(\varnothing) = 0.64, \quad A(\{a\}) = 0.82, \quad A(\{b\}) = 0.91,$$

$$A(\{c\}) = 0.45, \quad A(\{a,b,c\}) = 0.85.$$

则

$$A(x) \wedge A(y) \mapsto A(x \wedge y) \wedge A(x \vee y)$$

$$= \begin{cases} 1, & \text{当 } x = y \text{ 时,} \\ 1, & \text{当 } x,y \text{ 中有一个是 } \varnothing \text{ 或者 } \{a,b,c\} \text{ 时,} \\ 1, & \text{当 } x,y \text{ 中有一个是 } \varnothing \text{ 或者 } \{a,b,c\} \text{ 时,} \\ 1, & \text{当 } x,y \text{ 中有一个是 } \{c\} \text{ 时,} \\ 0.64, & \text{当 } x,y \text{ 分别是 } \{a\}, \{b\} \text{ 时.} \end{cases}$$

因此 $\mathscr{L}(A) = 0.64$.

再定义 $B \in L^M$ 为

$$B(\varnothing) = 0.35, \quad B(\{a\}) = 0.32, \quad B(\{b\}) = 0.32,$$

$$B(\{c\}) = 0.48, \quad B(\{a,b,c\}) = 0.32.$$

则

$$A(x) \wedge A(y) \mapsto A(x \wedge y) \wedge A(x \vee y)$$

$$= \begin{cases} 1, & \text{当 } x = y \text{ 时,} \\ 1, & \text{当 } x,y \text{ 中有一个是 } \varnothing \text{ 或者 } \{a,b,c\} \text{ 时,} \\ 1, & \text{当 } x,y \text{ 中有一个是 } \varnothing \text{ 或者 } \{a,b,c\} \text{ 时,} \\ 1, & \text{当 } x,y \text{ 分别是 } \{a\}, \{b\}, \{c\} \text{ 时.} \end{cases}$$

因此 $\mathscr{L}(A) = 1$, 也就是说 B 就是一个 L-模糊凸子格.

不难验证下面事实.

$$\mathscr{L}(A) = \bigwedge_{x,y \in X} ((A(x) \wedge A(y) \to A(x \wedge y)) \wedge (A(x) \wedge A(y) \to A(x \vee y))).$$

由蕴含算子的性质, 可得下面的引理.

引理 7.1.6 设 $(M, \vee, \wedge)$ 是一个格，$A \in L^M$. 则 $\mathscr{L}(A) \geqslant a$ 当且仅当任意的 $x, y \in M$,
$$A(x) \wedge A(y) \wedge a \leqslant A(x \wedge y) \wedge A(x \vee y)$$
或
$$A(x) \wedge A(y) \wedge a \leqslant A(x \wedge y) \quad \text{且} \quad A(x) \wedge A(y) \wedge a \leqslant A(x \vee y).$$

由引理 7.1.6, 我们不难得到下面定理.

定理 7.1.7 设 $(M, \vee, \wedge)$ 是一个格，$A \in L^M$. 则
$$\mathscr{L}(A) = \bigvee \{a \in L \mid A(x) \wedge A(y) \wedge a \leqslant A(x \wedge y) \wedge A(x \vee y), \forall x, y \in M\}$$
$$= \bigvee \{a \in J(L) \mid A(x) \wedge A(y) \wedge a \leqslant A(x \wedge y) \wedge A(x \vee y), \forall x, y \in M\}.$$

下面我们利用 L-模糊集的截集来刻画 L-模糊子格度.

定理 7.1.8 设 $(M, \vee, \wedge)$ 是一个格，$A \in L^M$. 则
(1) $\mathscr{L}(A) = \bigvee \{a \in L \mid \forall b \leqslant a, A_{[b]} \text{ 是 } M \text{ 的一个子格}\}$;
(2) $\mathscr{L}(A) = \bigvee \{a \in J(L) \mid \forall b \in J(L), b \leqslant a, A_{[b]} \text{ 是 } M \text{ 的一个子格}\}$;
(3) $\mathscr{L}(A) = \bigvee \{a \in L \mid \forall b \in P(L), b \ngeqslant a, A^{(b)} \text{ 是 } M \text{ 的一个子格}\}$;
(4) $\mathscr{L}(A) = \bigvee \{a \in L \mid \forall b \notin \alpha(a), A^{[b]} \text{ 是 } M \text{ 的一个子格}\}$;
(5) 若对任意的 $a, b \in L$, $\beta(a \wedge b) = \beta(a) \cap \beta(b)$, 则
$$\mathscr{L}(A) = \bigvee \{a \in L \mid \forall b \in \beta(a), A_{(b)} \text{ 是 } M \text{ 的一个子格}\}.$$

证明 (1) 假定 $a \in L$ 使得对任意的 $x, y \in M$,
$$A(x) \wedge A(y) \wedge a \leqslant A(x \wedge y) \wedge A(x \vee y).$$

任取 $b \leqslant a$, 则对任意的 $x, y \in M$, 由
$$x, y \in A_{[b]} \Rightarrow A(x) \geqslant b, A(y) \geqslant b$$
$$\Rightarrow A(x \wedge y) \geqslant b, A(x \vee y) \geqslant b$$
$$\Rightarrow x \wedge y, x \vee y \in A_{[b]}$$

可知 $A_{[b]}$ 是 M 的一个子格, 这意味着
$$\mathscr{L}(A) \leqslant \bigvee \{a \in L \mid \forall b \leqslant a, A_{[b]} \text{ 是 } M \text{ 的一个子格}\}.$$

反之, 假定 $a \in L$ 使得对任意的 $b \leqslant a$, $A_{[b]}$ 是 M 的一个子格. 对 $x, y \in X$, 令 $b = A(x) \wedge A(y) \wedge a$, 则 $b \leqslant a$, 且 $x, y \in A_{[b]}$, 这意味着 $x \vee y \in A_{[b]}$ 且 $x \wedge y \in A_{[b]}$, 从而

7.1 L-模糊子格度

$$A(x) \wedge A(y) \wedge a \leqslant A(x \wedge y) \wedge A(x \vee y).$$

这证明了 $\mathscr{L}(A) \geqslant \bigvee \{a \in L \mid \forall b \leqslant a, A_{[b]}$ 是 M 的一个子格$\}$.

(2) 的证明是类似的.

(3) 假定 $a \in L$ 使得对任意的 $x, y \in M$,

$$A(x) \wedge A(y) \wedge a \leqslant A(x \wedge y) \wedge A(x \vee y).$$

任取 $b \in P(L)$ 且 $b \not\geqslant a$, 则对任意的 $x, y \in M$, 由

$$x, y \in A^{(b)} \Rightarrow A(x) \not\leqslant b, A(y) \not\leqslant b$$
$$\Rightarrow A(x) \wedge A(y) \wedge a \not\leqslant b$$
$$\Rightarrow A(x \wedge y) \not\leqslant b, A(x \vee y) \not\leqslant b$$
$$\Rightarrow x \wedge y \in A^{(b)}, x \vee y \in A^{(b)}$$

可知 $A^{(b)}$ 是 M 的子格. 这表明

$$\mathscr{L}(A) \leqslant \bigvee \{a \in L \mid \forall b \in P(L), b \not\geqslant a, A^{(b)} \text{是 } M \text{ 的一个子格}\}.$$

反之, 假定 $a \in L$ 使得对任意的 $b \in P(L)$ 且 $b \not\geqslant a$, $A^{(b)}$ 是 M 的一个子格. 任取 $x, y \in M$, 则对任意的 $b \in P(L)$, 由

$$A(x) \wedge A(y) \wedge a \not\leqslant b \Rightarrow x, y \in A^{(b)}, a \not\leqslant b$$
$$\Rightarrow x \wedge y \in A^{(b)}, x \vee y \in A^{(b)}$$
$$\Rightarrow A(x \wedge y) \not\leqslant b, A(x \vee y) \not\leqslant b$$

可知

$$A(x) \wedge A(y) \wedge a \leqslant A(x \wedge y) \wedge A(x \vee y).$$

这表明

$$\mathscr{L}(A) \geqslant \bigvee \{a \in L \mid \forall b \in P(L), b \not\geqslant a, A^{(b)} \text{ 是 } M \text{ 的一个子格}\}.$$

(4) 假定 $a \in L$ 使得对任意的 $x, y \in M$,

$$A(x) \wedge A(y) \wedge a \leqslant A(x \wedge y) \wedge A(x \vee y).$$

任取 $b \notin \alpha(a)$, 则对任意的 $x, y \in M$, 由

$$x, y \in A^{[b]} \Rightarrow b \notin \alpha(A(x)), b \notin \alpha(A(y))$$

$$\Rightarrow b \notin \alpha(A(x)) \cup \alpha(A(y)) \cup \alpha(a)$$
$$\Rightarrow b \notin \alpha(A(x) \wedge A(y) \wedge a)$$
$$\Rightarrow b \notin \alpha(A(x \vee y) \wedge A(x \wedge y))$$
$$\Rightarrow b \notin \alpha(A(x \vee y)), b \notin \alpha(A(x \wedge y))$$
$$\Rightarrow x \vee y \in A^{[b]}, x \wedge y \in A^{[b]}$$

可知 $A^{[b]}$ 是 M 的一个子格. 这意味着

$$\mathscr{L}(A) \leqslant \bigvee \{a \in L \mid \forall b \notin \alpha(a), A^{[b]} \text{ 是 } M \text{ 的一个子格}\}.$$

反之, 假定 $a \in L$ 使得对任意的 $b \notin \alpha(a)$, $A^{[b]}$ 是 M 的一个子格. 任取 $x, y \in M$, 则对任意的 $b \in L$, 由

$$b \notin \alpha(A(x) \wedge A(y) \wedge a) = \alpha(A(x)) \cup \alpha(A(y)) \cup \alpha(a)$$
$$\Rightarrow x, y \in A^{[b]}, b \notin \alpha(a)$$
$$\Rightarrow x \vee y \in A^{[b]}, x \wedge y \in A^{[b]}$$
$$\Rightarrow b \notin \alpha(A(x \wedge y)), b \notin \alpha(A(x \wedge y))$$
$$\Rightarrow b \notin \alpha(A(x \wedge y)) \cup \alpha(A(x \wedge y))$$
$$\Rightarrow b \notin \alpha(A(x \wedge y) \wedge A(x \wedge y))$$

可知
$$\alpha(A(x) \wedge A(y) \wedge a) \supseteq \alpha(A(x \wedge y) \wedge A(x \vee y)).$$

因此
$$A(x) \wedge A(y) \wedge a = \bigwedge \alpha(A(x) \wedge A(y) \wedge a)$$
$$\leqslant \bigwedge \alpha(A(x \wedge y) \wedge A(x \vee y))$$
$$= A(x \wedge y) \wedge A(x \vee y).$$

这表明
$$\mathscr{L}(A) \geqslant \bigvee \{a \in L \mid \forall b \notin \alpha(a), A^{[b]} \text{ 是 } M \text{ 的一个子格}\}.$$

(5) 假定 $a \in L$ 使得对任意的 $x, y \in M$,
$$A(x) \wedge A(y) \wedge a \leqslant A(x \wedge y) \wedge A(x \vee y).$$

7.1 L-模糊子格度

任取 $b \in \beta(a)$, 则对任意的 $x, y \in M$, 由

$$x, y \in A_{(b)} \Rightarrow b \in \beta(A(x)), b \in \beta(A(y))$$
$$\Rightarrow b \in \beta(A(x)) \cap \beta(A(y)) \cap \beta(a)$$
$$\Rightarrow b \in \beta(A(x) \wedge A(y) \wedge a)$$
$$\Rightarrow b \in \beta(A(x \wedge y) \wedge A(x \vee y))$$
$$\Rightarrow b \in \beta(A(x \wedge y)), b \in \beta(A(x \vee y))$$
$$\Rightarrow x \wedge y \in A_{(b)}, x \vee y \in A_{(b)}$$

可知 $A_{(b)}$ 是 M 的一个子格. 这表明

$$\mathscr{L}(A) \leqslant \bigvee \{a \in L \mid \forall b \in \beta(a), A_{(b)} \text{ 是 } M \text{ 的一个子格}\}.$$

反之, 假定 $a \in L$ 使得对任意的 $b \in \beta(a)$, $A_{(b)}$ 是 M 的一个子格. 任取 $x, y \in X$, 则对任意的 $b \in L$, 由

$$b \in \beta(A(x) \wedge A(y) \wedge a) = \beta(A(x)) \cap \beta(A(y)) \cap \beta(a)$$
$$\Rightarrow x, y \in A_{(b)}, b \in \beta(a)$$
$$\Rightarrow x \wedge y \in A_{(b)}, x \vee y \in A_{(b)}$$
$$\Rightarrow b \in \beta(A(x \wedge y)), b \in \beta(A(x \vee y))$$
$$\Rightarrow b \in \beta(A(x \wedge y) \wedge A(x \vee y))$$

可知

$$\beta(A(x) \wedge A(y) \wedge a) \subseteq \beta(A(x \wedge y) \wedge A(x \vee y)).$$

因此

$$A(x) \wedge A(y) \wedge a = \bigvee \beta(A(x) \wedge A(y) \wedge a)$$
$$\leqslant \bigvee \beta(A(x \wedge y) \wedge A(x \vee y))$$
$$= A(x \wedge y) \wedge A(x \vee y).$$

这表明

$$\mathscr{L}(A) \geqslant \bigvee \{a \in L \mid \forall b \in \beta(a), A_{(b)} \text{ 是 } M \text{ 的一个子格}\}. \qquad \square$$

从定理 7.1.8, 我们能够得到下面推论.

推论 7.1.9 设 $(M, \vee, \wedge)$ 是一个格, $A \in L^M$. 则下列条件等价:

(1) A 是 M 的一个 L-模糊子格;

(2) $\forall a \in L$, $A_{[a]}$ 是 M 的一个子格;

(3) $\forall a \in J(L)$, $A_{[a]}$ 是 M 的一个子格;

(4) $\forall a \in L$, $A^{[a]}$ 是 M 的一个子格;

(5) $\forall a \in P(L)$, $A^{(a)}$ 是 M 的一个子格;

(6) 若对任意的 $a, b \in L$, $\beta(a \wedge b) = \beta(a) \cap \beta(b)$, 则 $\forall a \in L$, $A_{(a)}$ 是 M 的一个子格.

定理 7.1.10 设 $f : K \to M$ 是一个格同态, $A \in L^K$, $B \in L^M$. 则

(1) $\mathscr{L}(f_L^{\leftarrow}(B)) \geqslant \mathscr{L}(B)$; 若 f 是满射, 则 $\mathscr{L}(f_L^{\leftarrow}(B)) = \mathscr{L}(B)$.

(2) $\mathscr{L}(A) \leqslant \mathscr{L}(f_L^{\rightarrow}(A))$; 若 f 是单射, 则 $\mathscr{L}(A) = \mathscr{L}(f_L^{\rightarrow}(A))$.

证明 (1) 任取 $a \in L$ 使得对任意的 $x, y \in M$,

$$B(x) \wedge B(y) \wedge a \leqslant B(x \wedge y) \wedge B(x \vee y).$$

则对任意的 $s, t \in K$, 由

$$\begin{aligned} f_L^{\leftarrow}(B)(s) \wedge f_L^{\leftarrow}(B)(t) \wedge a &= B(f(s)) \wedge B(f(t)) \wedge a \\ &\leqslant B(f(s) \wedge f(t)) \wedge B(f(s) \vee f(t)) \\ &= B(f(s \wedge t)) \wedge B(f(s \vee t)) \\ &= f_L^{\leftarrow}(B)(s \wedge t) \wedge f_L^{\leftarrow}(B)(s \vee t) \end{aligned}$$

可知 $\mathscr{L}(f_L^{\leftarrow}(B)) \geqslant \mathscr{L}(B)$. 若 f 是满的, 则易验证 $\mathscr{L}(f_L^{\leftarrow}(B)) \leqslant \mathscr{L}(B)$. 因此此时有 $\mathscr{L}(f_L^{\leftarrow}(B)) = \mathscr{L}(B)$.

(2) 设 $a \in L$ 使得对任意的 $s, t \in K$,

$$A(s) \wedge A(t) \wedge a \leqslant A(s \wedge t) \wedge A(s \vee t).$$

则对任意的 $x, y \in M$, 由

$$\begin{aligned} &f_L^{\rightarrow}(A)(x) \wedge f_L^{\rightarrow}(A)(y) \wedge a \\ &= \bigvee_{f(s)=x} A(s) \wedge \bigvee_{f(t)=y} A(t) \wedge a \\ &= \bigvee \{A(s) \wedge A(t) \wedge a \mid f(s) = x, f(t) = y\} \\ &\leqslant \bigvee \{A(s \wedge t) \wedge A(s \vee t) \mid f(s) = x, f(t) = y\} \end{aligned}$$

7.1 L-模糊子格度

$$= \bigvee\{A(s\wedge t)\wedge A(s\vee t) \mid f(s\wedge t) = x\wedge y, f(s\vee t) = x\vee y\}$$

$$= \bigvee\{A(w)\wedge A(u) \mid f(w) = x\wedge y, f(u) = x\vee y\}$$

$$= f_L^{\rightarrow}(A)(x\wedge y)\wedge f_L^{\rightarrow}(A)(x\vee y).$$

可证 $\mathscr{L}(A) \leqslant \mathscr{L}(f_L^{\rightarrow}(A))$.

假设 f 是单射, 我们需证 $\mathscr{L}(f_L^{\rightarrow}(A)) \leqslant \mathscr{L}(A)$. 设 $a \in L$ 使得对任意的 $x, y \in M$,

$$f_L^{\rightarrow}(A)(x)\wedge f_L^{\rightarrow}(A)(y)\wedge a \leqslant f_L^{\rightarrow}(A)(x\wedge y)\wedge f_L^{\rightarrow}(A)(x\vee y).$$

任取 $s, t \in K$. 因为 f 是一个单同态, 所以有

$$f_L^{\rightarrow}(A)(f(s)) = A(s), \quad f_L^{\rightarrow}(A)(f(s)\wedge f(t)) = A(s\wedge t),$$
$$f_L^{\rightarrow}(A)(f(t)) = A(t), \quad f_L^{\rightarrow}(A)(f(s)\vee f(t)) = A(s\vee t).$$

于是

$$A(s)\wedge A(t)\wedge a = f_L^{\rightarrow}(A)(f(s))\wedge f_L^{\rightarrow}(A)(f(t))\wedge a$$
$$\leqslant f_L^{\rightarrow}(A)(f(s)\wedge f(t))\wedge f_L^{\rightarrow}(A)(f(s)\vee f(t))$$
$$= A(s\wedge t)\wedge A(s\vee t).$$

这表明 $\mathscr{L}(f_L^{\rightarrow}(A)) \leqslant \mathscr{L}(A)$. □

推论 7.1.11 设 $f: K \to M$ 是一个格同态, $A \in L^K$, $B \in L^M$. 则
(1) 若 B 是 M 的一个 L-模糊子格, 则 $f_L^{\leftarrow}(B)$ 是 K 的一个 L-模糊子格;
(2) 若 A 是 K 的一个 L-模糊子格, 则 $f_L^{\rightarrow}(A)$ 是 M 的一个 L-模糊子格.

定理 7.1.12 设 K 和 M 是两个格, $A \in L^K$ 且 $B \in L^M$. 则 $\mathscr{L}(A\times B) \geqslant \mathscr{L}(A)\wedge \mathscr{L}(B)$.

证明 对任意的 $a \in L$, 由下面蕴含式

$$a \prec \mathscr{L}(A)\wedge \mathscr{L}(B)$$

$$\Rightarrow a \prec \mathscr{L}(A), \ a \prec \mathscr{L}(B)$$

$$\Rightarrow \exists b_1 \in L \text{ 使得 } \forall b \leqslant b_1, A_{[b]} \text{ 是 } K \text{ 的一个子格且 } a \leqslant b_1,$$

同时

$$\exists b_2 \in L \text{ 使得 } \forall b \leqslant b_2, B_{[b]} \text{ 是 } M \text{ 的一个子格且 } a \leqslant b_2$$

$\Rightarrow \forall b \leqslant b_1 \wedge b_2, A_{[b]}, B_{[b]}$ 分别是 K, M 的子格且 $a \leqslant b_1 \wedge b_2$

$\Rightarrow \forall b \leqslant b_1 \wedge b_2, A_{[b]} \times B_{[b]}$ 是 $K \times M$ 的一个子格且 $a \leqslant b_1 \wedge b_2$

$\Rightarrow a \leqslant \mathscr{L}(A \times B)$.

可以证明 $\mathscr{L}(A \times B) \geqslant \mathscr{L}(A) \wedge \mathscr{L}(B)$. □

由定理 7.1.12 可得下面推论.

推论 7.1.13 设 $A \in L^K$ 和 $B \in L^M$ 分别是 K 和 M 的 L-模糊子格. 则 $A \times B$ 是 $K \times M$ 的 L-模糊子格.

7.2 L-模糊凸子格度

本节我们引入 L-模糊凸子格度的概念, 并给出它的截集刻画; 还讨论一个 L-子集的 L-模糊凸子格度分别与它的同态像和同态原像的 L-模糊凸子格度的关系以及 L-子集乘积的 L-模糊凸子格度.

定义 7.2.1 设 $(M, \vee, \wedge)$ 是一个格, $A \in M^X$. 令

$$\mathscr{L}_c(A) = \mathscr{L}(A) \wedge \bigwedge_{x,y,z \in M} \{A(x \wedge y) \wedge A(x \vee y) \mapsto A(z) \mid x \wedge y \leqslant z \leqslant x \vee y\},$$

则称 $\mathscr{L}_c(A)$ 为 A 的 L-模糊凸子格度.

注 7.2.2 显然 A 是一个 L-模糊凸子格当且仅当 $\mathscr{L}_c(A) = 1$. 一个 L-模糊子集 A 是一个 L-模糊凸子格的程度可以看作 $\mathscr{L}_c(A)$.

例 7.2.3 设 $M = \{\varnothing, \{a\}, \{b\}, \{c\}, \{a,b,c\}\}$ 是幂集格 $\mathbf{2}^{\{a,b,c\}}$ 的子格, $L = [0,1]$. 定义 $A \in L^X$ 为

$$A(\varnothing) = 0.64, \quad A(\{a\}) = 0.82, \quad A(\{b\}) = 0.91,$$

$$A(\{c\}) = 0.45, \quad A(\{a,b,c\}) = 0.85.$$

则

$$\bigwedge \{A(x \wedge y) \wedge A(x \vee y) \mapsto A(z) \mid x \wedge y \leqslant z \leqslant x \vee y\}$$

$$= \begin{cases} 1, & \text{当 } x = y \text{ 时}, \\ 0.45, & \text{当 } x, y \text{ 分别是 } \varnothing, \{a,b,c\} \text{ 时}, \\ 0.45, & \text{当 } x, y \text{ 分别是 } \{a\}, \{b\}, \{c\} \text{ 时}. \end{cases}$$

因此根据例 7.1.5 我们知道 $\mathscr{L}_c(A) = \mathscr{L}(A) \wedge 0.45 = 0.64 \wedge 0.45 = 0.45$.

7.2　L-模糊凸子格度

再定义 $B \in L^M$ 为

$$B(\varnothing) = 0.35, \quad B(\{a\}) = 0.32, \quad B(\{b\}) = 0.32,$$

$$B(\{c\}) = 0.48, \quad B(\{a,b,c\}) = 0.32.$$

易求得 $\mathscr{L}_c(A) = 1$, 也就是说 B 是一个 L-模糊凸子格.

类似于引理 7.1.6, 我们有下面的引理.

引理 7.2.4　设 $(M, \vee, \wedge)$ 是一个格, $A \in L^M$. 则 $\mathscr{L}(A) \geqslant a$ 当且仅当 $\mathscr{L}(A) \geqslant a$ 且对任意的 $x, y, z \in M$, $x \wedge y \leqslant z \leqslant x \vee y$, 都有

$$A(x \wedge y) \wedge A(x \vee y) \wedge a \leqslant A(z).$$

由引理 7.2.4, 可得下面的定理.

定理 7.2.5　设 $(M, \vee, \wedge)$ 是一个格, $A \in L^M$. 则

$$\mathscr{L}_c(A) = \mathscr{L}(A) \wedge \bigvee_{x,y,z \in M} \{a \in L \mid A(x \wedge y) \wedge A(x \vee y) \wedge a \leqslant A(z), x \wedge y \leqslant z \leqslant x \vee y\}.$$

定理 7.2.6　设 $(M, \vee, \wedge)$ 是一个格, $A \in L^M$. 则

$$\mathscr{L}_c(A) = \bigvee\{a \in L \mid \forall b \leqslant a, A_{[b]} \text{ 是一个凸子格}\}.$$

证明　假定 $a \in L$ 使得 $\mathscr{L}(A) \geqslant a$ 且对任意的 $x, y, z \in X$, $x \wedge y \leqslant z \leqslant x \vee y$ 蕴含 $A(x \wedge y) \wedge A(x \vee y) \wedge a \leqslant A(z)$. 由 $\mathscr{L}(A) \geqslant a$, 可知对任意的 $b \leqslant a$, $A_{[b]}$ 是一个子格. 只剩证明 $A_{[b]}$ 是凸的. 对任意的 $x, y, z \in X$, 若 $x \wedge y \leqslant z \leqslant x \vee y$, 则

$$x, y \in A_{[b]} \Rightarrow x \wedge y \in A_{[b]}, x \vee y \in A_{[b]}$$

$$\Rightarrow A(x \wedge y) \geqslant b, A(x \vee y) \geqslant b$$

$$\Rightarrow A(z) \geqslant A(x \wedge y) \wedge A(x \vee y) \wedge a \geqslant b$$

$$\Rightarrow z \in A_{[b]}.$$

因此 $A_{[b]}$ 是一个凸子格. 这证明了 $\mathscr{L}_c(A) \leqslant \bigvee\{a \in L \mid \forall b \leqslant a, A_{[b]} \text{ 是一个}$ 凸子格$\}$.

反之, 假定 $a \in L$ 使得任意的 $b \leqslant a$, $A_{[b]}$ 是凸子格. 则 $\mathscr{L}(A) \geqslant a$. 对任意的 $x, y, z \in X$, 若 $x \wedge y \leqslant z \leqslant x \vee y$, 令 $A(x \wedge y) \wedge A(x \vee y) \wedge a = b$, 则

$$A(x \wedge y) \wedge A(x \vee y) \wedge a = b \Rightarrow A(x \wedge y) \geqslant b, A(x \vee y) \geqslant b, a \geqslant b$$

$$\Rightarrow x \vee y \in A_{[b]}, x \wedge y \in A_{[b]}$$

$$\Rightarrow z \in A_{[b]}$$

$$\Rightarrow A(z) \geqslant b$$

$$\Rightarrow A(x \wedge y) \wedge A(x \vee y) \wedge a \leqslant A(z).$$

这蕴含 $\mathscr{L}_c(A) \geqslant \bigvee \{a \in L \mid \forall b \leqslant a, A_{[b]} \text{是凸子格}\}$. □

定理 7.2.7 设 $(X, \vee, \wedge)$ 是一个格, $A \in L^X$. 则

$$\mathscr{L}_c(A) = \bigvee \{a \in L \mid \forall b \in P(L), b \not\geqslant a, A^{(b)} \text{是一个凸子格}\}.$$

证明 假定 $a \in L$ 使得 $\mathscr{L}(A) \geqslant a$ 且对任意的 $x, y, z \in X$, $x \wedge y \leqslant z \leqslant x \vee y$ 蕴含 $A(x \wedge y) \wedge A(x \vee y) \wedge a \leqslant A(z)$. 由 $\mathscr{L}(A) \geqslant a$, 可知对任意的 $b \in P(L), b \not\geqslant a$, $A^{(b)}$ 是一个子格. 只剩证明 $A^{(b)}$ 是凸的. 对任意的 $x, y, z \in X$, 若 $x \wedge y \leqslant z \leqslant x \vee y$, 则

$$x, y \in A^{(b)} \Rightarrow A(x \wedge y) \not\leqslant b, A(x \vee y) \not\leqslant b$$

$$\Rightarrow A(x \wedge y) \wedge A(x \vee y) \wedge a \not\leqslant b$$

$$\Rightarrow A(z) \not\leqslant b$$

$$\Rightarrow z \in A^{(b)}.$$

因此 $A^{(b)}$ 是一个凸子格. 这蕴含 $\mathscr{L}_c(A) \leqslant \bigvee \{a \in L \mid \forall b \in P(L), b \not\geqslant a, A^{(b)} \text{是凸子格}\}$.

反之, 假定 $a \in L$ 使得对任意的 $b \in P(L)$ 满足 $b \not\geqslant a$, $A^{(b)}$ 是凸子格. 则 $\mathscr{L}(A) \geqslant a$. 对任意的 $x, y, z \in X$, 若 $x \wedge y \leqslant z \leqslant x \vee y$, 设 $A(x \wedge y) \wedge A(x \vee y) \wedge a \not\leqslant b$, 则

$$A(x \wedge y) \wedge A(x \vee y) \wedge a \not\leqslant b \Rightarrow x \vee y \in A^{(b)}, x \wedge y \in A^{(b)}, a \not\leqslant b$$

$$\Rightarrow z \in A^{(b)}$$

$$\Rightarrow A(z) \not\leqslant b.$$

因此 $A(x \wedge y) \wedge A(x \vee y) \wedge a \leqslant A(z)$. 这蕴含 $\mathscr{L}_c(A) \geqslant \bigvee \{a \in L \mid \forall b \in P(L), b \not\geqslant a, A^{(b)} \text{是一个凸子格}\}$. □

定理 7.2.8 设 $(X, \vee, \wedge)$ 是一个格, $A \in L^X$. 则

$$\mathscr{L}_c(A) = \bigvee \{a \in L \mid \forall b \notin \alpha(a), A^{[b]} \text{是一个凸子格}\}.$$

证明 假定 $a \in L$ 使得 $\mathscr{L}(A) \geqslant a$ 且对任意的 $x, y, z \in X$, $x \wedge y \leqslant z \leqslant x \vee y$ 蕴含 $A(x \wedge y) \wedge A(x \vee y) \wedge a \leqslant A(z)$. 由 $\mathscr{L}(A) \geqslant a$, 可知对任意的 $b \notin \alpha(a)$, $A^{[b]}$

7.2 L-模糊凸子格度

是一个子格. 只剩证明 $A^{[b]}$ 是凸的. 对任意的 $x,y,z \in X$, 若 $x \wedge y \leqslant z \leqslant x \vee y$, 则 $\alpha(A(z)) \subseteq \alpha(A(x \wedge y) \wedge A(x \vee y) \wedge a)$. 于是,

$$x, y \in A^{[b]} \Rightarrow x \wedge y \in A^{[b]}, x \vee y \in A^{[b]}$$
$$\Rightarrow b \notin \alpha(A(x \wedge y)), b \notin \alpha(A(x \vee y))$$
$$\Rightarrow b \notin \alpha(A(x \wedge y)) \cup \alpha(A(x \vee y)) \cup \alpha(a)$$
$$\Rightarrow b \notin \alpha(A(x \wedge y) \wedge A(x \vee y) \wedge a)$$
$$\Rightarrow b \notin \alpha(A(z))$$
$$\Rightarrow z \in A^{[b]}.$$

因此 $A^{[b]}$ 是凸子格. 这蕴含 $\mathscr{L}_c(A) \leqslant \bigvee\{a \in L \mid \forall b \notin \alpha(a), A^{[b]}$是一个凸子格$\}$.

反之, 假定 $a \in L$ 使得对任意的 $b \notin \alpha(a)$, $A^{[b]}$ 是一个凸子格. 则 $\mathscr{L}(A) \geqslant a$. 对任意的 $x, y, z \in X$, 若 $x \wedge y \leqslant z \leqslant x \vee y$, 设 $b \notin \alpha(A(x \wedge y) \wedge A(x \vee y) \wedge a)$. 则

$$b \notin \alpha(A(x \wedge y) \wedge A(x \vee y) \wedge a) \Rightarrow b \notin \alpha(A(x \wedge y)) \cup \alpha(A(x \vee y)) \cup \alpha(a)$$
$$\Rightarrow b \notin \alpha(A(x \wedge y)), b \notin \alpha(A(x \vee y)), b \notin \alpha(a)$$
$$\Rightarrow x \wedge y, x \vee y \in A^{[b]}, A^{[b]}\text{是凸子格}$$
$$\Rightarrow z \in A^{[b]}$$
$$\Rightarrow b \notin \alpha(A(z)).$$

这蕴含 $\alpha(A(x \wedge y) \wedge A(x \vee y) \wedge a) \supseteq \alpha(A(z))$. 因此

$$A(x \wedge y) \wedge A(x \vee y) \wedge a = \bigwedge \alpha(A(x \wedge y) \wedge A(x \vee y) \wedge a)$$
$$\leqslant \bigwedge \alpha(A(z))$$
$$= A(z).$$

这蕴含 $\mathscr{L}_c(A) \geqslant \bigvee\{a \in L \mid \forall b \notin \alpha(a), A^{[b]}$是一个凸子格$\}$. □

定理 7.2.9 设 $(X, \vee, \wedge)$ 是一个格, $A \in L^X$. 若 $\beta(a \wedge b) = \beta(a) \cap \beta(b)$, 则

$$\mathscr{L}_c(A) = \bigvee\{a \in L \mid \forall b \in \beta(a), A_{(b)} \text{ 是一个凸子格}\}.$$

证明 假定 $a \in L$ 使得 $\mathscr{L}(A) \geqslant a$ 且对任意的 $x, y, z \in X$, $x \wedge y \leqslant z \leqslant x \vee y$ 蕴含 $A(x \wedge y) \wedge A(x \vee y) \wedge a \leqslant A(z)$. 由 $\mathscr{L}(A) \geqslant a$, 可知对任意的 $b \in \beta(a)$, $A_{(b)}$

是一个子格. 只剩证明 $A_{(b)}$ 是凸的. 对任意的 $x, y, z \in X$, 若 $x \wedge y \leqslant z \leqslant x \vee y$, 则 $\beta(A(x \wedge y) \wedge A(x \vee y) \wedge a) \subseteq \beta(A(z))$. 于是,

$$x, y \in A_{(b)} \Rightarrow x \wedge y \in A_{(b)}, x \vee y \in A_{(b)}$$
$$\Rightarrow b \in \beta(A(x \wedge y)), b \in \beta(A(x \vee y))$$
$$\Rightarrow b \in \beta(A(x \wedge y)) \cap \beta(A(x \vee y)) \cap \beta(a)$$
$$\Rightarrow b \in \beta(A(x \wedge y) \wedge A(x \vee y) \wedge a)$$
$$\Rightarrow b \in \beta(A(z))$$
$$\Rightarrow z \in A_{(b)}.$$

因此 $A_{(b)}$ 是凸子格. 这蕴含 $\mathscr{L}_c(A) \leqslant \bigvee\{a \in L \mid \forall b \in \beta(a), A_{(b)}$ 是一个凸子格$\}$.

反之, 假定 $a \in L$ 使得对任意的 $b \in \beta(a)$, $A_{(b)}$ 是凸子格. 则 $\mathscr{L}(A) \geqslant a$. 对任意的 $x, y, z \in X$, 若 $x \wedge y \leqslant z \leqslant x \vee y$, 则对任意的 $b \in L$,

$$b \in \beta(A(x \wedge y) \wedge A(x \vee y) \wedge a) \Rightarrow b \in \beta(A(x \wedge y)) \cap \beta(A(x \vee y)) \cap \beta(a)$$
$$\Rightarrow b \in \beta(A(x \wedge y)) b \in \beta(A(x \vee y)), b \in \beta(a)$$
$$\Rightarrow x \vee y, x \wedge y \in A_{(b)}, A_{(b)} 是凸子格$$
$$\Rightarrow z \in A_{(b)}$$
$$\Rightarrow b \in \beta(A(z)).$$

这蕴含 $\beta(A(x \wedge y) \wedge A(x \vee y) \wedge a) \subseteq \beta(A(z))$. 因此

$$A(x \wedge y) \wedge A(x \vee y) \wedge a = \bigvee \beta(A(x \wedge y) \wedge A(x \vee y) \wedge a)$$
$$\leqslant \bigvee \beta(A(z))$$
$$= A(z).$$

这蕴含 $\mathscr{L}_c(A) \geqslant \bigvee\{a \in L \mid \forall b \in \beta(a), A_{(b)}$ 是一个凸子格$\}$. □

接下来, 我们将 f-不变模糊子集的概念以及结论推广到 L-值情形.

定义 7.2.10 设 $f : X \to Y$ 是一个格映射, $A \in L^X$. 若 $\forall x, y \in X$, $f(x) = f(y)$ 蕴含 $A(x) = A(y)$, 则 A 称为 f-不变的.

引理 7.2.11 设 $f : X \to Y$ 是一个格同态, A 是 X 的一个 f-不变的 L-子集. 则 $f(x) \leqslant f(y)$ 蕴含 $A(x) = A(x \wedge y)$ 和 $A(y) = A(x \vee y)$.

7.2 L-模糊凸子格度

定理 7.2.12 设 $f: X \to Y$ 是一个格同态, $A \in L^X, B \in L^Y$.
(1) $\mathscr{L}_X(f_L^\leftarrow(B)) \geqslant \mathscr{L}_Y(B)$, 且若 f 是满射, 则 $\mathscr{L}_X(f_L^\leftarrow(B)) = \mathscr{L}_Y(B)$;
(2) 若 f 是单射, A 是 f-不变的, 则 $\mathscr{L}_X(A) \leqslant \mathscr{L}_Y(f_L^\rightarrow(A))$.

证明 (1) 由定理 7.2.5, 可知 $\mathscr{L}_X(f_L^\leftarrow(B)) \geqslant \mathscr{L}_Y(B)$. 设 $a \in L$ 使得 $\mathscr{L}_X(f_L^\leftarrow(B)) \geqslant a$ 且对任意的 $x, y, z \in Y$, $x \wedge y \leqslant z \leqslant x \vee y$ 蕴含 $B(x \wedge y) \wedge B(x \vee y) \wedge a \leqslant B(z)$. 则 $\mathscr{L}_X(f_L^\leftarrow(B)) \geqslant a$. 对任意的 $s, t, w \in X$, 若 $s \wedge t \leqslant w \leqslant s \vee t$, 因为 f 保序的, 所以 $f(s \wedge t) \leqslant f(w) \leqslant f(s \vee t)$. 因此

$$f_L^\leftarrow(B)(s \wedge t) \wedge f_L^\leftarrow(B)(s \vee t) \wedge a$$
$$= A(f(s \wedge t)) \wedge A(f(s \vee t)) \wedge a$$
$$\leqslant B(f(w))$$
$$= f^\leftarrow(B)(w).$$

这蕴含 $\mathscr{L}_X(f_L^\leftarrow(B)) \geqslant \mathscr{L}_Y(B)$. 若 f 是满射, 则易验证 $\mathscr{L}_X(f_L^\leftarrow(B)) \leqslant \mathscr{L}_Y(B)$. 因此 $\mathscr{L}_X(f_L^\leftarrow(B)) = \mathscr{L}_Y(B)$.

(2) 设 $a \in L$ 使得 $a \leqslant \mathscr{L}_X(A)$ 且对任意的 $s, t, w \in X$, $s \wedge t \leqslant w \leqslant s \vee t$ 蕴含 $\mu(s \wedge t) \wedge \mu(s \vee t) \wedge a \leqslant A(w)$. 由定理 7.2.5 可得 $a \leqslant \mathscr{L}_Y(f_L^\rightarrow(A))$. 对任意的 $x, y, z \in Y$, 满足 $x \wedge y \leqslant z \leqslant x \vee y$. 则存在 $s, t, w \in X$ 使得 $f(s) = x \wedge y$, $f(t) = x \vee y$ 且 $f(w) = z$. 于是 $f(s) \leqslant f(w) \leqslant f(t)$. 因为 A 是 f-不变的, 由引理 7.2.11, 可知 $A(s) = A(s \wedge w)$ 且 $A(t) = A(w \vee t)$. 因为 $s \wedge w \leqslant w \vee t$, 所以 $A(s \wedge w) \wedge A(w \vee t) \wedge a \leqslant A(w)$, 即 $A(s) \wedge A(t) \wedge a \leqslant A(w)$. 因此

$$f_L^\rightarrow(A)(x) \wedge f_L^\rightarrow(A)(y) \wedge a$$
$$= \bigvee_{f(s) = x \wedge y} A(s) \wedge \bigvee_{f(t) = x \vee y} A(t) \wedge a$$
$$= \bigvee \{A(s) \wedge A(t) \wedge a \mid f(s) = x \wedge y, f(t) = x \vee y\}$$
$$\leqslant \bigvee \{A(w) \mid f(w) = z\}$$
$$= f_L^\rightarrow(A)(z).$$

这蕴含 $\mathscr{L}_X(A) \leqslant \mathscr{L}_Y(f_L^\rightarrow(A))$. $\square$

定理 7.2.13 设 K 和 M 是两个格, $A \in L^K$ 且 $B \in L^M$. 则 $\mathscr{L}_c(A \times B) \geqslant \mathscr{L}_c(A) \wedge \mathscr{L}_c(B)$.

7.3 由 L-模糊子格度确定的 L-模糊凸结构

在一个给定的格 M 中,对任意的 $A \in L^M$,都有它的 L-模糊子格度 $\mathscr{L}(A)$ 与之对应,这自然就确定了一个从 L^M 到 L 的映射 $\mathscr{L}$. 下面我们证明这个映射就是格 M 上的一个 L-模糊凸结构.

定理 7.3.1 设 $(M, \vee, \wedge)$ 是一个格,则 $\mathscr{L}$ 是 M 上的一个 L-模糊凸结构,称为由 L-模糊子格度确定的 L-模糊凸结构.

证明 (1) 显然 $\mathscr{L}(\chi_\varnothing) = \mathscr{L}(\chi_M) = 1$.

(2) 设 $\{A_i \mid i \in \Omega\} \subseteq L^M$ 是非空的. 我们来证明 $\bigwedge_{i \in \Omega} \mathscr{L}(A_i) \leqslant \mathscr{L}\left(\bigwedge_{i \in \Omega} A_i\right)$.

对任意的 $a \in L$,由

$$a \leqslant \bigwedge_{i \in \Omega} \mathscr{L}(A_i)$$

$$\Rightarrow a \leqslant \mathscr{L}(A_i), \forall i \in \Omega$$

$$\Rightarrow A_i(x) \wedge A_i(y) \wedge a \leqslant A_i(x \wedge y) \wedge A_i(x \vee y), \forall i \in \Omega, \forall x, y \in M$$

$$\Rightarrow \bigwedge_{i \in \Omega} A_i(x) \wedge \bigwedge_{i \in \Omega} A_i(y) \wedge a \leqslant \bigwedge_{i \in \Omega} A_i(x \wedge y) \wedge \bigwedge_{i \in \Omega} A_i(x \vee y)$$

$$\Rightarrow a \leqslant \mathscr{L}\left(\bigwedge_{i \in \Omega} A_i\right)$$

可证 $\bigwedge_{i \in \Omega} \mathscr{L}(A_i) \leqslant \mathscr{L}\left(\bigwedge_{i \in \Omega} A_i\right)$.

(3) 设 $\{A_i \mid i \in \Omega\} \subseteq L^M$ 是非空的且是全序的. 为证 $\mathscr{L}\left(\bigvee_{i \in \Omega} A_i\right) \geqslant \bigwedge_{i \in \Omega} \mathscr{L}(A_i)$,设 $a \in L$ 使得 $a \leqslant \bigwedge_{i \in \Omega} \mathscr{L}(A_i)$. 则对任意的 $i \in \Omega, x, y \in X$,有

$$A_i(x) \wedge A_i(y) \wedge a \leqslant A_i(x \wedge y) \wedge A_i(x \vee y).$$

接下来证明

$$\bigvee_{i \in \Omega} A_i(x) \wedge \bigvee_{i \in \Omega} A_i(y) \wedge a \leqslant \bigvee_{i \in \Omega} A_i(x \wedge y) \wedge \bigvee_{i \in \Omega} A_i(x \vee y).$$

任取 $b \in L$ 使得 $b \prec \bigvee_{i \in \Omega} A_i(x) \wedge \bigvee_{i \in \Omega} A_i(y) \wedge a$,则存在 $i, j \in \Omega$ 使得 $b \leqslant A_i(x)$,$b \leqslant A_j(y)$ 且 $b \leqslant a$. 因为 $\{A_i \mid i \in \Omega\}$ 是全序的,不妨假设 $A_j \leqslant A_i$. 于是

$b \leqslant A_i(x) \wedge A_i(y) \wedge a$. 由 $A_i(x) \wedge A_i(y) \wedge a \leqslant A_i(x \wedge y)$, 可得 $b \leqslant A_i(x \wedge y)$. 因此 $b \leqslant \bigvee_{i \in \Omega} A_i(x \wedge y)$. 这意味着

$$\bigvee_{i \in \Omega} A_i(x) \wedge \bigvee_{i \in \Omega} A_i(y) \wedge a \leqslant \bigvee_{i \in \Omega} A_i(x \wedge y).$$

同理可证

$$\bigvee_{i \in \Omega} A_i(x) \wedge \bigvee_{i \in \Omega} A_i(y) \wedge a \leqslant \bigvee_{i \in \Omega} A_i(x \vee y).$$

这表明 $a \leqslant \mathscr{L}\left(\bigvee_{i \in \Omega} A_i\right)$. 因此 $\mathscr{L}\left(\bigvee_{i \in \Omega} A_i\right) \geqslant \bigwedge_{i \in \Omega} \mathscr{L}(A_i)$.

综上可知 $\mathscr{L}$ 是 M 上的 L-模糊凸结构. □

由定理 7.1.10, 可得下面的定理.

定理 7.3.2 设 $f: K \to M$ 是一个格同态, $\mathscr{L}_K, \mathscr{L}_M$ 分别是 K 和 M 上由 L-模糊子格度确定的 L-模糊凸结构, 则 $f: (K, \mathscr{L}_K) \to (M, \mathscr{L}_M)$ 是一个 L-模糊凸保持映射和 L-模糊凸到凸映射.

若 $f: K \to M$ 是一个格同构, 则 $f: (K, \mathscr{L}_K) \to (M, \mathscr{L}_M)$ 是一个 L-模糊凸空间同构.

类似于定理 7.3.1, 可证下面定理.

定理 7.3.3 设 $(M, \vee, \wedge)$ 是一个格, 则 $\mathscr{L}_c$ 是 M 上的一个 L-模糊凸结构, 称为由 L-模糊凸子格度确定的 L-模糊凸结构.

习 题 7

1. 请将定理 7.1.12 推广到任意乘积.
2. 请将定理 7.2.13 推广到任意乘积.
3. 试证明定理 7.3.3.
4. 设 $f: K \to M$ 是一个格同态, $\mathscr{L}_c{}^K, \mathscr{L}_c{}^M$ 分别是 K 和 M 上由 L-模糊凸子格度确定的 L-模糊凸结构, 则 $f: (K, \mathscr{L}_c{}^K) \to (M, \mathscr{L}_c{}^M)$ 是否为一个 L-模糊凸保持映射和 L-模糊凸到凸映射?

若 $f: K \to M$ 是一个格同构, 则 $f: (K, \mathscr{L}_c{}^K) \to (M, \mathscr{L}_c{}^M)$ 是否为一个 L-模糊凸空间同构?

5. 问题: 格中理想和滤子的概念都可以推广到模糊情形, 同样的道理可以引入格中理想度和滤子度的概念, 有兴趣的读者可以讨论这些问题.
6. 在文献 [13, 15] 中, 提出了格的模糊素理想和模糊半素理想度的概念, 读者可以讨论它们和凸结构之间的关系.

第 8 章 L-模糊效应代数

效应代数在模型化量子逻辑方面有着重要的应用,它是相对广义的代数结构.在这一章中,我们首先给出效应代数上的 L-模糊子代数度和 L-子代数的定义,并利用四种截集给出一些相应的刻画. 通过 L-模糊子代数度可确定效应代数上的一个 L-模糊凸结构,并证明两个效应代数之间的同态是一个 L-模糊凸保持映射,且一个单态映射是 L-模糊凸到凸映射. 其次给出效应代数上的 L-模糊理想度的定义,并利用四种截集给出一些相应的刻画. 通过 L-模糊理想度同样可以确定效应代数上的一个 L-模糊凸结构.

8.1 效应代数基础

在这一节中,我们首先给出效应代数的一些基本概念.

定义 8.1.1 一个效应代数 $(E, 0, 1, \oplus)$ 包含一个集合 E,两个特殊元素 $0, 1 \in E$(称为零元和单位元),并含有一个部分二元运算 $\oplus$,其满足下列条件:对任意的 $x, y, z \in E$,

(E1) (交换律) 如果 $x \oplus y$ 有定义,则 $y \oplus x$ 有定义且 $x \oplus y = y \oplus x$;

(E2) (结合律) 如果 $x \oplus y$ 有定义且 $(x \oplus y) \oplus z$ 有定义,则 $y \oplus z$ 和 $x \oplus (y \oplus z)$ 有定义且 $(x \oplus y) \oplus z = x \oplus (y \oplus z)$;

(E3) (正交律) 对任意的 $x \in E$ 存在唯一的 $x' \in E$ 使得 $x \oplus x'$ 有定义且 $x \oplus x' = 1$,$x' \in E$ 称为 x 的正交补;

(E4) 如果 $x \oplus 1$ 有定义,则 $x = 0$.

为了简单起见,一个效应代数 $(E, 0, 1, \oplus)$ 也可以简单记为 E.

定义 8.1.2 如果效应代数 E 上的一个集合 S 满足下列条件:

(1) $0, 1 \in S$;

(2) $x \in S$ 能推出 $x' \in S$;

(3) 如果 $x, y \in S$ 且 $x \oplus y$ 有定义,则 $x \oplus y \in S$,

则称 S 为 E 上的一个子代数.

定义 8.1.3 设 E 和 F 是效应代数. 称映射 $g: E \to F$ 为

(1) 一个同态映射,如果 $g(1_E) = 1_F$,并且当 $x, y \in E$,$x \perp y$ 时,有 $g(x) \perp g(y)$ 且 $g(x \oplus y) = g(x) \oplus g(y)$;

(2) 一个单同态,如果 g 是一个同态映射,并且 $g(x) \perp g(y)$ 当且仅当 $x \perp y$.

定义 8.1.4 设 $\{E_i\}_{i\in\Lambda}$ 是一族效应代数, 定义 $\prod\limits_{i\in\Lambda} E_i$ 为

$$\prod_{i\in\Lambda} E_i = \left\{x \,\Big|\, x:\Lambda \to \bigcup_{i\in\Lambda} E_i \text{ s.t. } \forall i\in\Lambda, x(i)=x_i\in E_i\right\}.$$

定义 $\prod\limits_{i\in\Lambda} E_i$ 上的运算 $\oplus$ 如下: 对任意的 $x,y\in\prod\limits_{i\in\Lambda} E_i$, $x\perp y$ 当且仅当对所有的 $i\in\Lambda$, $x_i\perp y_i$. 此时, $(x\oplus y)(i)=x_i\oplus y_i$ 且 $x'(i)=x_i'$. 于是有 $\mathbf{0}_i=0_i, \mathbf{1}_i=1_i$, 其中 0_i 和 1_i 分别是 E_i 的最小元和最大元. 则称 $\left(\prod\limits_{i\in\Lambda} E_i, \oplus, \mathbf{0}, \mathbf{1}\right)$ 为效应代数的直积.

容易验证 $\left(\prod\limits_{i\in\Lambda} E_i, \oplus, \mathbf{0}, \mathbf{1}\right)$ 是一个效应代数.

定义 8.1.5 设 $\{E_i\}_{i\in\Lambda}$ 是一族效应代数且对任意的 $i\in\Lambda$, A_i 是 E_i 的一个 L-子集, 则 $\prod\limits_{i\in\Lambda} E_i$ 的 L-子集 $\prod\limits_{i\in\Lambda} A_i$ 定义为

$$\left(\prod_{i\in\Lambda} A_i\right)(x) = \bigwedge_{i\in\Lambda} A_i(x_i).$$

8.2 效应代数上的 L-模糊子代数度

在这一节中, 我们给出 L-模糊子代数度和 L-子代数的定义, 并利用四种截集刻画 L-模糊子代数度.

定义 8.2.1 设 $(E,\oplus,0,1)$ 是一个效应代数且 $A\in L^E$. 定义 A 的 L-模糊效应子代数度 $\mathscr{E}(A)$ 为

$$\mathscr{E}(A) \triangleq \bigwedge_{x,y\in E, x\perp y} \{(A(x)\to A(0)\wedge A(x'))\wedge((A(x)\wedge A(y))\to A(x\oplus y))\}.$$

当 $\mathscr{E}(A)=1$ 时, 称 A 为 E 的 L-模糊效应子代数. 我们约定 $\varnothing$ 是 E 的效应子代数.

例 8.2.2 设 $E=\{0,x,y,1\}$ 且 $\oplus$ 定义为

$\oplus$	0	x	y	1
0	0	x	y	1
x	x	1	$*$	$*$
y	y	$*$	1	$*$
1	1	$*$	$*$	$*$

则 $(E, \oplus, 0, 1)$ 是一个效应代数. 令 $L = [0,1]$, 并且定义 E 的 L-子集 A_1, A_2 和 A_3 如下:

(1) $A_1(0) = 0.8$, $A_1(x) = 0.5$, $A_1(y) = 0.5$, $A_1(1) = 0.8$.

由定义 8.2.1, 可以得到 $\mathscr{E}(A_1) = 1$. 事实上, 因为对任意的 $z \in E$, $A_1(z) \leqslant A_1(0)$, 所以对任意的 $z \in E$, 有 $A_1(z) \to A_1(0) = 1$. 容易验证 $x' = x$, $y' = y$, $0' = 1$ 和 $1' = 0$, 从而由 $A_1(0) = A_1(1)$ 可以得到 $A_1(z) = A_1(z')$, 所以对任意的 $z \in E$, 容易得到 $A_1(z) \to A_1(z') = 1$. 另外, 比较容易证明对所有满足 $z \perp w$ 的 $z, w \in E$, 都有 $(A(z) \wedge A(w)) \to A(z \oplus w) = 1$. 因此, $\mathscr{E}(A_1) = 1$, 那就是说 A_1 是 E 的 L-模糊效应子代数.

(2) $A_2(0) = 0.3$, $A_2(x) = 1$, $A_2(y) = 1$, $A_2(1) = 1$.

由定义 8.2.1, 可以得到 $\mathscr{E}(A_2) = 0.3$.

事实上, 因为 $A_2(x) \to A_2(0) = A_2(1) \to A_2(0) = A_2(y) \to A_2(0) = 0.3$, 所以 $\mathscr{E}(A_2) = 0.3$.

(3) $A_3(0) = 0$, $A_2(x) = 0.5$, $A_3(y) = 1$, $A_3(1) = 0.6$.

由定义 8.2.1, 可以计算出 $\mathscr{E}(A_3) = 0$.

事实上, 因为 $A_3(y) \to A_3(0) = 1 \to 0 = 0$, 所以 $\mathscr{E}(A_3) = 0$.

由蕴含算子的性质, 容易得到下面引理.

引理 8.2.3 设 $(E, \oplus, 0, 1)$ 是一个效应代数且 $A \in L^E$. 则 $\forall a \in L$, $\mathscr{E}(A) \geqslant a$ 当且仅当对满足 $x \perp y$ 的任意 $x, y \in E$, 都有

$$A(x) \wedge A(y) \wedge a \leqslant A(x \oplus y), \quad A(x) \wedge a \leqslant A(x') \wedge A(0).$$

由引理 8.2.3, 可以得到下面定理.

定理 8.2.4 设 $(E, \oplus, 0, 1)$ 是一个效应代数且 $A \in L^E$. 则

$$\mathscr{E}(A) = \bigvee \{a \in L \mid A(x) \wedge A(y) \wedge a \leqslant A(x \oplus y), A(x) \wedge a \leqslant A(x') \wedge A(0), x \perp y\}.$$

下面我们利用 L-模糊子集的四种截集去刻画 L-模糊效应子代数度.

定理 8.2.5 设 $(E, \oplus, 0, 1)$ 是一个效应代数且 $A \in L^E$. 则

(1) $\mathscr{E}(A) = \bigvee \{a \in L \mid \forall b \leqslant a, A_{[b]}$ 是 E 的一个效应子代数$\}$;

(2) $\mathscr{E}(A) = \bigvee \{a \in L \mid \forall b \in P(L), b \not\geqslant a, A^{(b)}$ 是 E 的一个效应子代数$\}$;

(3) $\mathscr{E}(A) = \bigvee \{a \in L \mid \forall b \notin \alpha(a), A^{[b]}$ 是 E 的一个效应子代数$\}$;

(4) $\mathscr{E}(A) = \bigvee \{a \in L \mid \forall b \in P(L), b \notin \alpha(a), A^{[b]}$ 是 E 的一个效应子代数$\}$;

(5) 如果对所有的 $a, b \in L$, $\beta(a \wedge b) = \beta(a) \cap \beta(b)$, 那么

$$\mathscr{E}(A) = \bigvee \{a \in L \mid \forall b \in \beta(a), A_{(b)}$ 是 E 的一个效应子代数$\}.$$

证明 (1) 对任意的 $x,y \in E$ 使得 $x \perp y$, 假设 $a \in L$ 满足

$$A(x) \wedge A(y) \wedge a \leqslant A(x \oplus y), \quad A(x) \wedge a \leqslant A(x') \wedge A(0).$$

对任意的 $b \leqslant a$, 假设 $A_{[b]} \neq \varnothing$, 令 $x,y \in A_{[b]}$ 满足 $x \perp y$, 则

$$b \leqslant A(x) \wedge A(y) \wedge a \leqslant A(x \oplus y), \quad b \leqslant A(x) \wedge a \leqslant A(x') \wedge A(0),$$

这意味着 $x \oplus y \in A_{[b]}$ 且 $x', 0 \in A_{[b]}$. 所以 $A_{[b]}$ 是 E 的一个效应子代数. 因此

$$\mathscr{E}(A) \leqslant \bigvee \{a \in L \mid \forall b \leqslant a, A_{[b]} \text{ 是 } E \text{ 的一个效应子代数}\}.$$

反之, 假设 $a \in L$ 使得对每一个 $b \leqslant a$, $A_{[b]}$ 是 E 的一个效应子代数. 对满足 $x \perp y$ 的任意 $x,y \in E$, 令

$$b = A(x) \wedge A(y) \wedge a \text{ 和 } c = A(x) \wedge a,$$

则 $b \leqslant a$ 和 $c \leqslant a$ 成立. 进而有 $x, y \in A_{[b]}$ 和 $x \in A_{[c]}$. 因为 $A_{[b]}$ 和 $A_{[c]}$ 都是 E 的效应子代数, 所以有 $x \oplus y \in A_{[b]}$ 和 $x', 0 \in A_{[c]}$, 即

$$A(x) \wedge A(y) \wedge a \leqslant A(x \oplus y) \quad \text{和} \quad A(x) \wedge a \leqslant A(x') \wedge A(0).$$

所以

$$\mathscr{E}(A) \geqslant \bigvee \{a \in L \mid \forall b \leqslant a, A_{[b]} \text{ 是 } E \text{ 的一个效应子代数}\}.$$

(2) 对满足 $x \perp y$ 的任意 $x,y \in E$, 假设 $a \in L$ 满足

$$A(x) \wedge A(y) \wedge a \leqslant A(x \oplus y) \quad \text{和} \quad A(x) \wedge a \leqslant A(x') \wedge A(0).$$

对满足 $b \not\geqslant a$ 的任意 $b \in P(L)$, 假设 $x,y \in A^{(b)}$ 使得 $x \perp y$, 则 $A(x) \not\leqslant b$ 和 $A(y) \not\leqslant b$, 又因为 b 是素的, 所以 $A(x) \wedge A(y) \wedge a \not\leqslant b$, $A(x) \wedge a \not\leqslant b$. 由

$$A(x) \wedge A(y) \wedge a \leqslant A(x \oplus y) \quad \text{和} \quad A(x) \wedge a \leqslant A(x') \wedge A(0)$$

可以得到 $A(x \oplus y) \not\leqslant b$ 和 $A(x') \wedge A(0) \not\leqslant b$, 即, $x \oplus y, x', 0 \in A^{(b)}$. 这表明 $A^{(b)}$ 是 E 的一个效应子代数. 因此

$$\mathscr{E}(A) \leqslant \bigvee \{a \in L \mid \forall b \in P(L), b \not\geqslant a, A^{(b)} \text{ 是 } E \text{ 的一个效应子代数}\}.$$

反之, 假设 $a \in L$ 且对满足 $b \not\geqslant a$ 的任意 $b \in P(L)$, $A^{(b)}$ 是 E 的一个效应子代数. 对满足 $x \perp y$ 的任意 $x,y \in E$, 令 $b \in P(L)$ 使得 $A(x) \wedge A(y) \wedge a \not\leqslant b$, 则 $x,y \in A^{(b)}$ 且 $a \not\leqslant b$ 成立. 因为 $A^{(b)}$ 是 E 的一个效应子代数, 所以 $x \oplus y \in A^{(b)}$,

即 $A(x \oplus y) \not\leqslant b$. 于是我们可以得到 $A(x) \wedge A(y) \wedge a \leqslant A(x \oplus y)$. 类似地, 可以证明 $A(x) \wedge a \leqslant A(x') \wedge A(0)$. 这表明

$$\mathscr{E}(A) \geqslant \bigvee\{a \in L \mid \forall b \in P(L), b \not\geqslant a, A^{(b)} \text{ 是 } E \text{ 的一个效应子代数}\}.$$

(3) 设 $x, y \in E$ 使得 $x \perp y$, $a \in L$ 满足

$$A(x) \wedge A(y) \wedge a \leqslant A(x \oplus y) \quad \text{和} \quad A(x) \wedge a \leqslant A(x') \wedge A(0).$$

对任意的 $b \notin \alpha(a)$, 假设 $x, y \in A^{[b]}$ 满足 $x \perp y$, 则 $b \notin \alpha(A(x))$ 且 $b \notin \alpha(A(y))$. 由 $b \notin \alpha(a)$ 可以得到 $b \notin \alpha(A(x)) \cup \alpha(A(y)) \cup \alpha(a)$. 因为

$$\alpha(A(x)) \cup \alpha(A(y)) \cup \alpha(a) = \alpha(A(x) \wedge A(y) \wedge a),$$

$$\alpha(A(x)) \cup \alpha(a) = \alpha(A(x) \wedge a),$$

所以 $b \notin \alpha(A(x) \wedge A(y) \wedge a)$ 和 $b \notin \alpha(A(x) \wedge a)$. 由

$$A(x) \wedge A(y) \wedge a \leqslant A(x \oplus y) \quad \text{和} \quad A(x) \wedge a \leqslant A(x') \wedge A(0),$$

我们知道有 $b \notin \alpha(A(x \oplus y))$, $b \notin \alpha(A(x'))$ 和 $b \notin \alpha(A(0))$, 即有 $x \oplus y, x', 0 \in A^{[b]}$ 成立. 所以 $A^{[b]}$ 是 E 的一个效应子代数. 这表明

$$\mathscr{E}(A) \leqslant \bigvee\{a \in L \mid \forall b \notin \alpha(a), A^{[b]} \text{ 是 } E \text{ 的一个效应子代数}\}.$$

反之, 假设 $a \in L$ 使得对于所有的 $b \notin \alpha(a)$, $A^{[b]}$ 是 E 的一个效应子代数. 现在证明, 对满足 $x \perp y$ 的所有 $x, y \in E$, 都有

$$A(x) \wedge A(y) \wedge a \leqslant A(x \oplus y) \quad \text{和} \quad A(x) \wedge a \leqslant A(x') \wedge A(0).$$

为此设 $b \in L$ 且 $b \notin \alpha(A(x) \wedge A(y) \wedge a)$, 则 $b \notin \alpha(A(x)) \cup \alpha(A(y)) \cup \alpha(a)$, 这意味着 $x, y \in A^{[b]}$ 和 $b \notin \alpha(a)$. 因为 $A^{[b]}$ 是 E 的一个效应子代数, 所以有 $x \oplus y \in A^{[b]}$, 即 $b \notin \alpha(A(x \oplus y))$, 这表明 $\alpha(A(x) \wedge A(y) \wedge a) \supseteq \alpha(A(x \oplus y))$. 因此

$$A(x) \wedge A(y) \wedge a = \bigwedge \alpha(A(x) \wedge A(y) \wedge a) \leqslant \bigwedge \alpha(A(x \oplus y)) = A(x \oplus y).$$

类似地, 可以证明 $A(x) \wedge a \leqslant A(x') \wedge A(0)$. 故有

$$\mathscr{E}(A) \geqslant \bigvee\{a \in L \mid \forall b \notin \alpha(a), A^{[b]} \text{ 是 } E \text{ 的一个效应子代数}\}.$$

(4) 由 (3) 的等式易见

$$\mathscr{E}(A) \leqslant \bigvee\{a \in L \mid \forall b \in P(L), b \notin \alpha(a), A^{[b]} \text{ 是 } E \text{ 的一个效应子代数}\}.$$

8.2 效应代数上的 L-模糊子代数度

反之,假设 $a \in L$ 使得对满足 $b \notin \alpha(a)$ 的所有 $b \in P(L)$, $A^{[b]}$ 是 E 的一个效应子代数. 现在证明, 对满足 $x \perp y$ 的所有 $x, y \in E$, 都有

$$A(x) \wedge A(y) \wedge a \leqslant A(x \oplus y) \quad \text{和} \quad A(x) \wedge a \leqslant A(x') \wedge A(0).$$

为此设 $b \in P(L)$ 且 $b \notin \alpha^*(A(x) \wedge A(y) \wedge a)$, 则有

$$b \notin \alpha(A(x) \wedge A(y) \wedge a) = \alpha(A(x)) \cup \alpha(A(y)) \cup \alpha(a).$$

这意味着 $x, y \in A^{[b]}$ 且 $b \notin \alpha(a)$. 因为 $A^{[b]}$ 是 E 的一个效应子代数, 所以 $x \oplus y \in A^{[b]}$, 即, $b \notin \alpha(A(x \oplus y))$. 于是有 $b \notin \alpha^*(A(x \oplus y))$, 这表明

$$\alpha^*(A(x) \wedge A(y) \wedge a) \supseteq \alpha^*(A(x \oplus y)).$$

因此可得

$$A(x) \wedge A(y) \wedge a = \bigwedge \alpha^*(A(x) \wedge A(y) \wedge a) \leqslant \bigwedge \alpha^*(A(x \oplus y)) = A(x \oplus y).$$

类似地, 可以证明 $A(x) \wedge a \leqslant A(x') \wedge A(0)$. 这样就证明了

$$\mathscr{E}(A) \geqslant \bigvee \{a \in L \mid \forall b \in P(L), b \notin \alpha(a), A^{[b]} \text{ 是 } E \text{ 的一个效应子代数}\}.$$

(5) 对满足 $x \perp y$ 的任意 $x, y \in E$, 假设 $0 \neq a \in L$ 满足

$$A(x) \wedge A(y) \wedge a \leqslant A(x \oplus y) \quad \text{和} \quad A(x) \wedge a \leqslant A(x') \wedge A(0).$$

对任意的 $b \in \beta(a)$, 假设 $A_{(b)} \neq \varnothing$, 令 $x, y \in A_{(b)}$ 满足 $x \perp y$, 则 $b \in \beta(A(x))$ 和 $b \in \beta(A(y))$. 由 $b \in \beta(a)$ 可以得到 $b \in \beta(A(x)) \cap \beta(A(y)) \cap \beta(a)$. 从下面两个等式

$$\beta(A(x)) \cap \beta(A(y)) \cap \beta(a) = \beta(A(x) \wedge A(y) \wedge a),$$

$$\beta(A(x)) \cap \beta(a) = \beta(A(x) \wedge a),$$

可知 $b \in \beta(A(x) \wedge A(y) \wedge a)$ 和 $b \in \beta(A(x) \wedge a)$ 成立. 于是由

$$A(x) \wedge A(y) \wedge a \leqslant A(x \oplus y) \quad \text{和} \quad A(x) \wedge a \leqslant A(x') \wedge A(0)$$

可以得到 $b \in \beta(A(x \oplus y)), b \in \beta(A(x'))$ 和 $b \in \beta(A(0))$ 成立, 即 $x \oplus y, x', 0 \in A_{(b)}$. 所以 $A_{(b)}$ 是 E 的一个效应子代数. 这样就证明了

$$\mathscr{E}(A) \leqslant \bigvee \{a \in L \mid \forall b \in \beta(a), A_{(b)} \text{ 是 } E \text{ 的一个效应子代数}\}.$$

反之，假设 $a \in L$ 使得对所有的 $b \in \beta(a)$，$A_{(b)}$ 是 E 的一个效应子代数. 现在证明对满足 $x \perp y$ 的所有 $x, y \in E$，均有

$$A(x) \wedge A(y) \wedge a \leqslant A(x \oplus y) \quad \text{和} \quad A(x) \wedge a \leqslant A(x') \wedge A(0).$$

当 $A(x) \wedge A(y) \wedge a = 0$ 且 $A(x) \wedge a = 0$ 时，上面两个不等式显然都成立. 现在假设 $A(x) \wedge A(y) \wedge a \neq 0$，于是有 $A(x) \neq 0, A(y) \neq 0$ 和 $a \neq 0$. 取 $b \in L$ 使得 $b \in \beta(A(x) \wedge A(y) \wedge a)$，则 $b \in \beta(A(x)) \cap \beta(A(y)) \cap \beta(a)$，从而 $x, y \in A_{(b)}$ 且 $b \in \beta(a)$. 因为 $A_{(b)}$ 是 E 的一个效应子代数，所以有 $x \oplus y \in A_{(b)}$，即 $b \in \beta(A(x \oplus y))$，这意味着 $\beta(A(x) \wedge A(y) \wedge a) \subseteq \beta(A(x \oplus y))$. 进一步我们可得

$$A(x) \wedge A(y) \wedge a = \bigvee \beta(A(x) \wedge A(y) \wedge a) \leqslant \bigvee \beta(A(x \oplus y)) = A(x \oplus y).$$

类似地，可以证明 $A(x) \wedge a \leqslant A(x') \wedge A(0)$. 这样我们就证明了

$$\mathscr{E}(A) \geqslant \bigvee \{a \in L \mid \forall b \in \beta(a), A_{(b)} \text{ 是 } E \text{ 的一个效应子代数}\}. \qquad \square$$

从定义 8.2.1 和引理 8.2.3，下列命题是显然的.

推论 8.2.6 设 E 是一个效应代数且 $A \in L^E$. 则 A 是一个 L-效应子代数当且仅当对满足 $x \perp y$ 的任意 $x, y \in E$，都有

$$A(0) \geqslant A(x), \quad A(x') \geqslant A(x), \quad A(x \oplus y) \geqslant A(x) \wedge A(y).$$

推论 8.2.7 设 A 是效应代数 E 的一个 L-效应子代数. 则
(1) 对任意的 $x \in E$, $A(1) \geqslant A(x)$;
(2) 对任意的 $x \in E$, $A(x) = A(x')$;
(3) $A(1) = A(0)$.

证明 (1) 因为 $x \perp x'$ 且 A 是 E 的一个 L-效应子代数，所以

$$A(1) = A(x \oplus x') \geqslant A(x) \wedge A(x') = A(x).$$

(2) 首先由推论 8.2.6 可以得到 $A(x') \geqslant A(x)$. 另一方面，由定义 8.1.1，可得 $x'' = x$，进而有 $A(x) = A(x'') \geqslant A(x')$.

(3) 由 (1) 可得 $A(1) \geqslant A(0)$. 由推论 8.2.6，可得 $A(1) \leqslant A(0)$，于是有 $A(1) = A(0)$. $\qquad \square$

由定义 8.2.1 和定理 8.2.5，可以得到下列两个结论.

推论 8.2.8 设 A 是效应代数 E 的一个 L-模糊子集. 则下列条件是等价的:
(1) A 是一个 L-效应子代数;
(2) 对任意的 $a \in J(L)$, $A_{[a]}$ 是 E 的一个效应子代数;

(3) 对任意的 $a \in L$, $A_{[a]}$ 是 E 的一个效应子代数;
(4) 对任意的 $a \in L$, $A^{[a]}$ 是 E 的一个效应子代数;
(5) 对任意的 $a \in P(L)$, $A^{[a]}$ 是 E 的一个效应子代数;
(6) 对任意的 $a \in P(L)$, $A^{(a)}$ 是 E 的一个效应子代数.

推论 8.2.9 设 A 是效应代数 E 的一个 L-模糊子集. 如果对所有的 $a, b \in L$, $\beta(a \wedge b) = \beta(a) \cap \beta(b)$, 则下列条件是等价的:
(1) A 是 E 的一个 L-效应子代数;
(2) 对任意的 $a \in L$, $A_{(a)}$ 是 E 的一个效应子代数;
(3) 对任意的 $a \in J(L)$, $A_{(a)}$ 是 E 的一个效应子代数.

8.3 由模糊效应子代数确定的 L-模糊凸结构

在这一节中, 我们研究效应代数上的 L-模糊效应子代数度与 L-模糊凸结构之间的关系. 并证明两个效应代数之间的同态是一个 L-模糊凸保持映射, 一个单态映射是 L-模糊凸到凸映射. 最后证明效应代数上的所有 L-效应子代数之集是一个 L-凸结构.

对任意的 $A \in L^E$, $\mathscr{E}$ 可以看作一个映射 $\mathscr{E} : L^E \to L$, 它定义为 $A \longmapsto \mathscr{E}(A)$. 下列定理表明 $\mathscr{E}$ 是效应代数 E 上的一个 L-模糊凸结构.

定理 8.3.1 设 E 是一个效应代数. 则 $\mathscr{E} : L^E \to L$ 是 E 上的一个 L-模糊凸结构, 称为由 L-模糊效应子代数度确定的 L-模糊凸结构.

证明 (1) 由定义 8.2.1 易知 $\mathscr{E}(\chi_\varnothing) = \mathscr{E}(\chi_E) = 1$.

(2) 设 $\{A_i\}_{i \in \Omega} \subseteq L^E$ 非空. 现在证明 $\mathscr{E}\left(\bigwedge_{i \in \Omega} A_i\right) \geqslant \bigwedge_{i \in \Omega} \mathscr{E}(A_i)$. 设 $a \in L$ 使得其满足 $a \leqslant \bigwedge_{i \in \Omega} \mathscr{E}(A_i)$. 则对任意的 $i \in \Omega$, 可以得到 $a \leqslant \mathscr{E}(A_i)$, 这意味着对满足 $x \perp y$ 的任意 $x, y \in E$, 都有

$$A_i(x) \wedge A_i(y) \wedge a \leqslant A_i(x \oplus y) \quad \text{和} \quad A_i(x) \wedge a \leqslant A_i(x') \wedge A_i(0).$$

进而有

$$\bigwedge_{i \in \Omega} A_i(x) \wedge \bigwedge_{i \in \Omega} A_i(y) \wedge a \leqslant \bigwedge_{i \in \Omega} A_i(x \oplus y),$$

$$\bigwedge_{i \in \Omega} A_i(x) \wedge a \leqslant \bigwedge_{i \in \Omega} A_i(x') \wedge \bigwedge_{i \in \Omega} A_i(0).$$

这表明 $a \leqslant \mathscr{E}\left(\bigwedge_{i \in \Omega} A_i\right)$. 于是我们得到 $\mathscr{E}\left(\bigwedge_{i \in \Omega} A_i\right) \geqslant \bigwedge_{i \in \Omega} \mathscr{E}(A_i)$.

(3) 设 $\{A_i\}_{i\in\Omega} \subseteq L^E$ 是非空且定向的. 现在证明 $\mathscr{E}\left(\bigvee_{i\in\Omega} A_i\right) \geqslant \bigwedge_{i\in\Omega} \mathscr{E}(A_i)$. 为此设 $a \in L$ 满足 $a \leqslant \bigwedge_{i\in\Omega} \mathscr{E}(A_i)$, 则对任意的 $i \in \Omega, a \leqslant \mathscr{E}(A_i)$. 这意味着对满足 $x \perp y$ 的任意 $x, y \in E$, 都有

$$A_i(x) \wedge A_i(y) \wedge a \leqslant A_i(x \oplus y) \quad \text{和} \quad A_i(x) \wedge a \leqslant A_i(x') \wedge A_i(0).$$

下面我们来证明

$$\bigvee_{i\in\Omega} A_i(x) \wedge \bigvee_{i\in\Omega} A_i(y) \wedge a \leqslant \bigvee_{i\in\Omega} A_i(x \oplus y).$$

取 $b \in L$ 使得

$$b \prec \bigvee_{i\in\Omega} A_i(x) \wedge \bigvee_{i\in\Omega} A_i(y) \wedge a,$$

则存在 $i, j \in \Omega$ 使得 $b \leqslant A_i(x), b \leqslant A_j(y)$ 和 $b \leqslant a$. 因为 $\{A_i\}_{i\in\Omega}$ 是定向的, 所以存在 $i_0 \in \Omega$ 使得 $A_i \leqslant A_{i_0}$ 和 $A_j \leqslant A_{i_0}$, 于是有 $b \leqslant A_{i_0}(x) \wedge A_{i_0}(y) \wedge a$. 由 $A_{i_0}(x) \wedge A_{i_0}(y) \wedge a \leqslant A_{i_0}(x \oplus y)$ 可得 $b \leqslant A_{i_0}(x \oplus y) \leqslant \bigvee_{i\in\Omega} A_i(x \oplus y)$. 于是我们可以得到

$$\bigvee_{i\in\Omega} A_i(x) \wedge \bigvee_{i\in\Omega} A_i(y) \wedge a \leqslant \bigvee_{i\in\Omega} A_i(x \oplus y).$$

类似地, 可以证明

$$\bigvee_{i\in\Omega} A_i(x) \wedge a \leqslant \bigvee_{i\in\Omega} A_i(x') \wedge \bigvee_{i\in\Omega} A_i(0).$$

综上可知 $a \leqslant \mathscr{E}\left(\bigvee_{i\in\Omega} A_i\right)$, 这表明 $\mathscr{E}\left(\bigvee_{i\in\Omega} A_i\right) \geqslant \bigwedge_{i\in\Omega} \mathscr{E}(A_i)$.

这就证明了映射 $\mathscr{E}: L^E \to L$ 是一个 L-模糊凸结构. □

下面我们来研究效应代数的同态性质. 首先给出下列引理.

引理 8.3.2 设 E 和 F 是效应代数且 $g: E \to F$ 是一个同态映射. 则 $g(0_E) = 0_F$ 且对任意的 $x \in E, g(x') = g(x)'$.

证明 由定义 8.1.3, 可知 $g(0_E \oplus 1_E) = g(0_E) \oplus g(1_E) = g(0_E) \oplus 1_F$, 由定义 8.1.3 知 $g(0_E) = 0_F$. 因为对任意的 $x \in E, x \perp x'$, 所以 $g(1_E) = g(x \oplus x') = g(x) \oplus g(x')$, 因此 $g(x') = g(x)'$. □

定理 8.3.3 设 $g: E \to F$ 为两个效应代数之间的一个同态映射. 则

(1) $g: (E, \mathscr{E}_E) \to (F, \mathscr{E}_F)$ 是 L-模糊凸保持的;

8.3 由模糊效应子代数确定的 L-模糊凸结构

(2) 如果 $g: E \to F$ 是一个单同态, 则 $g: (E, \mathscr{E}_E) \to (F, \mathscr{E}_F)$ 是 L-模糊凸到凸的.

证明 (1) 为了证明 $g: (E, \mathscr{E}_E) \to (F, \mathscr{E}_F)$ 是 L-模糊凸保持的, 只需证明对任意的 $B \in L^F$, $\mathscr{E}_E(g_L^\leftarrow(B)) \geqslant \mathscr{E}_F(B)$.

设 $a \in L$ 使得其满足 $a \leqslant \mathscr{E}_F(B)$. 由引理 8.2.3 可知, 对满足 $x \perp y$ 的任意 $x, y \in F$, 都有

$$B(x) \wedge B(y) \wedge a \leqslant B(x \oplus y) \quad \text{和} \quad B(x) \wedge a \leqslant B(x') \wedge B(0_F).$$

于是对满足 $s \perp t$ 的任意 $s, t \in E$, 我们能够得到

$$g_L^\leftarrow(B)(s) \wedge g_L^\leftarrow(B)(t) \wedge a = B(g(s)) \wedge B(g(t)) \wedge a$$
$$\leqslant B(g(s) \oplus g(t))$$
$$= B(g(s \oplus t))$$
$$= g_L^\leftarrow(B)(s \oplus t).$$

类似地, 可以得到

$$g_L^\leftarrow(B)(s) \wedge a \leqslant g_L^\leftarrow(B)(s') \wedge g_L^\leftarrow(B)(0_E),$$

这表明 $a \leqslant \mathscr{E}_E(g_L^\leftarrow(B))$. 这样我们就证明了 $\mathscr{E}_E(g_L^\leftarrow(B)) \geqslant \mathscr{E}_F(B)$.

(2) 为了证明 $g: (E, \mathscr{E}_E) \to (F, \mathscr{E}_F)$ 是 L-模糊凸到凸的, 只需证明对任意的 $A \in L^E$, $\mathscr{E}_E(A) \leqslant \mathscr{E}_F(g_L^\to(A))$.

为此设 $a \in L$ 使得其满足 $a \leqslant \mathscr{E}_E(A)$, 则对满足 $s \perp t$ 的任意 $s, t \in E$, 都有

$$A(s) \wedge A(t) \wedge a \leqslant A(s \oplus t) \quad \text{和} \quad A(s) \wedge a \leqslant A(s') \wedge A(0_E).$$

这样, 对满足 $x \perp y$ 的任意 $x, y \in F$, 我们能够得到

$$g_L^\to(A)(x) \wedge g_L^\to(A)(y) \wedge a$$
$$= \bigvee_{g(s)=x} A(s) \wedge \bigvee_{g(t)=y} A(t) \wedge a$$
$$= \bigvee \{A(s) \wedge A(t) \wedge a \mid g(s) = x, g(t) = y\}$$
$$\leqslant \bigvee \{A(s \oplus t) \mid g(s) = x, g(t) = y\}$$
$$\leqslant \bigvee \{A(s \oplus t) \mid g(s \oplus t) = x \oplus y\}$$

$$\leqslant \bigvee\{A(r) \mid g(r) = x \oplus y\}$$
$$= (g_L^{\to}(A))(x \oplus y).$$

类似地, 可以证明
$$(g_L^{\to}(A))(x) \wedge a \leqslant (g_L^{\to}(A))(x') \wedge (g_L^{\to}(A))(0_F).$$

这表明 $a \leqslant \mathscr{E}_F(g_L^{\to}(A))$. 因此 $\mathscr{E}_E(A) \leqslant \mathscr{E}_F(g_L^{\to}(A))$. 这样就证明了 $g : (E, \mathscr{E}_E) \to (F, \mathscr{E}_F)$ 是 L-模糊凸到凸的. □

推论 8.3.4 设 $\{E_i\}_{i \in \Omega}$ 是一族效应代数且 $\prod_{i \in \Omega} A_i$ 是 $\{A_i\}_{i \in \Omega}$ 的乘积, 其中 $A_i \in L^{E_i}$, $P_i : \prod_{i \in \Omega} E_i \to E_i$ 是投射, 让 $\mathscr{E}$ 和 $\mathscr{E}_i$ 分别表示 $\prod_{i \in \Omega} E_i$ 与 E_i 上的由 L-模糊效应子代数度确定的 L-模糊凸结构, 则对任意的 $i \in \Omega$, $P_i : \left(\prod_{i \in \Omega} E_i, \mathscr{E}\right) \to (E_i, \mathscr{E}_i)$ 是 L-模糊凸保持映射.

证明 容易证明 $P_i : \prod_{i \in \Omega} E_i \to E_i$ 是效应代数之间的一个同态映射, 由定理 8.3.3, 可知对任意的 $i \in \Omega$, P_i 是 L-模糊凸保持映射. □

定理 8.3.5 设 $\{E_i\}_{i \in \Omega}$ 是一族效应代数且 $\prod_{i \in \Omega} A_i$ 是 $\{A_i\}_{i \in \Omega}$ 的直积, 其中 $A_i \in L^{E_i}$. 则 $\mathscr{E}\left(\prod_{i \in \Omega} A_i\right) \geqslant \bigwedge_{i \in \Omega} \mathscr{E}_i(A_i)$.

证明 设 $P_i : \prod_{i \in \Omega} E_i \to E_i$ 是投射. 容易证明 $\prod_{i \in \Omega} A_i = \bigwedge_{i \in \Omega} P_i^{\leftarrow}(A_i)$. 由定理 8.3.1 中 (2) 的证明以及推论 8.3.4, 可得
$$\mathscr{E}\left(\prod_{i \in \Omega} A_i\right) = \mathscr{E}\left(\bigwedge_{i \in \Omega} P_i^{\leftarrow}(A_i)\right) \geqslant \bigwedge_{i \in \Omega} \mathscr{E}(P_i^{\leftarrow}(A_i)) \geqslant \bigwedge_{i \in \Omega} \mathscr{E}_i(A_i).$$

因此 P_i 是 L-模糊凸保持映射. □

8.4 效应代数中的 L-模糊理想度

在这一节中, 我们引入效应代数中的 L-模糊理想度的概念, 并利用截集给出它的刻画. 进一步研究 L-模糊理想度与 L-模糊凸结构的关系.

定义 8.4.1 设 E 是一个效应代数, A 是 E 的一个 L-模糊集. 则 A 是 L-模糊理想的程度 $\Im(A)$ 被定义为
$$\Im(A) = \bigwedge_{\substack{x,y,z,w \in E, \\ z \perp w, x \leqslant y}} (A(y) \to A(x)) \wedge (A(z) \wedge A(w) \to A(z+w)).$$

8.4 效应代数中的 L-模糊理想度

当 $\mathfrak{I}(A) = 1$ 时, 称 A 是 E 的一个 L-模糊理想. 当 $L = [0,1]$ 时, L-模糊理想也称为模糊理想.

当 $\mathfrak{I}(A) = 1$ 时, L-模糊理想的定义可以如下表述.

定义 8.4.2 设 E 是一个效应代数, A 是 E 的一个 L-模糊集. A 称为 E 的一个 L-模糊理想, 如果它满足下面两个条件: 对于任意的 $x, y \in E$,

(II1) 如果 $x \leqslant y$, 那么 $A(y) \leqslant A(x)$;

(II2) 如果 $x \perp y$, 那么 $A(x) \wedge A(y) \leqslant A(x+y)$.

下面我们给出一些 L-模糊理想度的例子.

例 8.4.3 设 $E = \{0, x, x', 1\}$ 满足 $0 \leqslant x \leqslant x' \leqslant 1$, $x + x' = 1$. 那么 E 是一个效应代数.

(1) 设 $A : E \to [0,1]$ 是一个常值映射, 对于任意的 $x, y, z, w \in E$, 当 $x \leqslant y$ 且 $z \perp w$ 时, 一定有

$$A(y) \leqslant A(x) \quad \text{和} \quad A(z) \wedge A(w) \leqslant A(z+w).$$

也就是 $\mathfrak{I}(A) = 1$. 所以 A 是一个模糊理想.

(2) 设 $A : E \to [0,1]$ 是一个模糊集, 其取值如下:

$$A(t) = \begin{cases} 0.1, & t = 0, \\ 0.3, & t = x, \\ 0.7, & t = x', \\ 1, & t = 1, \end{cases}$$

那么当 $z \perp w$ 时, 有 $A(z) \wedge A(w) \leqslant A(z+w)$. 但是因为 $x \leqslant x'$, 而

$$A(x') = 0.7 \nleqslant 0.3 = A(x),$$

所以 A 就不是一个模糊理想, 不过此时我们可以看出 $\mathfrak{I}(A) = 0.1$, 因此可以认为它是模糊理想的程度是 0.1.

(3) 设 $A : E \to [0,1]$ 是一个模糊集, 其取值如下:

$$A(t) = \begin{cases} 0.1, & t = 1, \\ 0.7, & t = x, \\ 0.3, & t = x', \\ 1, & t = 0. \end{cases}$$

我们可以计算出当 $z \leqslant y$ 时, 有 $A(y) \leqslant A(z)$. 但是因为 $x + x' = 1$, 而

$$A(x) \wedge A(x') = 0.3 \wedge 0.7 \nleqslant 0.1 = A(1),$$

所以 A 也不是一个模糊理想, 不过此时我们可以计算出 $\mathfrak{I}(A) = 0.1$, 因此 A 在 0.1 程度上是一个模糊理想.

(4) 设 $A : E \to [0, 1]$ 是一个模糊集, 其取值为

$$A(t) = \begin{cases} 0, & t = 0, \\ 0.7, & t = x, \\ 1, & t = x', \\ 0.3, & t = 1. \end{cases}$$

可以计算出 $\mathfrak{I}(A) = 0$, 所以 A 是一个模糊理想的程度是 0.

引理 8.4.4 设 E 是一个效应代数且 $A \in L^E$. 对于任意的 $a \in L$, $a \leqslant \mathfrak{I}(A)$ 当且仅当对于满足 $x \leqslant y$ 和 $z \perp w$ 的任意 $x, y, z, w \in E$, 都有

$$a \wedge A(y) \leqslant A(x) \quad \text{和} \quad a \wedge A(z) \wedge A(w) \leqslant A(z + w).$$

证明 ($\Rightarrow$) 任取 $a \in L$ 使得 $a \leqslant \mathfrak{I}(A)$, 则

$$a \leqslant \bigwedge_{\substack{x,y,z,w \in E, \\ z \perp w, x \leqslant y}} (A(y) \to A(x)) \wedge (A(z) \wedge A(w) \to A(z+w)).$$

这意味着对于任意的 $x, y, z, w \in E$(这里 $x \leqslant y$ 且 $z \perp w$), 都有

$$a \leqslant A(y) \to A(x) \quad \text{和} \quad a \leqslant A(z) \wedge A(w) \to A(z+w).$$

进一步可得到

$$a \wedge A(y) \leqslant A(x) \quad \text{和} \quad a \wedge A(z) \wedge A(w) \leqslant A(z+w).$$

($\Leftarrow$) 假设 $a \in L$ 使得对于满足 $x \leqslant y$ 和 $z \perp w$ 的任意 $x, y, z, w \in E$, 都有

$$a \wedge A(y) \leqslant A(x) \quad \text{和} \quad a \wedge A(z) \wedge A(w) \leqslant A(z+w).$$

那么

$$a \leqslant A(y) \to A(x) \quad \text{且} \quad a \leqslant A(z) \wedge A(w) \to A(z+w),$$

即

$$a \leqslant (A(y) \to A(x)) \wedge (A(z) \wedge A(w) \to A(z+w)).$$

8.4 效应代数中的 L-模糊理想度

于是就得到了

$$a \leqslant \bigwedge_{\substack{x,y,z,w \in E, \\ z \perp w, x \leqslant y}} (A(y) \to A(x)) \wedge (A(z) \wedge A(w) \to A(z+w)).$$

这样就完成了引理的证明. □

从引理 8.4.4 可直接得到下面定理.

定理 8.4.5 设 E 是一个效应代数且 $A \in L^E$, 则

$$\Im(A) = \bigvee \left\{ a \in L \mid a \wedge A(y) \leqslant A(x), a \wedge A(z) \wedge A(w) \leqslant A(z+w), x \leqslant y, z \perp w \right\}.$$

下面我们利用各种截集给出 L-模糊理想度的刻画.

定理 8.4.6 设 E 是一个效应代数且 $A \in L^E$. 则

$$\Im(A) = \bigvee \left\{ a \in L \mid \forall b \leqslant a, A_{[b]} \text{ 是 } E \text{ 的一个理想} \right\}.$$

证明 假设 $a \in L$ 使得对于任意的 $b \leqslant a$, $A_{[b]}$ 是 E 的一个理想. 对于满足 $x \leqslant y$ 的任意 $x, y \in E$, 设 $c = a \wedge A(y)$, 则 $c \leqslant a$ 且 $c \leqslant A(y)$, 这意味着 $y \in A_{[c]}$. 由假设可知 $A_{[c]}$ 是 E 的一个理想, 于是便有 $x \in A_{[c]}$, 进而可以得到 $c \leqslant A(x)$. 因此我们有

$$a \wedge A(y) = c \leqslant A(x).$$

类似地, 对于满足 $z \perp w$ 的任意 $z, w \in E$, 可以证明

$$a \wedge A(z) \wedge A(w) \leqslant A(z+w).$$

这样, 我们就证明了

$$\Im(A) = \bigvee \left\{ a \in L \mid a \wedge A(y) \leqslant A(x), a \wedge A(z) \wedge A(w) \leqslant A(z+w), x \leqslant y, z \perp w \right\}$$

$$\geqslant \bigvee \left\{ a \in L \mid \forall b \leqslant a, A_{[b]} \text{ 是 } E \text{ 的一个理想} \right\}.$$

反之, 假设 $x, y, z, w \in E$ 使得当 $z \perp w$ 和 $x \leqslant y$ 时, 有

$$a \wedge A(y) \leqslant A(x) \quad \text{和} \quad a \wedge A(z) \wedge A(w) \leqslant A(z+w).$$

对于任意的 $b \leqslant a$, 我们需要证明 $A_{[b]}$ 是 E 的一个理想.

(I1) 如果 $y \in A_{[b]}$ 且 $x \leqslant y$, 那么由 $b \leqslant A(y)$ 可知

$$b \leqslant a \wedge A(y) \leqslant A(x),$$

这意味着 $x \in A_{[b]}$.

(I2) 如果 $z, w \in A_{[b]}$ 且 $z \perp w$, 那么
$$b \leqslant a \wedge A(z) \wedge A(w) \leqslant A(z+w).$$

于是可以得到 $z + w \in A_{[b]}$. 这表明 $A_{[b]}$ 是 E 的一个理想. 故得到
$$\mathfrak{I}(A) = \bigvee \Big\{ a \in L \mid a \wedge A(y) \leqslant A(x), a \wedge A(z) \wedge A(w) \leqslant A(z+w), x \leqslant y, z \perp w \Big\}$$
$$\leqslant \bigvee \Big\{ a \in L \mid \forall b \leqslant a, A_{[b]} \text{ 是 } E \text{ 的一个理想} \Big\}. \qquad \Box$$

定理 8.4.7 设 E 是一个效应代数且 $A \in L^E$, 则
$$\mathfrak{I}(A) = \bigvee \Big\{ a \in L \mid b \notin \alpha(a), A^{[b]} \text{ 是 } E \text{ 的一个理想} \Big\}.$$

证明 假设 $a \in L$ 使得对于满足 $x \leqslant y$ 和 $z \perp w$ 的任意 $x, y, z, w \in E$, 都有
$$a \wedge A(y) \leqslant A(x) \quad \text{和} \quad a \wedge A(z) \wedge A(w) \leqslant A(z+w).$$

对于 $b \notin \alpha(a)$, 我们需要证明 $A^{[b]}$ 是 E 的一个理想.

(I1) 如果 $x \leqslant y$ 和 $y \in A^{[b]}$, 那么 $b \notin \alpha(A(y))$. 由
$$a \wedge A(y) \leqslant A(x)$$

可以得到
$$\alpha(A(x)) \subseteq \alpha(a \wedge A(y)) = \alpha(a) \cup \alpha(A(y)),$$

这意味着 $b \notin \alpha(A(x))$. 于是可以得到 $x \in A^{[b]}$.

(I2) 如果 $z, w \in A^{[b]}$ 而且 $z \perp w$, 那么
$$b \notin \alpha(A(z)) \cup \alpha(A(w)) \cup \alpha(a) = \alpha(a \wedge A(z) \wedge A(w)).$$

由
$$a \wedge A(z) \wedge A(w) \leqslant A(z+w)$$

可以得到
$$\alpha(A(z+w)) \subseteq \alpha(a \wedge A(z) \wedge A(w)).$$

这意味着 $b \notin \alpha(A(z+w))$, 也就是 $z + w \in A^{[b]}$. 这表明
$$\mathfrak{I}(A) = \bigvee \Big\{ a \in L \mid a \wedge A(y) \leqslant A(x), a \wedge A(z) \wedge A(w) \leqslant A(z+w), x \leqslant y, z \perp w \Big\}$$

8.4 效应代数中的 L-模糊理想度

$$\leqslant \bigvee \left\{ a \in L \mid b \notin \alpha(a), A^{[b]} \text{ 是 } E \text{ 的一个理想} \right\}.$$

相反地, 假设 $a \in L$ 使得对任意 $b \notin \alpha(a)$, 都有 $A^{[b]}$ 是 E 的一个理想. 下面我们证明对于任意的 $x, y, z, w \in E$, 当 $x \leqslant y$ 且 $z \perp w$ 时, 都有

$$a \wedge A(y) \leqslant A(x) \quad \text{且} \quad a \wedge A(z) \wedge A(w) \leqslant A(z+w).$$

为此我们假设 $b \notin \alpha(a \wedge A(y))$. 由

$$\alpha(a \wedge A(y)) = \alpha(a) \cup \alpha(A(y))$$

可知

$$b \notin \alpha(a) \quad \text{且} \quad b \notin \alpha(A(y)).$$

这说明 $y \in A^{[b]}$. 从开始的假设条件可以知道 $A^{[b]}$ 是 E 的一个理想, 所以可得 $x \in A^{[b]}$, 这意味着 $b \notin \alpha(A(x))$. 于是我们就证明了 $a \wedge A(y) \leqslant A(x)$.

类似地, 对于满足 $z \perp w$ 的任意 $z, w \in E$, 可以证明

$$a \wedge A(z) \wedge A(w) \leqslant A(z+w).$$

因此我们得到所要证明的下述不等式:

$$\mathfrak{I}(A) = \bigvee \left\{ a \in L \mid a \wedge A(y) \leqslant A(x), a \wedge A(z) \wedge A(w) \leqslant A(z+w), x \leqslant y, z \perp w \right\}$$

$$\geqslant \bigvee \left\{ a \in L \mid b \notin \alpha(a), A^{[b]} \text{ 是 } E \text{ 的一个理想} \right\}. \qquad \square$$

定理 8.4.8 设 E 是一个效应代数且 A 是 E 的一个 L-模糊集. 则

$$\mathfrak{I}(A) = \bigvee \left\{ a \in L \mid \forall b \in P(L), a \not\leqslant b, A^{(b)} \text{ 是 } E \text{ 的一个理想} \right\}.$$

证明 假设 $a \in L$ 使得对于满足 $x \leqslant y$ 和 $z \perp w$ 的任意 $x, y, z, w \in E$, 都有

$$a \wedge A(y) \leqslant A(x) \quad \text{和} \quad a \wedge A(z) \wedge A(w) \leqslant A(z+w).$$

我们需要证明对任意的 $b \in P(L)$, 当 $a \not\leqslant b$ 时, $A^{(b)}$ 都是 E 的理想.

为了这个目的, 我们假设 $y \in A^{(b)}$. 如果 $x \notin A^{(b)}$, 那么 $A(x) \leqslant b$. 由

$$a \wedge A(y) \leqslant A(x)$$

可知

$$a \wedge A(y) \leqslant b.$$

因为 $b \in P(L)$ 且 $y \in A^{(b)}$, 也就是 $A(y) \not\leqslant b$, 所以可以得到 $a \leqslant b$, 这是一个矛盾. 因此我们必有 $x \in A^{(b)}$.

类似地, 对于任意的 $z, w \in E$, 当 $z \perp w$ 时, 可以证明

$$z, w \in A^{(b)} \Rightarrow z + w \in A^{(b)}.$$

因此我们证明了下面不等式成立:

$$\mathfrak{I}(A) = \bigvee \{a \in L \mid a \wedge A(y) \leqslant A(x), a \wedge A(z) \wedge A(w) \leqslant A(z+w), x \leqslant y, z \perp w\}$$
$$\leqslant \bigvee \{a \in L \mid \forall b \in P(L), a \not\leqslant b, A^{(b)} \text{ 是 } E \text{ 的一个理想}\}.$$

相反地, 假设 $a \in L$ 使得对于任意的 $b \in P(L)$, 当 $a \not\leqslant b$ 时, $A^{(b)}$ 都是 E 的理想. 下面我们来证明: 对于满足 $x \leqslant y$ 和 $z \perp w$ 的任意 $x, y, z, w \in E$, 都有下面不等式成立:

$$a \wedge A(y) \leqslant A(x) \quad \text{且} \quad a \wedge A(z) \wedge A(w) \leqslant A(z+w).$$

为此我们假设 $b \in P(L)$ 使得对于满足 $x \leqslant y$ 的任意 $x, y \in E$, 都有 $a \wedge A(y) \not\leqslant b$. 此时必有

$$a \not\leqslant b \quad \text{且} \quad A(y) \not\leqslant b.$$

于是得到 $y \in A^{(b)}$. 因为 $A^{(b)}$ 是 E 的一个理想, 所以 $x \in A^{(b)}$. 这表明 $A(x) \not\leqslant b$. 这恰好证明了 $a \wedge A(y) \leqslant A(x)$.

类似地, 对于满足 $z \perp w$ 的任意 $z, w \in E$, 可以证明

$$a \wedge A(z) \wedge A(w) \leqslant A(z+w).$$

于是我们证明了下面不等式:

$$\mathfrak{I}(A) = \bigvee \{a \in L \mid a \wedge A(y) \leqslant A(x), a \wedge A(z) \wedge A(w) \leqslant A(z+w), x \leqslant y, z \perp w\}$$
$$\geqslant \bigvee \{a \in L \mid \forall b \in P(L), a \not\leqslant b, A^{(b)} \text{ 是 } E \text{ 的一个理想}\}. \qquad \square$$

定理 8.4.9 设 E 是一个效应代数且 A 是 E 的一个 L-模糊子集. 如果对于任意的 $a, b \in L$, 都有 $\beta(a \wedge b) = \beta(a) \cap \beta(b)$, 那么

$$\mathfrak{I}(A) = \bigvee \{a \in L \mid \forall b \in \beta(a), A_{(b)} \text{ 是 } E \text{ 的一个理想}\}.$$

证明 假设 $a \in L$ 使得对于满足 $x \leqslant y$ 和 $z \perp w$ 的任意 $x, y, z, w \in E$, 都有

$$a \wedge A(y) \leqslant A(x) \quad \text{和} \quad a \wedge A(z) \wedge A(w) \leqslant A(z+w).$$

下面我们来证明: 对于任意的 $b \in \beta(a)$, $A_{(b)}$ 是 E 的一个理想.

(I1) 如果 $y \in A_{(b)}$ 和 $x \leqslant y$, 那么由

$$b \in \beta(A(y)) \cap \beta(a) = \beta(A(y) \wedge a) \subseteq \beta(A(x))$$

可以得到 $x \in A_{(b)}$.

(I2) 如果 $z, w \in A_{(b)}$ 且 $z \perp w$, 那么

$$b \in \beta(A(z)) \cap \beta(A(w)) \cap \beta(a) = \beta(A(z) \wedge A(w) \wedge a) \subseteq \beta(A(z+w)).$$

因此可得 $z + w \in A_{(\mu)}$. 于是我们得到了

$$\mathfrak{I}(A) = \bigvee \{a \in L \mid a \wedge A(y) \leqslant A(x), a \wedge A(z) \wedge A(w) \leqslant A(z+w), x \leqslant y, z \perp w\}$$
$$\leqslant \bigvee \{a \in L \mid \forall b \in \beta(a), A_{(b)} \text{ 是 } E \text{ 的一个理想}\}.$$

相反地, 假设 $a \in L$ 使得对于任意 $b \in \beta(a)$, $A_{(b)}$ 是 E 的一个理想. 下面我们需要证明: 对于满足 $x \leqslant y$ 和 $z \perp w$ 的任意 $x, y, z, w \in E$, 都有

$$a \wedge A(y) \leqslant A(x) \quad \text{和} \quad a \wedge A(z) \wedge A(w) \leqslant A(z+w).$$

(i) 对于 $x, y \in E$ 和 $x \leqslant y$, 假设 $b \in \beta(a \wedge A(y))$. 根据

$$\beta(a \wedge A(y)) = \beta(A(y)) \cap \beta(a)$$

可以得到

$$b \in \beta(a) \quad \text{和} \quad b \in \beta(A(y)),$$

这说明 $y \in A_{(b)}$. 通过假设条件我们知道 $A_{(b)}$ 是 E 的一个理想, 所以 $x \in A_{(b)}$. 于是可得 $b \in \beta(A(x))$. 这样就证明了 $a \wedge A(y) \leqslant A(x)$.

(ii) 对于满足 $z \perp w$ 的 $z, w \in E$, 假设 $b \in \beta(a \wedge A(z) \wedge A(w))$. 根据

$$\beta(a \wedge A(z) \wedge A(w)) = \beta(a) \cap \beta(A(z)) \cap \beta(A(w))$$

可以得到

$$b \in \beta(a), \quad b \in \beta(A(z)) \quad \text{和} \quad b \in \beta(A(w)).$$

这说明 $z, w \in A_{(b)}$. 通过假设条件我们知道 $A_{(b)}$ 是 E 的一个理想且 $z \perp w$, 因此可以得到 $z + w \in A_{(b)}$, 也就是 $b \in \beta(A(z+w))$, 进而得到

$$a \wedge A(z) \wedge A(w) \leqslant A(z+w).$$

这就证明了

$$\mathfrak{I}(A) = \bigvee \{a \in L \mid a \wedge A(y) \leqslant A(x), a \wedge A(z) \wedge A(w) \leqslant A(z+w), x \leqslant y, z \perp w\}$$

$$\geqslant \bigvee \{a \in L \mid \forall b \in \beta(a), A_{(b)} \text{ 是 } E \text{ 的一个理想}\}. \qquad \square$$

从上面证明的几个截集刻画可以自然地得到下面几个推论.

推论 8.4.10 设 E 是一个效应代数且 A 是 E 的一个 L-模糊子集. 则下面的陈述等价:

(1) A 是 E 的 L-模糊理想;

(2) 对于任意的 $a \in L$, $A_{[a]}$ 是一个理想;

(3) 对于任意的 $a \in J(L)$, $A_{[a]}$ 是一个理想;

(4) 对于任意的 $a \in L$, $A^{[a]}$ 是一个理想;

(5) 对于任意的 $a \in P(L)$, $A^{[a]}$ 是一个理想;

(6) 对于任意的 $a \in P(L)$, $A^{(a)}$ 是一个理想.

推论 8.4.11 设 E 是一个效应代数且 A 是 E 的一个 L-模糊子集. 如果对于任意的 $a, b \in L$, 都有 $\beta(a \wedge b) = \beta(a) \cap \beta(b)$, 那么下面的陈述等价:

(1) A 是 E 的 L-模糊理想;

(2) 对于任意的 $a \in J(L)$, $A_{(a)}$ 是一个理想;

(3) 对于任意的 $a \in L$, $A_{(a)}$ 是一个理想.

8.5　由 L-模糊理想度确定的 L-模糊凸结构

在这一节中, 我们借助于 L-模糊理想度来确定一个 L-模糊凸结构, 从而得到二者之间的对应关系, 通过效应代数之间的同态映射, 可以得到 L-模糊凸空间之间的一种凸保持映射和凸到凸映射.

设 E 是一个效应代数, 对每个 $A \in L^E$, 让 $\mathfrak{I}(A)$ 与之对应, 则得到一个映射 $\mathfrak{I}: L^E \to L$, 于是我们有下面定理.

定理 8.5.1 设 E 是一个效应代数, 则 $\mathfrak{I}$ 是 E 上的一个 L-模糊凸结构, 称为由 L-模糊理想度确定的 L-模糊凸结构.

证明 (1) 显然 $\mathfrak{I}(\chi_\varnothing) = \mathfrak{I}(\chi_E) = 1$.

(2) 设 $\{A_i\}_{i \in \Omega}$ 是 E 的一族 L-模糊子集. 则由定义 8.4.1 可得下面不等式:

$$\mathfrak{I}\left(\bigwedge_{i \in \Omega} A_i\right)$$

8.5 由 L-模糊理想度确定的 L-模糊凸结构

$$= \bigwedge_{\substack{x,y,z,w\in E,\\ z\perp w, x\leqslant y}} \left(\bigwedge_{i\in\Omega}A_i(y) \to \bigwedge_{i\in\Omega}A_i(x)\right) \wedge \left(\bigwedge_{i\in\Omega}A_i(z)\wedge\bigwedge_{i\in\Omega}A_i(w) \to \bigwedge_{i\in\Omega}A_i(z+w)\right)$$

$$= \bigwedge_{\substack{x,y,z,w\in E,\\ z\perp w, x\leqslant y}} \bigwedge_{i\in\Omega} \left(\bigwedge_{j\in I}A_j(y) \to A_i(x)\right) \wedge \bigwedge_{i\in\Omega} \left(\bigwedge_{j\in I}A_j(z)\wedge\bigwedge_{j\in I}A_j(w) \to A_i(z+w)\right)$$

$$= \bigwedge_{\substack{x,y,z,w\in E,\\ z\perp w, x\leqslant y}} \bigwedge_{i\in\Omega} \left[\left(\bigwedge_{j\in I}A_j(y) \to A_i(x)\right) \wedge \left(\bigwedge_{j\in I}A_j(z)\wedge\bigwedge_{j\in I}A_j(w) \to A_i(z+w)\right)\right]$$

$$\geqslant \bigwedge_{i\in\Omega} \bigwedge_{\substack{x,y,z,w\in E,\\ z\perp w, x\leqslant y}} \left(A_i(y) \to A_i(x)\right) \wedge \left(A_i(z)\wedge A_i(w) \to A_i(z+w)\right)$$

$$= \bigwedge_{i\in\Omega} \mathfrak{I}(A_i).$$

(3) 设 $\{A_i\}_{i\in\Omega}$ 是 E 的一族上定向的 L-模糊子集. 为了证明

$$\mathfrak{I}\left(\bigvee_{i\in\Omega}A_i\right) \geqslant \bigwedge_{i\in\Omega}\mathfrak{I}(A_i),$$

取 $a\in L$ 使得 $a\leqslant \bigwedge_{i\in\Omega}\mathfrak{I}(A_i)$. 则对于任意的 $i\in\Omega$, 有 $a\leqslant\mathfrak{I}(A_i)$ 成立. 由引理 8.4.4 可知, 对于满足 $x\leqslant y, z\perp w$ 的任意的 $x,y,z,w\in E$ 和对任意的 $i\in\Omega$, 都有

$$a\wedge A_i(y) \leqslant A_i(x) \quad \text{和} \quad a\wedge A_i(z)\wedge A_i(w) \leqslant A_i(z+w).$$

下面我们只需要证明: 对于满足 $x\leqslant y$ 和 $z\perp w$ 的任意 $x,y,z,w\in E$, 都有

$$a\wedge\left(\bigvee_{i\in\Omega}A_i(y)\right) \leqslant \bigvee_{i\in\Omega}A_i(x)$$

和

$$a\wedge\left(\bigvee_{i\in\Omega}A_i(z)\right)\wedge\left(\bigvee_{i\in\Omega}A_i(w)\right) \leqslant \bigvee_{i\in\Omega}A_i(z+w).$$

为此, 我们任取 $b\prec a\wedge\left(\bigvee_{i\in\Omega}A_i(z)\right)\wedge\left(\bigvee_{i\in\Omega}A_i(w)\right)$, 则存在 $i,j\in\Omega$ 使得

$$b\leqslant A_i(z), \quad b\leqslant A_j(w) \quad \text{且} \quad b\leqslant a.$$

因为 $\{A_i\}_{i\in\Omega}$ 是上定向的, 所以存在 $k \in \Omega$ 使得 $A_i \leqslant A_k$ 和 $A_j \leqslant A_k$ 成立. 这样进一步就可以得到

$$b \leqslant a \wedge A_k(z) \wedge A_k(w) \leqslant A_k(z+w) \leqslant \bigvee_{i\in\Omega} A_i(z+w).$$

这表明

$$a \wedge \left(\bigvee_{i\in\Omega} A_i(z)\right) \wedge \left(\bigvee_{i\in\Omega} A_i(w)\right) \leqslant \bigvee_{i\in\Omega} A_i(z+w).$$

类似地, 可以证明对于满足 $x \leqslant y$ 的任意 $x, y \in E$, 都有

$$a \wedge \left(\bigvee_{i\in\Omega} A_i(y)\right) \leqslant \bigvee_{i\in\Omega} A_i(x).$$

根据引理 8.4.4 可以得到 $a \leqslant \mathfrak{I}\left(\bigvee_{i\in\Omega} A_i\right)$. 这样, 我们就证明了

$$\mathfrak{I}\left(\bigvee_{i\in\Omega} A_i\right) \geqslant \bigwedge_{i\in\Omega} \mathfrak{I}(A_i).$$

因此, $\mathfrak{I}$ 是 E 上的一个 L-模糊凸结构. □

定理 8.5.2 设 $\mathfrak{I}_E, \mathfrak{I}_F$ 分别是效应代数 E 和 F 上由 L-模糊理想度确定的 L-模糊凸结构, 且 $f: E \to F$ 是效应代数之间的一个同态映射. 则 $f: (E, \mathfrak{I}_E) \to (F, \mathfrak{I}_F)$ 是一个凸保持映射.

证明 对 E 的任意一个 L-模糊子集 A 来说, 因为

$$\begin{aligned}
& \mathfrak{I}_E(f_L^{\leftarrow}(A)) \\
=\ & \bigwedge_{\substack{x,y,z,w\in E,\\ z\perp w, x\leqslant y}} \left(f_L^{\leftarrow}(A)(y) \to f_L^{\leftarrow}(A)(x)\right) \wedge \left(f_L^{\leftarrow}(A)(z) \wedge f_L^{\leftarrow}(A)(w) \to f_L^{\leftarrow}(A)(z+w)\right) \\
=\ & \bigwedge_{\substack{x,y,z,w\in E,\\ z\perp w, x\leqslant y}} \left(A(f(y)) \to A(f(x))\right) \wedge \left(A(f(z)) \wedge A(f(w)) \to A(f(z)+f(w))\right) \\
\geqslant\ & \bigwedge_{\substack{x_1,y_1,z_1,w_1\in F,\\ z_1\perp w_1, x_1\leqslant y_1}} \left(A(y_1) \to A(x_1)\right) \wedge \left(A(z_1) \wedge A(w_1) \to A(z_1+w_1)\right) \\
=\ & \mathfrak{I}_F(A),
\end{aligned}$$

所以 $f: (E, \mathfrak{I}_E) \to (F, \mathfrak{I}_F)$ 是凸保持的. □

由定理 8.5.2, 我们易得下面推论.

推论 8.5.3 设 E 和 F 是两个效应代数且 $f: E \to F$ 是单同态. 如果 B 是 F 的 L-模糊理想, 那么 $f_L^{\leftarrow}(B)$ 是 E 的 L-模糊理想.

习 题 8

1. 计算例 8.2.2 中 $A_1 \wedge A_2$ 的 L-模糊效应子代数度.
2. 请利用推论 8.2.6 直接证明推论 8.2.8 和推论 8.2.9 成立.
3. 设 E 是一个效应代数, 请举出一个 L-子代数 A 的例子, 使得 L 不满足下列条件:
$$\beta(a \wedge b) = \beta(a) \cap \beta(b), \quad \forall a, b \in L,$$
但存在 $c \in L$ 使得 $A_{(c)}$ 不是一个效应子代数.

4. 设 E 是一个效应代数, 请举出一个 L-模糊理想 A 的例子, 使得 L 不满足下列条件:
$$\beta(a \wedge b) = \beta(a) \cap \beta(b), \quad \forall a, b \in L,$$
但存在 $c \in L$ 使得 $A_{(c)}$ 不是一个理想.

5. 如果 $L = [0,1]$, 那么在定理 8.4.9 中, 是否要求 $\beta(a \wedge b) = \beta(a) \cap \beta(b)$ 成立才能证明定理结论?

6. 证明推论 8.4.10.
7. 证明推论 8.4.11.
8. 证明推论 8.5.3.
9. 理想和滤子是对应的两个概念, 效应代数中也有滤子的定义, 读者可以研究效应代数中 L-模糊滤子度及其刻画.
10. 给出效应代数上的 L-模糊滤子度的定义后, 再研究其与 L-模糊凸结构之间的关系.

参 考 文 献

[1] 李洪兴, 罗承忠, 汪培庄. 如何定义模糊集的映射. 北京师范大学学报 (自然科学版), 1993, 29: 1–9.

[2] 吴广庆, 杜萍, 张国铭. L-fuzzy 环与L-fuzzy 理想的分解定理. 河北师范大学学报 (自然科学版), 1996, 20: 20–22.

[3] 孟晗, 程本肃, 史福贵. L-fuzzy 拓扑群的直积. 烟台师范学院学报 (自然科学版), 1997, 13: 29–32.

[4] 张成, 刚家泰, 张广济. 关于模糊向量子空间的研究. 大连大学学报, 2002, 23: 26–31.

[5] 魏晓伟, 岳跃利, 黄春娥. L-集上的模糊泛代数. 四川师范大学学报 (自然科学版), 2019, 42: 63–68.

[6] 欧阳军, 潘庆年. Fuzzy 商环及其 Fuzzy 同态同构基本定理. 阜阳师范学院学报 (自然科学版), 1993, (2): 19–22.

[7] 张德学, 刘应明. L-fuzzy 拓扑空间的弱诱导化. 数学学报, 1993, 36: 68–73.

[8] 史福贵, 王国民. L-fuzzy 子群与L-fuzzy 正规子群的表现定理. 烟台师范学院学报 (自然科学版), 1995, 11: 16–19.

[9] 聂灵沼, 丁石孙. 代数学引论. 2 版. 北京: 高等教育出版社, 2000.

[10] 吴广庆, 史福贵. Hutton 一致结构的点式处理. 模糊系统与数学, 2000, 14: 38–42.

[11] 郝瑶媛, 王绪柱. 模糊子环的度量. 模糊系统与数学, 2016, 30: 22–31.

[12] 沈冲, 史福贵. L-凸系统. 模糊系统与数学, 2017, 31: 18–23.

[13] 黄飞, 廖祖华. 格的模糊素理想的度量. 模糊系统与数学, 2019, 33: 63–74.

[14] 朱翔, 廖祖华. 格的模糊子集的模糊理想度. 郑州大学学报 (理学版), 2019, 51: 107–112.

[15] 黄飞, 廖祖华. 格的模糊半素理想及其度量. 模糊系统与数学, 2020, 34: 38–52.

[16] 罗承忠. Fuzzy 集与集合层. 模糊数学, 1983, 4: 113–126.

[17] 张文修. 模糊数学基础. 西安: 西安交通大学出版社, 1984.

[18] 梁基华. 关于不分明度量空间的几个问题. 数学年刊 A 辑 (中文版), 1985, 6: 59–67.

[19] 彭先图. Fuzzy 群范畴. 模糊数学, 1987, 1: 61–68.

[20] 王国俊. L-fuzzy 拓扑空间论. 西安: 陕西师范大学出版社, 1988.

[21] 王国俊. 拓扑分子格理论. 西安: 陕西师范大学出版社, 1990.

[22] 史福贵. 由普通导算子诱导的 L 导算子及由 L 导算子诱导的 L 拓扑. 模糊系统与数学, 1991, 1: 32–37.

[23] 彭育威. 关于格值诱导空间的两个公开问题. 数学学报, 1993, 35: 751–757.

[24] 谭宜家. 半群上的 L-fuzzy 同余关系. 福州大学学报 (自然科学版), 1994, 22: 8–13.

[25] 史福贵. L_β-集合套与L_α-集合套理论及其应用. 模糊系统与数学, 1995, 9: 65–72.

[26] 史福贵. 分子集合套理论及其应用. 烟台师范学院学报 (自然科学版), 1996, 12: 33–36.

[27] 史福贵. L-fuzzy 集与素元集合套. 数学研究与评论, 1996, 16: 398–402.
[28] 姚炳学. LF 商环的同态. 聊城师院学报 (自然科学版), 1998, 11: 11–14.
[29] 赵立军. L-fuzzy 子环的 L-fuzzy 理想. 烟台师范学院学报 (自然科学版), 2000, 16: 91–94.
[30] 史福贵. L-fuzzy mappings of L-fuzzy sets. 模糊系统与数学, 2000, 14: 16–24.
[31] 何凤兰. L-fuzzy 素理想的再定义 (英文). 模糊系统与数学, 2001, 15: 36–39.
[32] 赵立军. L-fuzzy 子半群的 L-fuzzy 双 (内) 理想. 首都师范大学学报 (自然科学版), 2002, 23: 5–8.
[33] 乔丙武. L-fuzzy 环与 L-fuzzy 商环. 河北师范大学学报 (自然科学版), 2003, 27: 241–243.
[34] 赵立军. L-fuzzy 子群的 L-fuzzy 直积. 模糊系统与数学, 2003, 17: 37–40.
[35] 张可铭. L-fuzzy 子域与 L-fuzzy 线性空间. 数学杂志, 2004, 24: 53–58.
[36] 罗承忠. 模糊集引论: 上册. 2 版. 北京: 北京师范大学出版社, 2005.
[37] 刘东利. 伪效应代数中的模糊滤子与模糊理想. 计算机工程与应用, 2011, 47: 50–52.
[38] 方进明. 剩余格与模糊集. 北京: 科学出版社, 2012.
[39] 修振宇. (L, M)-模糊凸空间及其相关理论研究. 北京: 北京理工大学, 2015.
[40] 冯娟. 关于 t-模糊子半群度的一些结论. 成都: 四川师范大学, 2016.
[41] 张禾瑞. 近世代数基础. 修订版. 北京: 高等教育出版社, 2010.
[42] Ajmal N. Homomorphism of fuzzy groups, correspondence theorem and fuzzy quotient groups. Fuzzy Sets and Systems, 1994, 61: 329–339.
[43] Ajmal N. Fuzzy groups with sup property. Information Sciences, 1996, 93: 247–264.
[44] Ajmal N. Fuzzy group theory: A comparison of different notions of product of fuzzy sets. Fuzzy Sets and Systems, 2000, 110: 437–446.
[45] Ajmal N, Thomas K. Fuzzy lattices. Information Sciences, 1994, 79: 271–291.
[46] Aktaş H, Çağman N. Generalized product of fuzzy subgroups and t-level subgroups. Mathematical Communications, 2006, 11: 121–128.
[47] An Y Y, Shi F G, Wang L. A generalized definition of fuzzy subrings. Journal of Mathematics, 2022, 5341207: 1–11.
[48] Anthony J, Sherwood H. Fuzzy groups redefined. Journal of Mathematical Analysis and Applications, 1979, 69: 124–130.
[49] Anthony J, Sherwood H. A characterization of fuzzy subgroups. Fuzzy Sets and Systems, 1982, 7: 297–305.
[50] Bakhshi M. On fuzzy convex lattice-ordered subgroups. Iranian Journal of Fuzzy Systems, 2013, 10: 159–172.
[51] Bandelt H J, Hedlíková J. Median algebras. Discrete Mathematics, 1983, 45: 1–30.
[52] Berger M. Convexity. The American Mathematical Monthly, 1990, 97: 650–678.
[53] Mukherjee N, Bhattacharya P. Fuzzy normal subgroups and fuzzy cosets. Information Sciences, 1984, 34: 225–239.

[54] Birkhoff G. On the combination of subalgebras. Mathematical Proceedings of the Cambridge Philosophical Society. Cambridge: Cambridge University Press, 1933: 441–464.

[55] Birkhoff G. Abstract linear dependence and lattices. American Journal of Mathematics, 1935, 57: 800–804.

[56] Birkhoff G. Lattice Theory. New York: American Mathematical Society Colloquium Publications, 1948.

[57] Biswas R. Fuzzy fields and fuzzy linear spaces redefined. Fuzzy Sets and Systems, 1989, 33: 257–259.

[58] Chang C L. Fuzzy topological spaces. Journal of Mathematical Analysis and Applications, 1968, 24: 182–190.

[59] Che Y H, Liu Q. A novel concept of fuzzy subsemigroup (ideal). Journal of Mathematics, 2022, Article ID 5830590, 10 pages.

[60] Chon I. Fuzzy subgroups as products. Fuzzy Sets and Systems, 2004, 141: 505–508.

[61] Das P S. Fuzzy groups and level subgroups. Journal of Mathematical Analysis and Applications, 1981, 84: 264–269.

[62] Davey B A, Priestley H A. Introduction to Lattices and Order. Cambridge: Cambridge University Press, 2002.

[63] Dilworth R P. Dependence relations in a semi-modular lattice. Duke Mathematical Journal, 1944, 11: 575–587.

[64] Dixit V, Bhambri S, Kumar P. Union of fuzzy subgroups. Fuzzy Sets and Systems, 1996, 78: 121–123.

[65] Dixit V, Kumar R, Ajmal N. Level subgroups and union of fuzzy subgroups. Fuzzy Sets and Systems, 1990, 37: 359–371.

[66] Dixit V, Kumar R, Ajmal N. Fuzzy ideals and fuzzy prime ideals of a ring. Fuzzy Sets and Systems, 1991, 44: 127–138.

[67] Dong Y Y, Li J. Fuzzy convex structure and prime fuzzy ideal space on residuated lattices. Journal of Nonlinear and Convex Analysis, 2020, 21: 2725–2735.

[68] Duchet P. Convexity in combinatorial structures. Proceedings of the 14th Winter School on Abstract Analysis, Circolo Matematico di Palermo, 1987: 261–293.

[69] Dvurečenskij A, Pulmannová S. New Trends in Quantum Structures. Dordrecht: Springer Science & Business Media, 2000.

[70] Dwinger P. Characterization of the complete homomorphic images of a completely distributive complete lattice. I. Indagationes Mathematicae (Proceedings), 1982, 85: 403–414.

[71] Eckhoff J. Helly, radon, and carathéodory type theorems// Handbook of Convex Geometry. Amsterdam: North-Holland, 1993: 389–448.

[72] Eroğlu M S. The homomorphic image of a fuzzy subgroup is always a fuzzy subgroup. Fuzzy Sets and Systems, 1989, 33: 255–256.

[73] Fan C, Shi F G, Mehmood F. M-hazy Γ-semigroup. Journal of Nonlinear and Convex Analysis, 2020, 21: 2659–2669.

[74] Foulis D J, Bennett M K. Effect algebras and unsharp quantum logics. Foundations of Physics, 1994, 24: 1331–1352.

[75] Gierz G, Hofmann K H, Keimel K, et al. A Compendium of Continuous Lattices. Berlin: Springer Science & Business Media, 1980.

[76] Gierz G, Hofmann K H, Keimel K, et al. Continuous Lattices and Domains. Cambridge: Cambridge University Press, 2003.

[77] Goguen J A. L-fuzzy sets. Journal of Mathematical Analysis and Applications, 1967, 18: 145–174.

[78] Gu W, Lu T. Fuzzy algebras over fuzzy fields redefined. Fuzzy Sets and Systems, 1993, 53: 105–107.

[79] Han Y L, Shi F G. L-fuzzy convexity induced by L-convex fuzzy ideal degree. Journal of Intelligent & Fuzzy Systems, 2019, 36: 1705–1714.

[80] Höhle U, Šostak A. Axiomatic foundations of fixed-basis fuzzy topology//Höhle U, Rodabaugh S E, eds. Mathematics of Fuzzy Sets: Logic, Topology, and Measure Theory. Berlin: Springer, 1999: 123-272.

[81] Hua X J, Xin X L, Zhu X. Generalized (convex) fuzzy sublattices. Computers & Mathematics with Applications, 2011, 62: 699–708.

[82] Huang H L, Shi F G. L-fuzzy numbers and their properties. Information Sciences, 2008, 178: 1141–1151.

[83] Shan J. New characterizations of L-fuzzy ideals. Fuzzy Systems and Mathematics, 2022, 16: 22–26.

[84] Katsaras A, Liu D B. Fuzzy vector spaces and fuzzy topological vector spaces. Journal of Mathematical Analysis and Applications, 1977, 58: 135–146.

[85] Kubiak T. On fuzzy topologies. Poznan: Adam Mickiewicz University, 1985.

[86] Kubiś W. Abstract convex structures in topology and set theory. Katowice: University of Silesia, 1999.

[87] Kuraoka T, Kuroki N. On fuzzy quotient rings induced by fuzzy ideals. Fuzzy Sets and Systems, 1992, 47: 381–386.

[88] Lassak M. On metric B-convexity for which diameters of any set and its hull are equal. Bull. Acad. Polon. Sci., 1977, 25: 969–975.

[89] Levi F W. On Helly's theorem and the axioms of convexity. J. Indian Math. Soc., 1951, 15: 65–76.

[90] Li J, Shi F G. L-fuzzy convexity induced by L-convex fuzzy sublattice degree. Iranian Journal of Fuzzy Systems, 2017, 14: 83–102.

[91] Liu D, Wang G. Fuzzy filters in effect algebras. Fuzzy Systems and Mathematics, 2009, 23: 6–17.

[92] Liu Q, Shi F G. M-fuzzifying median algebras and its induced convexities. Journal of Intelligent & Fuzzy Systems, 2019, 36: 1–9.

[93] Liu Q, Shi F G. M-hazy lattices and its induced fuzzifying convexities. Journal of Intelligent & Fuzzy Systems, 2019, 37: 2419–2433.

[94] Liu Q, Shi F G. A new approach to the fuzzification of groups. Journal of Intelligent & Fuzzy Systems, 2019, 37: 6429–6442.

[95] Liu W J. Fuzzy invariant subgroups and fuzzy ideals. Fuzzy Sets and Systems, 1982, 8: 133–139.

[96] Liu W J. Operations on fuzzy ideals. Fuzzy Sets and Systems, 1983, 11: 19–29.

[97] Liu Y, Luo M. Fuzzy Topology. Singapore: World Scientific Publishing, 1997.

[98] Lowen R. convex fuzzy sets. Fuzzy Sets and Systems, 1980, 3: 291–310.

[99] Lubczonok P. Fuzzy vector spaces. Fuzzy Sets and Systems, 1990, 38: 329–343.

[100] Ma Z. Note on ideals of effect algebras. Information Sciences, 2009, 179: 505–507.

[101] MacLane S. Some interpretations of abstract linear dependence in terms of projective geometry. American Journal of Mathematics, 1936, 58: 236–240.

[102] Malik D S, Mordeson J N. Fuzzy subfields. Fuzzy Sets and Systems, 1990, 37: 383–388.

[103] Malik D S, Mordeson J N. Fuzzy direct sums of fuzzy rings. Fuzzy Sets and Systems, 1992, 45: 83–91.

[104] Maruyama Y. Lattice-valued fuzzy convex geometry. 数理解析研究所講究録, 2009, 1641: 22–37.

[105] Mehmood F, Shi F G. M-hazy vector spaces over M-hazy field. Mathematics, 2021, 9: 1118.

[106] Mehmood F, Shi F G, Hayat K. A new approach to the fuzzification of rings. Journal of Nonlinear and Convex Analysis, 2020, 21: 2637–2646.

[107] Mehmood F, Shi F G, Hayat K, et al. The homomorphism theorems of M-hazy rings and their induced fuzzifying convexities. Mathematics, 2020, 8: 411.

[108] Karl M. Untersuchungen über allgemeine metrik. Mathematische Annalen, 1928, 100: 75–163.

[109] Močkoř J. α-Cuts and models of fuzzy logic. International Journal of General Systems, 2013, 42: 67–78.

[110] Mordeson J N, Malik D S. Fuzzy Commutative Algebra. Singapore: World Scientific, 1998.

[111] Mordeson J N, Malik D S, Kuroki N. Fuzzy Semigroups. Berlin: Springer, 2003.

[112] Muhiuddin G, Mahboob A, Khan N M, et al. New types of fuzzy (m, n)-ideals ordered semigroups. Journal of Intelligent & Fuzzy Systems, 2021, 41: 6561–6574.

[113] Muhiuddina G, Alenzea E N, Mahboobb A, et al. Some new concepts on int-soft ideals in ordered semigroups. New Mathematics and Natural Computation, 2021, 17: 267–279.

[114] Nanda S. Fuzzy fields and fuzzy linear spaces. Fuzzy Sets and Systems, 1986, 19: 89–94.

[115] Negoiţă C V, Ralescu D A. Applications of Fuzzy Sets to Systems Analysis. New York: Birkhauser, 1975.

[116] Novák V. Fuzzy Sets and Their Applications. Bristol: Adam Hilger, 1989.

[117] Abu Osman M T. On t-fuzzy subfield and t-fuzzy vector subspaces. Fuzzy Sets and Systems, 1989, 33: 111–117.

[118] Öztürk M A, Jun Y B, Yazarli H. A new view of fuzzy gamma rings. Hacettepe Journal of Mathematics and Statistics, 2010, 39: 365–378.

[119] Pang B. Bases and subbases in (L, M)-fuzzy convex spaces. Computational and Applied Mathematics, 2020, 39: 1–2.

[120] Pang B. Convergence structures in M-fuzzifying convex spaces. Quaestiones Mathematicae, 2020, 43: 1541–1561.

[121] Pang B. L-fuzzifying convex structures as L-convex structures. Journal of Nonlinear and Convex Analysis, 2020, 21: 2831–2841.

[122] Pang B, Shi F G. Strong inclusion orders between L-subsets and its applications in L-convex spaces. Quaestiones Mathematicae, 2018, 41: 1021–1043.

[123] Pang B, Zhao Y. Characterizations of L-convex spaces. Iranian Journal of Fuzzy Systems, 2016, 13: 51–61.

[124] Pei D, Fan T. On generalized fuzzy rough sets. International Journal of General Systems, 2009, 38: 255–271.

[125] Ray A K. On product of fuzzy subgroups. Fuzzy Sets and Systems, 1999, 105: 181–183.

[126] Rosa M. On fuzzy topology fuzzy convexity spaces and fuzzy local convexity. Fuzzy Sets and Systems, 1994, 62: 97–100.

[127] Rosenfeld A. Fuzzy groups. Journal of Mathematical Analysis and Applications, 1971, 35: 512–517.

[128] Shen C, Shi F G. L-convex systems and the categorical isomorphism to scott-hull operators. Iranian Journal of Fuzzy Systems, 2018, 15: 23–40.

[129] Shen C, Shi F G. Characterizations of L-convex spaces via domain theory. Fuzzy Sets and Systems, 2020, 380: 44–63.

[130] Sherwood H. Products of fuzzy subgroups. Fuzzy Sets and Systems, 1983, 11: 79–89.

[131] Shi F G. L-fuzzy relations and L-fuzzy subgroups. Journal of Fuzzy Mathematics, 2000, 8: 491–499.

[132] Shi F G. L-fuzzy interiors and L-fuzzy closures. Fuzzy Sets and Systems, 2009, 160: 1218–1232.

[133] Shi F G. (L, M)-fuzzy matroids. Fuzzy Sets and Systems, 2009, 160: 2387–2400.

[134] Shi F G. A new approach to the fuzzification of matroids. Fuzzy Sets and Systems, 2009, 160: 696–705.

[135] Shi F G. Novel characterizations of LM-fuzzy open operators. Transactions on Fuzzy Sets and Systems, 2022, 1: 6–20.

[136] Shi F G, Huang C E. Fuzzy bases and the fuzzy dimension of fuzzy vector spaces. Mathematical Communications, 2010, 15: 303–310.

[137] Shi F G, Xin X. L-fuzzy subgroup degrees and L-fuzzy normal subgroup degrees. Journal of Advanced Research in Pure Mathematics, 2011, 3: 92–108.

[138] Shi F G, Xiu Z Y. A new approach to the fuzzification of convex structures. Journal of Applied Mathematics, 2014, 2014: 249183.

[139] Shi F G, Xiu Z Y. (L,M)-fuzzy convex structures. Journal of Nonlinear Sciences and Applications, 2017, 10: 3655–3669.

[140] Shi S R. Measures of fuzzy subgroups. Proyecciones (Antofagasta), 2010, 29: 41–48.

[141] Shi Y, Huang H L. A characterization of strong Q-concave spaces. Journal of Nonlinear and Convex Analysis, 2020, 21: 2771–2781.

[142] Shi Y, Shi F G. Lattice-valued betweenness relations and its induced lattice-valued convex structures. Journal of Intelligent & Fuzzy Systems, 2019, 37: 1–11.

[143] Soltan V P. D-convexity in graphs. Soviet. Math. Dokl., 1983, 28: 419–421.

[144] Swamy U, Swamy K. Fuzzy prime ideals of rings. Journal of Mathematical Analysis and Applications, 1988, 134: 94–103.

[145] 谭宜家. Fuzzy 半群中的 fuzzy 理想. 模糊系统与数学, 1995, 9(2): 59–64.

[146] Tepavčević A, Trajkovski G. L-fuzzy lattices: an introduction. Fuzzy Sets and Systems, 2001, 123: 209–216.

[147] Van De Vel M L. Theory of Convex Structures. New York: Elsevier, 1993.

[148] Šostak A. On a fuzzy topological structure. Rend. Circ. Mat. Palermo (Suppl. Ser. II), 1985, 11: 89–103.

[149] Wang G J. Theory of topological molecular lattices. Fuzzy Sets and Systems, 1992, 47: 351–376.

[150] Wang K, Shi F G. M-fuzzifying topological convex spaces. Iranian Journal of Fuzzy Systems, 2018, 15: 159–174.

[151] Wang L, Wu X Y, Xiu Z Y. A degree approach to relationship among fuzzy convex structures, fuzzy closure systems and fuzzy Alexandrov topologies. Open Mathematics, 2019, 17: 913–928.

[152] Wei X, Pang B, Mi J S. Axiomatic characterizations of L-valued rough sets using a single axiom. Information Sciences, 2021, 580: 283–310.

[153] Wei X, Pang B, Mi J S. Axiomatic characterizations of L-fuzzy rough sets by L-fuzzy unions and L-fuzzy intersections. International Journal of General Systems, 2022, 51: 277–312.

[154] Wei X, Wang B. Fuzzy (restricted) hull operators and fuzzy convex structures on L-sets. Journal of Nonlinear and Convex Analysis, 2020, 21: 2805–2815.

[155] Wen Y F, Zhong Y, Shi F G. L-fuzzy convexity induced by L-convex degree on vector spaces. Journal of Intelligent & Fuzzy Systems, 2017, 33: 4031–4041.

[156] Williams D R P, Latha K. Fuzzy bi-Γ-ideals in Γ-semigroups. Hacettepe Journal of Mathematics and Statistics, 2009, 38: 1–15.

[157] Jing W. Ideals, filters and supports in pseudo effect algebras. International Journal of Theoretical Physics, 2004, 43: 349–358.

[158] Wu W. Normal fuzzy subgroups. Fuzzy Math., 1981, 1: 21–30.

[159] Xin X, Fu Y. Some results of convex fuzzy sublattices. Journal of Intelligent & Fuzzy Systems, 2014, 27: 287–298.

[160] Xiu Z Y, Pang B. Base axioms and subbase axioms in M-fuzzifying convex spaces. Iranian Journal of Fuzzy Systems, 2018, 15: 75–87.

[161] Yang H, Li E Q. A new approach to interval operators in L-convex spaces. Journal of Nonlinear and Convex Analysis, 2020, 21: 2705–2714.

[162] Yao W. An approach to the fuzzification of complete lattices. Journal of Intelligent & Fuzzy Systems, 2014, 26: 2239–2249.

[163] Yehia S E B. Fuzzy partitions and fuzzy-quotient rings. Fuzzy Sets and Systems, 1993, 54: 57–62.

[164] Ying M S. Fuzzifying topology based on complete residuated lattice-valued logic (i). Fuzzy Sets and Systems, 1993, 56: 337–373.

[165] Yuan B, Wu W. Fuzzy ideals on a distributive lattice. Fuzzy Sets and Systems, 1990, 35: 231–240.

[166] Zadeh L A. Fuzzy sets. Information and Control, 1965, 8: 338–353.

[167] Zahedi M. A characterization of L-fuzzy prime ideals. Fuzzy Sets and Systems, 1991, 44: 147–160.

[168] Zeng M Y, Wang L, Shi F G. A novel approach to the fuzzification of fields. Symmetry, 2022, 14: 1190.

[169] Zhan J, Jun Y B. Generalized fuzzy interior ideals of semigroups. Neural Computing and Applications, 2010, 19: 515–519.

[170] Zhang C. Fuzzy subfield and it's pointwise characterization. Fuzzy Theory and Application: Selected Papers of the 9th Annual Meeting of Fuzzy Mathematics and Systems Association of China, 1998.

[171] Zhang Q. Ideals and filters in dual effect algebras. Hennan Science, 2016, 34: 1211–1214.

[172] Zhang Q. Note on ideals in effect algebras. Hennan Science, 2017, 35: 1567–1569.

[173] Zhang Y. Some properties on fuzzy subgroups. Fuzzy Sets and Systems, 2001, 119: 427–438.

[174] Zhao F, Huang H L. The relationships among L-ordered hull operators, restricted L-hull operators and strong L-fuzzy convex structures. Journal of Nonlinear and Convex Analysis, 2020, 21: 2817–2829.

[175] Zhong Y, Shi F G. Characterizations of (L,M)-fuzzy topological degrees. Iranian Journal of Fuzzy Systems, 2018, 15: 129–149.

[176] Zhong Y, Yang S J, Shi F G. The relationship between L-subuniverses and L-convexities. Journal of Intelligent & Fuzzy Systems, 2017, 33: 3363–3372.

[177] Zhou X W, Shi F G. Some new results on six types mappings between L-convex spaces. Filomat, 2020, 34: 4767–4781.

[178] Zhou X W, Wang L. Four kinds of special mappings between M-fuzzifying convex spaces. Journal of Nonlinear and Convex Analysis, 2020, 21: 2683–2692.

《模糊数学与系统及其应用丛书》已出版书目
（按出版时间排序）

1 犹豫模糊集理论及应用　2018.2　徐泽水　赵　华　著
2 聚合函数及其应用　2019.2　覃　锋　著
3 逻辑代数上的非概率测度　2019.3　辛小龙　王军涛　杨　将　著
4 直觉模糊偏好关系群决策理论与方法　2019.11　万树平　王　枫　董九英　著
5 基于模糊逻辑代数的判断矩阵及其群体决策方法　2020.4　徐泽水　马振明　著
6 广义积分论　2021.12　张德利　著
7 一致模　2022.12　王住登　苏　勇　史雪荣　著
8 序与拓扑(第二版)　2022.12　徐晓泉　著
9 格值模糊凸结构与格值模糊代数　2024.11　史福贵　著